Springer Biographies

The books published in the Springer Biographies tell of the life and work of scholars, innovators, and pioneers in all fields of learning and throughout the ages. Prominent scientists and philosophers will feature, but so too will lesser known personalities whose significant contributions deserve greater recognition and whose remarkable life stories will stir and motivate readers. Authored by historians and other academic writers, the volumes describe and analyse the main achievements of their subjects in manner accessible to nonspecialists, interweaving these with salient aspects of the protagonists' personal lives. Autobiographies and memoirs also fall into the scope of the series.

Simine Short

Flight Not Improbable

Octave Chanute and the Worldwide Race Toward Flight

 Springer

Simine Short
National Soaring Museum
Elmira, NY, USA

ISSN 2365-0613 ISSN 2365-0621 (electronic)
Springer Biographies
ISBN 978-3-031-24429-2 ISBN 978-3-031-24430-8 (eBook)
https://doi.org/10.1007/978-3-031-24430-8

O. Chanute

We have no reason to believe that a practical flying machine will be the result of the happy thought of one or two persons. It will come rather by a process of evolution: one man accomplishing some promising results, but stopping short of success; the next carrying the investigation somewhat further, and thus on until a machine is produced which will be as practical as the "safety" bicycle, which took some eighty years for its development from the original despised velocipede.

Octave Chanute "Experiments in Flying," 1900

*This book is dedicated to my husband Jim,
my favorite glider pilot, who is just as interested in
the history of the sport as in flying.*

Foreword

I was pleased when asked to write the foreword for Simine Short's new book on Octave Chanute because I greatly appreciated her first book on that subject, *Locomotive to Aeromotive*. My delight in that book stemmed from my life-long interests in railroad *and* aviation history. I am glad to relate that her new book, "*Flight Not Improbable*", continues to build on her previous work in a very effective manner, by further examining this most-amazing individual. To say that Octave Chanute was a very accomplished man would be most appropriate. A highly successful civil engineer, working for various railroads, his career received an early boost when, in 1865, he designed and supervised the construction of the Chicago Stock Yards. This was a huge and complex project that brought Chanute's name to prominence. Before long he was designing bridges for such major railroads as the Chicago & Alton, AT&SF and Erie. The first bridge to successfully cross the "unbridgeable" Missouri River at Kansas City was designed by Chanute and completed under his supervision in 1869. This iron bridge withstood the forces of nature and was used for many years before being replaced by a more modern (steel) structure that was designed to handle the greater loads produced by increasingly heavy locomotives and rolling stock. A standing testament to his abilities as a *manager* came when the Erie Railroad, having declared bankruptcy in 1878, gave Chanute—who at the time was serving as Assistant General Superintendant—the very difficult task of bringing the company back to solvency. This was an amazing undertaking for a civil engineer/bridge builder! By the time he resigned from the Erie in 1883, the railroad was operating at a profit for the first time in years.

Working for the Erie during that period was quite stressful and in 1875, at the suggestion of his physician, Chanute traveled to England and France for a vacation. While in England, he chanced upon the 10th annual meeting of the *Aeronautical Society of Great Britain*. Established in 1866, this group was formed to pursue the science of manned flight. Please bear in mind that British aeronautical experimentation went back to the early 19th century with the experiments of Sir George Cayley and, for that reason, such activities enjoyed a significant following in England. Chanute was captivated by this group and went back home with a growing fascination with the prospect of developing a flying machine. Unlike in England, however, the subject of manned flight was taboo among civil engineers in the U.S. and it took him several years to get that subject into the mainstream of American engineering. That he was able to do this at all was due to the fact that he was an eminent member of the American engineering fraternity. He approached this challenge as he had all others—in a methodical and well-thought-out manner that could be understood—and appreciated—by his colleagues. In the spring of 1880, Chanute was asked to deliver the key-note message at the May meeting of the *American Society of Civil Engineers* (ASCE). He was a long-time member of this prestigious organization (founded in 1852) and at the time, was serving as vice president. The subject of his talk was to be "Progress of Engineering in the United States" and he was certainly well-equipped to deliver it. Railroads during this period were by far the predominate subject for such meetings and this one was to be no exception—except at the very end—when he introduced the subject of aerial navigation! As I mentioned above, this was a subject that was heretofore not discussed in these gatherings. Octave Chanute had the clout to do this and the fortitude to go ahead with it. He felt very strongly that, in order to achieve ultimate success in this new field of endeavor, a strong following among the engineering community would be required. He became the leader and, if you will, a "lightning rod"—for this new movement in the United States and, to a certain extent, abroad. As one progresses through this book, it becomes abundantly clear that, in regard to aerial experimentation, Chanute knew just about everyone worth knowing. That's probably an exaggeration—but only a small one when we consider that he corresponded regularly with aviation enthusiasts from the United States, Europe, Russia, Australia, and others. That he became a central person in this effort was due to the fact that Chanute firmly believed in the sharing of knowledge in order to further the science as a whole. He freely shared ideas based on his own experiences as a structural engineer and passed along any pertinent information, always giving credit where credit was due.

As you work your way through this book, if you are anything like I am, you will be amazed at the great number of individuals who were involved in aviation's infancy and by the fact that they were positively influenced by their association with Octave Chanute. I have been an aviation enthusiast pretty much my entire life—my father's father was taught to fly by Glenn Curtiss in 1911 and my mother's father attained the rank of Air Chief Marshal in the RAF—so I learned about Curtiss Pushers and Supermarine Spitfires at a very early age, and I considered myself quite well versed on the subject of aviation history. Well, it turns out that I was not as well versed as I thought—not by a long shot. There is so much more to be learned about the history of early aviation, and books such as this one provide the great service of presenting this lengthy period of innovation and development in a straightforward manner. This helps us to consider the ingenuity and outright bravery of the pioneers of aviation. Simine Short's efforts—past and present—have done much to raise my awareness and appreciation of this select group of individuals, many of whom have remained largely unknown to subsequent generations.

Enjoy this book.

Trafford L.-M. Doherty
Executive Director
National Soaring Museum
Harris Hill, Elmira, NY, USA

Preface

"Birth of Aviation"
Frieze of American History, United States Capitol, Washington, D.C.
Courtesy Architect of the Capital.

Gazing up the steps of the United States Capitol, thousands of visitors to Washington, D.C., are struck every day by the grandeur of this symbol of the American people. Nowhere is the pageant of America's early history so well showcased as inside the Capitol's great ceremonial rotunda. Sixty feet above its floor, the eight-foot-high frescoed Frieze of American History encircles the rotunda, giving the illusion of a sculptured relief and depicting nineteen scenes, from the "Discovery of America" to the "Birth of Aviation".

The final scene shows Orville and Wilbur Wright's first powered airplane flight in 1903 and three of their acknowledged precursors: the Italian Leonardo Da Vinci and two Americans, Samuel P. Langley, secretary of the Smithsonian Institution, and Octave Chanute, one of the renowned civil engineers of the nineteenth century.

Chanute developed a successful series of gliders and was a mentor to the Wright Brothers. In this scene, formally attired with long suit, starched collar, and trademark goatee, Chanute holds his Katydid multiplane glider. He stands closest to the Wrights, possibly because of the dates and depth of his aviation involvement but also possibly because of his close and continuing friendship with the brothers.

How did Chanute, a self-educated French immigrant, earn a place in the Frieze of American History among others whose names are better known? How did Octave Chanute rise from being a penniless immigrant to rank among the elite American engineers of the nineteenth century? How did he learn from others and then give back by mentoring his juniors? What inspired him to spend a lifetime in transportation and engineering, and how did he learn to work with others to achieve what one man alone usually could not accomplish? How did he capitalize on his early perception of the need to preserve natural resources? Finally, why, when others would have retired to a life of leisure, did he pursue his ultimate transportation passion—aviation—until his final days?

And finally, why did so many enthusiasts in the beginning of the twentieth century contact him to learn about his experiences building and flying gliders? How did Octave Chanute become the pivotal person in the Wright Brothers' quest for powered, controlled flight? Why, when the Wrights flew their powered flight of December 17, 1903, did their sister Katharine immediately wire the news to Chanute, making him the first, and for some time the only, person outside the Wright family to know of the epic event?

This book aims not only to answer these questions, but also to chronicle the amazingly productive decades that earned Octave Chanute a place in the *Frieze of American History*.

Elmira, NY, USA   Simine Short

Acknowledgements

A very big thank you goes to my husband Jim who provided not only help, but also constant support for the past two decades to first finish the "*Locomotive to Aeromotive*" book, published in 2011, and now this one. His patience in the various stages of writing, sharing thoughts for improvement, editing, trimming contents, and proofreading was extremely helpful and encouraging.

People Who Helped with this Book

Shortly before retiring from Argonne National Laboratory, I decided to write a book on Octave Chanute, highlighting his contribution to the development of aviation. After explaining my general idea to the acquisition editor of the University of Illinois Press, Laurie Matheson strongly suggested to first write a biography, as many people could not connect the name with a person. I did as suggested and *"Locomotive to Aeromotive"* was published in 2011. Hoping that the public is now more familiar with this eminent engineer, this book will discuss Chanute's aeronautical work, as I had envisioned originally. Dr. Leonard C. Bruno (retired) from the Library of Congress suggested to look for an international publisher, because Chanute worked with so many enthusiasts worldwide. Hearing this, Dr. Markus Raffel and Dr. Bernd Lukasch suggested contacting Springer Nature.

As my research stretched over so many years, please accept my apologies if I have missed listing someone who was kind in the past. I would be glad to hear from anyone I may have inadvertently missed.

Special thanks go to Joseph Hodges and his wife Jean of Denver, who allowed me to look through the Chanute family papers, collected and assembled by Joe Hodges's mother, Mrs. Elaine Chanute Hodges, and her brother Octave A. "Ox" Chanute, the two great-grandchildren of Octave Chanute.

Other notables who were so helpful in my research include (in no particular order): William Barry (retired), NASA, Chief Historian, who helped with translating articles written in the Russian language; Albion Bowers and Cam Martin (both now retired), NASA Dryden Flight Research Center, Edwards,

California; Gary Bradshaw, arguably the first person to establish a website focusing on early aeronautics in 1994 (the "Virtual Museum covering the invention of the airplane"); Leonard C. Bruno (retired) and Lewis Wyman, Manuscript Division, Library of Congress. Washington, D.C. When Len retired, Lewis introduced himself and usually helped find the needed answers; Jean Conklin, St. Joseph, Michigan, who researched Augustus M. Herring for her unpublished book; Tom Crouch, Peter Jakab, Russell Lee, Chris Moore, Brian Nicklas, and Elizabeth C. Borja, National Air and Space Museum, Washington, D.C.; Ellen Alers and Mary Markey, Smithsonian Institution Archives, Reference Services; Paul Dees, Seattle, Washington, an aeronautical engineer and hang glider pilot who shared his experiences of first building and then flying his Chanute-type reproduction in 1996; Trafford Doherty (now director of the National Soaring Museum) and Rick Leisenring, Glenn Curtiss Museum, Hammondsport, New York; Patricia Goitein, Peoria, Illinois, a longtime friend who shared her knowledge on the Civil War, slavery, ballooning and life in Peoria in the nineteenth century; Hayden Hamilton, editor of the American Aviation Historical Society Journal (a friend who loves flying gliders), Huntington Beach, California; Philip Jarrett, who shared his knowledge on Percy Pilcher; Reinhard Keimel (deceased), Technisches Museum Wien and Aviatika Museum, Vienna, Austria, drew the Chanute glider designs as they developed from the ladder kite to the biplane; Jean-Luc Claessens, free-lance photographer from Paris, and Laurent Rabier from the Musée de l'Air et de l'Espace, Le Bourget Aéroport de Paris, France; Barbara Kern and Urszula Kerkhoven (retired), John Crerar Library; thanks also go to Debra Levine and Jay Satterfield, Special Collections, Regenstein Library, Joe Gerdeman, Interlibrary Loan Division, Chicago, Illinois; Kevin Kochersberger, associate professor, College of Engineering, Virginia Tech, Blacksburg, Virginia; Bernd Lukasch, Otto-Lilienthal-Museum, Anklam, Germany; Markus Raffel, German Aerospace Center (DLR), Institute of Aerodynamics and Flow Technology, Head Department of Helicopters, Goettingen, Germany; Laurie Matheson, University of Illinois Press, still available for comments and advice; Stephan Nitsch (deceased), Langenhagen, Germany, who built and flew his Chanute-type glider in the early 1990s and again for television crews in 2003; Bruce Rowe, Indiana Dunes National Park, Porter, Indiana; Paul Nelson, creator of the documentary "Octave Chanute, Patron Saint of Flight;" Wm. Kevin Cawley, Joe Smith, and Robert J. Havlik (retired), University of Notre Dame Library and Archive, South Bend, Indiana; Trafford Doherty (who wrote the foreword for this book), National Soaring Museum, Elmira, New York; Jamie Seemiller and Marilyn Chang, Denver Public Library; Peter Selinger, Stuttgart, Germany;

George Rogge, Chanute Aquatorium Society, Gary, Indiana; Steve Spicer, Miller Beach, Gary, Indiana, who created a website to help publicize the one-hundredth anniversary of Chanute's gliding efforts in his hometown in 1995 and continues to add new material regularly; Ann and Jim Kepler, Maggie White, and Tim Seeden, Western Society of Engineers, Chicago, Illinois; Darrell Collins (retired), Wright Brothers National Memorial, Manteo, North Carolina; Bill Stolz and John Armstrong, Wright State University, Special Collections, Dayton, Ohio. Last, but not least, I would like to acknowledge the help I had received from Angela Lahee and Ashok Arumairaj and his layout and production team at Springer Nature, to make this book a worthwhile edition to the literature on early aviation.

If this book contributes something worthwhile to a better understanding of Octave Chanute and his contribution to the development of mechanical flight, then it is these people who deserve much credit for making it become reality.

Contents

1

How It All Began

Facility of locomotion, whether by land or by water, is one of the evidences of advanced civilization. On land, man could have scarcely emerged from barbarism without the subjection of the inferior animals to his rule. On water, he has at least partially accomplished the still more noble task of remembering the elements subservient to his will.

Campbell's Foreign Monthly Magazine; August 1843

Mississippi Delta. December 1838—After almost two months at sea the *Havre Paquet* entered the muddy Mississippi River delta; many of its passengers stood on deck, as they were anxious to start a new life in America where "the air is more free." [1]. One of the passengers traveling with his six-year-old son in the cabin section, was the cultivated 42-year-old Joseph Chanut; he too looked for a new and better life.

Seven years earlier, Joseph's maternal aunt had arranged a marriage between the then 35-year-old Joseph and the 19-year-old Elise Sophie de Bonnaire. They were married in the Catholic Parish Church of Saint Sulpice on April 19, 1831 [2]. Their first child, Octave Alexandre, was born on February 18, 1832 and christened the next day. The following year, the couple moved with Elise's mother to Sceaux, a southern suburb of Paris. A second boy, Emile Frédéric, was born there in May 1833. A third child, Leon, was born in May 1835 and two weeks later Elise requested a legal separation. Joseph moved back to the University section in Paris and concentrated on his academic life. In late summer 1838 he received an unsolicited offer to join Jefferson College, one of three major colleges in antebellum Louisiana, as a vice president and

© The Author(s), under exclusive license to Springer Nature Switzerland AG 2023
S. Short, *Flight Not Improbable*, Springer Biographies, https://doi.org/10.1007/978-3-031-24430-8_1

to teach French history, literature and geography [3]. Joseph was very ready to move to the New World, far away from Paris! He and his estranged wife agreed to share the responsibility for the three children. Joseph was to take his oldest boy Octave, while Elise was to rear the two younger boys in Paris, with the help of her mother. The six-year-old Octave was told to pack his belongings and go with his father, whom he barely knew, to a place far away from home.

Now the boy stood close to his father, watching the boat being pulled up river into New Orleans. Joseph most likely wondered about his future at Jefferson College, and young Octave probably wondered if he could make friends and have toys again.

The new vice president of Jefferson College, Joseph Chanut, arrived with his son in the small town of Convent, about 60 miles north of New Orleans, late in December 1838, starting a life that was not Parisian at all. Without input on how to bring up a boy, he used the same methods as his parents did some forty years earlier. When Joseph needed to leave his son alone at the house, he normally locked the door of their living quarters. This prevented young Octave from disappearing, when he, like other inquisitive lads, was motivated to discover his new world amid the surrounding sugar plantation or to mingle with other children of his age or any of the college's sixty slaves. Joseph home-schooled his son, and his French-speaking colleagues supplied a teaching curriculum according to their expertise, usually communicating in their mother tongue. They not only taught the youngster to read and write, but also to tell the truth and observe the general rules of etiquette. Manners of early-nineteenth-century French society formed an integral part of the boy's daily life.

The college, which provided a classic education to the boys of the surrounding plantations, was at its most successful level in the early 1840s [4], but then a fire destroyed most of the buildings. Not receiving paychecks regularly, Joseph resigned in the summer of 1844 and moved with his son to the Crescent City, or as the French called it, *La Nouvelle Orléans*. He accepted a job teaching at a local plantation, which was not exactly what Joseph considered a rewarding career; so he gave more thought to moving to New York City, where other French immigrants had started successful lives. The beginnings of the Mexican–American War in the spring of 1846, and the fear of a possible occupation of New Orleans, may have speeded his decision to move.

Joseph's financial situation did not permit the expensive mode of travel by steamer along the Atlantic Coast, but he could afford the month-long, 1,600-mile inland route via the Mississippi and Ohio Rivers at a cost of about $23 a person. Not only was this significantly lower in cost, but Joseph also thought that venturing into the heartland of the continent and meeting people of different cultures would prove a learning experience for both of them.

Early in August 1846, Joseph and Octave boarded the Mississippi packet for a slow trip upstream. They passed the town of Convent with Jefferson College on the eastern shore, perhaps recalling bittersweet memories. Villas of wealthy plantation owners, residences of common people and slave quarters were part of the ever-changing scenery as they traveled the 600 miles upriver to the mouth of the Ohio.

They disembarked at Cairo, Illinois. Their final destination was almost one thousand miles further to the northeast. Father and son now boarded a steamboat heading toward the smoky industrial center of Pittsburgh, a transportation hub with thousands of travelers passing through all year long [5]. The city at the confluence of the Allegheny and Monongahela rivers, which formed the Ohio River, had experienced a devasting fire a year earlier, and John A. Roebling had designed a wire suspension bridge with eight spans across the Monongahela river, which had opened just a few weeks earlier. The 15-year old Octave stood in awe seeing this novel feat of technology: a bridge held together by cables only! [6] Walking across he could feel the bridge swaying slightly in the wind, this was absolutely fascinating! Little did he know that he would discuss the various options in the design and construction of bridges twenty years later with the 36-year older civil engineer Roebling.

Joseph located the canal basin with its packets at 2nd Avenue; the eye-catching sign, "Through to Philadelphia in three days and a half" could not be overlooked. The agent told them proudly that the average travel speed, using canal boats and mountain railroads, was 4.34 miles per hour [7]. Early the next morning, the Chanuts stowed their belongings on the *Pioneer Canal Packet* Boat for their trip across the Allegheny mountains. Shortly after their departure, a stationary engine pulled the boat through a tunnel. On the other side, the ropes were attached to two horses, that pulled the boat on a one-mile-long aqueduct across the Allegheny River and then on a canal, dug parallel to the river [8] over the 104-mile stretch to Johnstown. Instead of sitting on the boat Octave usually joined other passengers walking briskly on the towpath.

At Johnstown the boats were hoisted out of the water and transferred onto flatcars of the Allegheny Portage Railroad to cross the densely forested Allegheny Mountains. As the locomotive could not travel faster than fifteen miles an hour, nor climb a steep grade, engineers introduced a rope-and-pulley-system. The cars were unhitched from the locomotive at the foot of incline No. 1 and fastened to a rope attached to a stationary steam-driven winch at the head of the incline that pulled the cars majestically up the steep slope. Between inclines No. 1 and 2, the tracks went through the nine-hundred-foot-long Staple Bend Mountain, the first railroad tunnel built in the United States. Travelers traversed the fourteen-mile-long second level in one hour, crossing the Conemaugh River via the Horseshoe Bend. This viaduct went in a perfect eighty-foot diameter semicircle, carrying the tracks seventy feet above the water surface and ascending over its whole course. The train travelled over the final incline and reached Blair's Gap Summit, 2,326 feet above sea level, during the early afternoon. A stone tavern, the Lemon House, invited passengers to relax during the one-hour wait while the crew reconnected the cars for the downhill trip. Travelers also visited the engine house with its thirty-horsepower steam engine and inspected the thick wire ropes that pulled the cars uphill. Octave was probably not the first to wonder how the descending and ascending trains could move at the same time in opposite directions on opposite sides of the mountain on the double-track roadbed.

In late afternoon, they reached the canal basin at Hollidaysburg on the Juniata River, about seven miles south of modern-day Altoona. The boat sections were now hoisted from the railroad cars and reassembled to glide eastward on the water for the next 172 miles to Columbia. Here passengers had to disembark and travel the final 82-mile trip on the Columbia Railroad, one of the earliest railroad lines in the United States, descending slightly into Philadelphia. For the final stretch Joseph selected the Philadelphia & Trenton Railroad. Arriving in Jersey City, New Jersey, passengers and freight were transferred across the Hudson River via a steam ferry into New York City.

In early September 1846 a new life began for father and son. Joseph boarded with a friend, while young Octave entered the Coudert Lyceum as an "indigent child" for an education quite different to what he had received in Louisiana. The lyceum was a boarding school for about thirty pupils, providing a classical French education. As the state required each school to accept a limited number of children whose parents could not pay the tuition, Octave was accepted and became part of the Coudert family, quickly starting a friendship with their oldest son Frédéric René.

The challenges of moving freight and passengers across the Alleghenies were fixed in Octave's mind. Could there be a more efficient way? Could he possibly help make a change? His father made it clear that there was no money to send him to college, so the teenager studied magazines to determine how to enter the high-tech world of transportation. An editorial in the

Merchant's Magazine described the benefits of railroads in early 1848: "The railroads are the chief means by which the whole commerce of the earth, its movement and its population, are to be connected together and the ends of the world literally united. No man can over-estimate their value." [9]. The future of transportation was clearly with the railroad. Reading reports by the chief engineer of the Hudson River Railroad, John B. Jervis, who had accomplished several marvelous engineering feats, totally impressed Octave. This railroad was now building tracks for almost 150 miles between New York City and Albany; once completed, this would be an unbelievable achievement.

After graduating from Coudert's Lyceum in August 1848 with a degree similar to a high school diploma, and being gifted in mathematics, the 16-year-old Octave decided that the Hudson River Railroad was to be his path into the future. He gave himself a haircut and put on his best clothes, all cleaned and pressed, to call on his future employer. As the teenager was leaving, his father reminded him to wear his black gloves. Octave realized too late that there was a big hole in one glove, so he took along thread to fix the glove on the long ride upriver—the best he knew how.

The 7 a.m. steamboat departed from the North River dock on time, arriving six hours later in Sing Sing, the headquarters of the Hudson River Railroad. The glove was repaired and the teenager was ready.

Without a formal letter of introduction, Octave simply knocked on the door of the office of the railroad and asked to speak with Chief Engineer Jervis. Introducing himself again, he asked for employment. Looking at the slender teenager, wearing his best clothes, Jervis did not think he should hire the young man. His reply was simple and discouraging: there were no vacancies. To prove that he was an industrious worker, Octave pleaded and offered to work without pay. Looking at the lad again, Jervis thought he recognized him as a fellow passenger who had traveled on the same boat in the morning. He had watched the teenager industriously fix his glove and was impressed [10]. When asked, Octave shyly admitted that he did indeed mend his glove on the ride up. "Well, I think you are careful and industrious and deserve a trial," were Jervis' comments. Success. He was allowed to come to work, but without pay, and was assigned to the survey party. Late in December 1848, his supervisor Henry Gardner recommended hiring Octave as assistant chainman for $1.12½ a day. Even though this was the lowest-paid job in the full surveying party, Octave felt like the richest man in the world. Soon, perhaps, he would be a real civil engineer and could add the title "C.E." to his name, just like Mr. Gardner and Mr. Jervis!

Twelve years after coming to the New World, Octave's father Joseph decided to move back to France, while the eighteen-year-old Octave decided to become a real American. Working far upriver in late December 1850, Octave wrote an emotional "Good Bye" letter. "… I have exactly the same opinion as you have about the parts played by energy und perseverance in achieving success. However, I often wonder how this happens when I see around me so many young men filled with energy and talents, having many more chances of success than I have, and they are no further advanced than I am. There must be some unknown circumstances that have prevented them from getting wealthy. I can see only one reason: it is not how worthy a man is, but rather how clever, how cunning he is. All one can do is meet events with fortitude; if they are favorable, they make you a wealthy man; if they are not, the effort was a waste of time. Although I hope to become wealthy in a certain, or rather an uncertain, number of years, I do not think that making money is the only goal a person should have. Money is only precious because of the pleasures it gives us. Consequently, I believe the only way to achieve happiness is to save all one can without depriving oneself of the many pleasures of one's age and position." [11].

When the Hudson River Railroad was fully operational, Gardner moved to his next job in Illinois and offered Octave the assistant engineer position on the Chicago, Alton & St. Louis Railroad laying-out the Joliet Cut-Off between Joliet, Illinois, and Lake Station, Indiana. Octave accepted and moved west in April 1853.

As soon as he could, Octave filed his intent to apply for citizenship. After all, he received a paycheck, was Americanized, and would soon be a real civil engineer. Almost three years to the day of filing and after one year of State residency, Octave Alexandre Chanut became a United States Citizen. On April 17, 1854, the 22-year-old engineer was sworn in at the McLean County Courthouse in Bloomington, Illinois, with one of his supervisors, Richard Morgan, as his witness. As part of the naturalization process he anglicized his name to Octave Chanute, dropping his middle name, Alexandre, and adding the letter "e" to his family name [12].

When the Joliet Cut-Off was completed, the 22-year-old Chanute moved to Peoria to accept the chief engineer position of the Peoria & Oquawka Railroad, Eastern Extension, laying tracks from Peoria to the Illinois-Indiana state line, and designing and building the first railroad bridge across the Illinois River at Peoria.

Living and working in Peoria was more social than Joliet; the budding civil engineer, standing 5-foot 6-inches tall, decided to grow an imperial or half beard with a tiny goatee, which was just as curly as his main hair. As a young bachelor, Octave was reportedly a dandy in dress and courtly in manners. No wonder he aroused the interest of Annie Riddell James when her older brother Charles took her to a Saturday evening dance in one of the town halls frequented by Octave and fellow rowing club members. Octave and Annie received the "Rites of Marriage" from the Clerk of Peoria County on March 11, 1857 [13]. They eventually raised two boys and three girls, with one girl dying a few months after she was born.

The Eastern Extension was sold to the Toledo, Wabash & Western Railroad in 1860. As Chanute did not always receive a paycheck, he arranged to take home pieces from his office in lieu of a paycheck, including a desk and a letterpress, which he thought would help in a business that he and his brother-in-law were running. Starting in early September 1860 Chanute copied almost all his outgoing letters to letterpress books, to keep a record for himself and for posterity.

When the road reached the Indiana state line, Chanute accepted the chief engineer position of the Chicago & Alton Railroad and moved his family to Chicago.

To keep up with the needs of the Civil War and the growing population, major slaughterhouses were erected in various parts of Chicago. The Union Stock Yard and Transit Company was chartered in February 1865, when ten railroads and eight meat packers joined to establish a single yard, to stop drovers from driving the cattle and hogs over crowded city streets from one yard to the other [14]. Boston capitalists purchased 320 acres of swampland west of Halsted Street and east of the south fork of the Chicago River. To attract the best engineering minds, incorporators advertised a competition, inviting engineers to submit plans for an efficient stock yard. The winner was to be appointed Chief Engineer.

To design an efficient yard with railroads entering and leaving on double tracks, a supply of fresh water for the animals and a sewage drainage system was challenging. Chanute knew how to build roads and drain swampland, but everything else was new and he accepted the challenge. When the entries were reviewed, the best proposal was judged to be that of Chanute who commented dryly to a friend: "This, in addition to my salary from the Chicago & Alton Road, will assist my income, but keep me dreadfully busy." [15] (Fig. 1.1).

With great fanfare, the Union Stock Yards opened on December 26, 1865, just six months after construction began. The formerly worthless swampland in South Chicago became a miniature city with banking accommodations, hotel, animal hospital, fire department, weighing facilities, seven miles of streets, sewer and water lines, and feeding pens for the livestock.

The Chanute family left Chicago in mid 1867 and moved to Kansas City, as he had accepted a new assignment with a salary of $5,000/year. Boston

Fig. 1.1 Octave Chanute, chief engineer of the Chicago & Alton Railrod (circa 1864). Courtesy Chanute Family

investors, the same group that backed the Chicago stock yard, wanted to build a "transcontinental railroad," and cross the "unbridgeable" Missouri on a bridge, and not like the Pacific Railroad on a ferry. After several technical set-backs, the bridge at Kansas City opened with a huge celebration on July 3, 1869, eight weeks after the golden spike was ceremoniously driven by the Pacific Railroad crew on their "Transcontinental Railroad."

The American Society of Civil Engineers (ASCE) was revitalized in 1867, and Chanute's former coworkers, Henry Gardner, John Jervis and Thomas Meyer, thought this society might play an important part in developing engineering as a profession in the United States. Every engineer knew that successful engineering required an accurate knowledge of science, careful observation, skillful deduction and logical reasoning; blunders and miscalculations were usually followed by failure and disaster. Agreeing with his friends,

Chanute applied for membership and submitted the required three reference letters. His membership was accepted in February 1868. Hearing of the "Civil Engineers' Club of the Northwest" being formed in Chicago, Chanute joined that group in July 1869 and became friends with several engineers, who would be part of his career growth. This club later changed its name to "Western Society of Engineers" (WSE), becoming one of the most prestigious engineering organizations outside the national society.

One of his new friends was the mechanical engineer **Matthias N. Forney** (1835–1908) who had moved to Chicago to join the *Railroad Gazette* as associate editor. After the Chicago Fire in 1871, the office was moved to New York City and so did the 36-year-old Forney. To be more influential in what the *Gazette* published, Forney purchased one-half ownership of the weekly paper and promoted the activities of the various engineering societies. Early in 1872, he explained in an editorial [16] that there were now two entirely distinct associations of civil engineers in this country. The ASCE, with headquarters in New York City, whose membership was composed largely of older engineers, could be called the "static" society. The other group, he opined, was the Civil Engineers' Club of the Northwest with headquarters in Chicago, whose members were mostly younger men. They were as much, or maybe even more, actively engaged in the practice of their profession, and should be called the "dynamic" society.

After much internal debate, members of the New York based ASCE accepted the invitation of the Civil Engineers' Club to hold the 4th Annual Convention in June 1872 in Chicago. Living in Lawrence, Kansas, an easy overnight train ride from Chicago, Chanute joined this, his first, ASCE meeting. He volunteered to serve on the newly formed ASCE Publications Committee, as this would provide an opportunity to learn and extend his personal base of contacts.

The annual ASCE Banquet was quite different than anything Chanute had ever attended. The reporter from the Chicago *Tribune* described it as a remarkable gathering. "There were men who had constructed mighty works that will live after them, great tunnels, bridges, railroads, docks, locks, and dams. They were rather a jolly whole-souled party, bent on unbending themselves, and their humor was enjoyable, although sines, cosines, tangents, cube root, the unknown quantity and other astronomical geometrical, trigonometrical, geological and algebraic materials entered into its conjunction." [17].

Back home in Lawrence, Chanute gave more thought to another career change, even though his wife Annie had just renovated their rented house to include a fireplace and a library for his books. Using his networking skills,

Chanute, who had just turned 41, received an interesting job offer from Erie Railway management to make their outdated road profitable again. He was appointed general manager in late February 1873, and joined the upper ranks of one of the largest eastern railroads with an office in New York City. Not knowing what the future would bring, they decided that Annie should move with their five children back to Peoria and stay with her family until further notice, while her husband moved east.

In mid-March Chanute traveled 1,200 miles over three days on three different railroads from Lawrence to New York. Having just purchased Jules Verne's book "*Around the World in Eighty Days*," he read it while traveling over the same route, by the same railroads, through the same towns, to his new job in New York, just like the book's hero, Mr. Phileas Fogg. "The train passed rapidly across the State of Iowa, by Council Bluffs, Des Moines, and Iowa City. During the night it crossed the Mississippi at Davenport, and by Rock Island entered Illinois. The next day at four in the evening, it reached Chicago, already risen from its ruins [of the Great Chicago fire], and more proudly seated than ever on the borders of its beautiful Lake Michigan. Nine hundred miles separated Chicago from New York; but trains are not wanting at Chicago. Mr. Fogg passed at once from one to the other, and the locomotive of the Pittsburg, Fort Wayne and Chicago Railway left at full speed, as if it fully comprehended that that gentleman had no time to lose. It traversed Indiana, Ohio, Pennsylvania, and New Jersey like a flash, rushing through towns with antique names, some of which had streets and car-tracks, but as yet no houses. At last, the Hudson came into view …"

Perhaps this paragraph from Chanute's favorite author, Jules Verne, provided a poignant transition, as Chanute left his old job of building a lofty rail system in the West to start a new one in New York City to rejuvenate an antiquated rail giant with financial and political problems in the east.

Moving from eastern Kansas, considered by many New Yorkers as the "wild west," to life in the big city required some adjusting. The regular day-to-day work with the Erie brought challenges, but nothing that stimulated him. The ASCE semi-monthly meetings were good, and Chanute quickly made friends with equally minded professionals for brainstorming. He also enjoyed discussions at the impromptu "smokers" on non-standard topics, including the perceived obstacles of commanding locomotion through the air. The student of mechanical science knew there was nothing contradictory within the principles of manflight; the flying machine would eventually join the other means of transportation by land and sea.

In early 1875, only two years after moving to New York, the Erie was taken over by a receiver and everything seemed to collapse around the 43-year-old

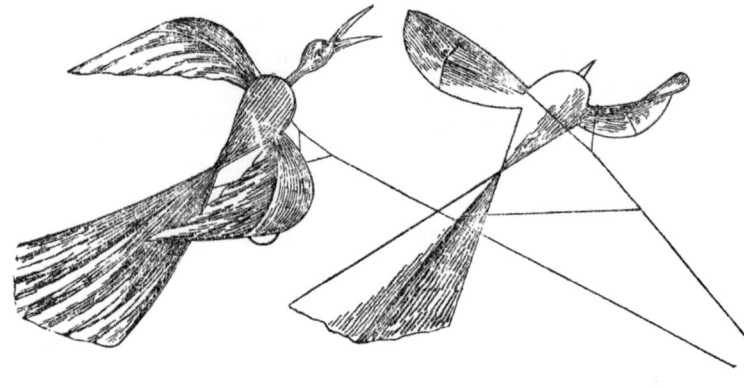

CHINESE BIRD KITE.

Fig. 1.2 Among ingenious fancies of the Chinese is their bird kite, which has an elastic frame. The thin paper, attached to the wings, moves under the action of the wind and simulates the flapping of the wings. This kite is sometimes 3 ft. in length. From *Progress in Flying Machines*

Chanute. His physician recommended a lengthy vacation far away from the hectic business environment, and to temporarily forget the anxiety of the world of railroads and telegraphs [18]. Following doctor's orders, Chanute left for Europe on June 6, 1875.

Not all passengers on this ocean voyage were to his liking. "Have been keeping very quiet during the whole voyage and so much that I am probably thought very uncivil by the other passengers, particularly the ladies. We have a large number of ill-bred persons on board. I never knew before how disagreeable Americans can make themselves" [19]. He became acquainted with members of the American Rifle Team, who entertained fellow passengers by demonstrating their Chinese bird kites, and explaining how to estimate the force and direction of the wind during shooting matches. The kite's construction was ingenious: Chanute watched how the thin paper, attached to the wings, moved by the action of the wind, simulating the flapping of bird wings. He needed to study this movement of the wind (Fig. 1.2).

After a brief stop in Liverpool, Chanute continued his trip to London, did some sightseeing and spent time on "other items of interest." The British Institution of Civil Engineers had no meetings scheduled during the summer months, but then he heard of the Aeronautical Society of Great Britain holding its tenth annual meeting on June 23, with James Glaisher presiding. This group had attracted scientists of stature and vision, people who wanted to solve the problems of mechanical flight in a logical fashion. Hearing that several engineers from other countries were to attend, Chanute joined as well. Thomas Moy read a paper on aeronautical progress and remarked in closing:

"I believe in work and in making use of the present time, and, as I told you in 1868, so I tell you now, that thought, work, and money can and will achieve aerial navigation" [20].

Next stop was Paris, spending some time with his mother and family. Chanute had hoped to meet members of the French aeronautic community, but they did not meet in the summer either. Toward the end of July, he finally started his "walking tour," as ordered by his physician, but soon realized that he needed to get back home to his family and a paycheck.

Back in New York, full of energy and new ideas, Chanute successfully picked up his job where he left off. The Erie Railway soon promoted him to Assistant General Superintendent, in addition to his duties as Chief Engineer, one step closer to the very top in management. As the road's financial situation seemed to be under control, he rented a house at 128 High Street and picked up his family in Peoria.

Technical educators and engineers held a joint meeting in Philadelphia, as part of the Centennial Exhibition in 1876. ASCE vice president Theodore G. Ellis discussed in his talk "Rise and Progress of Civil Engineering" the improvements made during the past year, but also some futuristic projects. He envisioned that "bridges of a mile span would be entirely possible, railway trains for passengers could be reduced to at least one quarter of their current weight, and the navigation of the air would become practicable." [21]. This talk, prepared with Chanute's help, appears to be the first to mention air travel by a presiding civil engineer in an annual address.

In April 1878 the Erie Railway declared bankruptcy, and Chanute was concerned about another stress-related breakdown. Not wanting to take another trip to Europe, he sought mental relaxation by delving deeper into the science of aeronautics. Reading reports on the flights of Charles F. Ritchel in his recently patented flying machine [22, 23] made Chanute compose his thoughts on mechanical flight: "Every few years a new flying machine is invented, tried and found a failure, each fresh attempt proving by the attention it excites how deep a hold the problem has upon the popular imagination. Inventors constantly retry old experiments, and in their isolated efforts they neither seek to ascertain what has already failed, nor the principles upon which success is possible." He left the second page blank, and the remaining pages discuss the possibility of artificial flight, the requisites of a flying machine, the sustaining power of air, and a discussion on how to operate and land a flying machine. "The fact that birds fly is by no means conclusive evidence that artificial flight is possible to man. It would be rash to say that it cannot be solved by man, but it may safely be assumed that the successful flying machine of the future will no more resemble a bird than a steamer resembles the whale, or the locomotive the hare." [24].

On rather short notice the ASCE president, Albert Fink, was called for consultation to England, so he asked for a volunteer to deliver the presidential address on "Progress of Engineering in the United States" in May 1880. Everyone pointed to vice president Chanute who accepted. Preparing for this talk was a turning point in his career and life, as it provided the opportunity for soul searching. He questioned:

- What had been accomplished in the United States, now that civil engineering was recognized as a distinct profession?
  - What had he, personally, accomplished as a professional civil engineer?
- What lay ahead in the engineering field?
  - What would the future bring in his personal engineering career?
- What could be changed and improved in the engineering field?
  - How could he contribute?
  - What could he change in his own career?

More than 100 members from every corner of the country were present at the 12th Annual ASCE Convention at Washington University in St. Louis. Chanute's address was a comprehensive review of engineering in the United States, stressing the importance to improve construction and lower the cost of railroad operation. To possibly highlight his own contribution, he prepared a table showing how the railroad system had expanded from 2,000 miles in operation in 1838 when he came to the United States, to almost 6,000 miles, when he began his career with the Hudson River Railroad in 1848, to almost 90,000 miles in 1880. He also stressed that associations like the ASCE needed to be administered by leaders with courage, practical wisdom, self-control and a strong commitment to justice. They could not be effective leaders if they could not persuade others to follow, and they could not be good leaders unless they were good people. Advocating further progress, Chanute also expressed hopes that the last mode of transportation, traveling through the air by "Flyer," borrowing a word from the railroad practice, could soon be solved. He stated, "There are signs that a new motive power will be invented, which shall be safer, of greater energy and less wasteful than steam." The gasoline engine had shown some success, and Chanute theorized that with this new motive power might come the solution of the last transportation problem. "I suppose you will smile when I say that the atmosphere yet remains to be conquered; but wildly improbable as my remarks may now seem, there may be engineers in this room who will yet see men safely sailing through the air." [25]. This is the first time that navigating through the air was discussed in detail in an annual address on engineering progress by an elected officer of the most prestigious American engineering society. As could be expected, fellow engineers scoffed at him at the next meetings and told him that it was undignified and unprofessional for an officer of a prestigious association to make a public statement hinting at the possibility of air travel or for man to fly [26].

Abstracts of his address were widely reprinted, however his predictions on air travel were usually not included. The ancient taboo of flying was still very real in the early 1880s, and an educated man who cared to stand well with good people found it safest to say nothing about it [27]. But Chanute was sure that only a civil engineer could solve this challenging engineering problem, thus an engineer needed to take the lead to get the job done. Others had to be pulled into his research circle so that a successful, practical flying machine could become reality. "How often is it that the imagined things of today become the accomplished things of tomorrow?"

Chanute had stated more than once that he was a Democrat of the "Jefferson School" and, like his father, believed in the Jeffersonian philosophy, which was one of reason, responsibility, individualism, liberty and limited government. Having watched the engineering profession grow, he also had watched some engineers not always act like professionals. He wrote an editorial on "The Ethics of Good Consulting Practice," because " 'a Code of Ethics,' or of professional good manners and fair dealings, is necessarily a matter of slow growth. General acceptance of any rule of conduct has to be universally accepted. Engineering is a new profession; it needs to have rigidly established rules of practice." Elementary rules of professional good manners, he noted, should be established in the engineering practice to be followed by every engineer [28].

Even though Chanute was born French, he dreamed American. His interest in engineering spurred him to a lifelong program of self-education; being an advocate for idea-sharing across disciplines and continents, he leveraged his skills to become a builder, a leader, a businessman, and even a dreamer who pursued his passions.

With his wide range of knowledge, he helped create the world in which we live today, setting a great example for multidisciplinary knowledge, collaboration, mentoring and out-reach. Taking an unofficial leadership, Chanute served as a powerful communication mechanism between experimenters around the world. He was not only the conduit for information and ideas, but he also acted as a mentor and cheerleader for those engaged in discovering the secrets of manflight. He was what we would call today the world-wide-web of aviation or the "Center of Knowledge on Flight" in Chicago. Enthusiasts contacted him via the postal mail to share what they had accomplished and at the same time learn everything they could about what other pioneers in aviation were doing to solve the mystery of flight. Chanute knew how to inspire people. Leadership, informal or formal, is a way of thinking, a way of acting and, most importantly, a way of communicating. He knew he could not change the world or do the work for others; he needed to help his correspondents figure out how to do things by themselves so that they could succeed beyond what they thought to be possible.

Chanute also believed in the philosophy of the utopian regime "Republic of Letters," which fostered communication among intellectuals, or *philosophes,* as he had learned from his father back in France. To him, communication, or letter writing, was a basic essence of science and research. He usually answered all letters systematically, one after the other, always copying each page into his letterpress book for record-keeping. And if needed, he shared the letters with other enthusiasts, so that new information could be passed along and studied by others. At the American Association for the Advancement of Science (AAAS) meeting in 1886, Chanute firmly stated his belief, "The busy men who are developing this country need to keep up with new discoveries and progress even before they are reduced to practical account. Engineers owe it to themselves to acquaint others with whatever new facts they may have acquired outside of the routine of their profession … provided that they always stick accurately to the facts."

As civil engineers usually do, Chanute saved all his incoming correspondence, including the envelopes, but also newspaper and magazine clippings, and any scrap of paper. The bulk of Chanute's engineering library, about 2,000 books and pamphlets, was donated to the John Crerar Library (now part of the University of Chicago Library system) in 1911. Almost twenty years later, Albert Francis Zahm, the new Daniel Guggenheim Chair in Aeronautics at the Library of Congress, Washington, DC, approached the Chanute family to have their father's personal papers and letters, especially the 24 letterpress books, transferred to the Library of Congress. And another twenty years later, Marvin W. McFarland joined the Library of Congress and noted in his diary, that the Wrights "had but one true predecessor, Lilienthal, and only one tutor, actually mentor, Chanute." [29]. He collected most of the letters between Chanute and the Wrights, and published them as the *Papers of Wilbur and Orville Wright, including the Chanute-Wright letters* in 1953.

Chanute's life-long obsession with detail and perhaps also his sense or vision of a nascent world-changing development is documented in these letters. They are virtually an insider's history of American engineering and aeronautics between 1860 and 1910, allowing one to assemble a remarkably accurate record of the development and growth of engineering with Chanute being part of this global progress.

2

Flight Is Not Improbable

If there be a domineering, tyrant thought, it is the conception that the problem of flight may be solved by man. When once this idea has invaded the brain, it possesses it exclusively. It is then a haunting thought, a walking nightmare, impossible to cast off.

Louis-Pierre Mouillard (1881)

America has long been known as a nation of inventive tinkerers. As the Honorable Samuel S. Fisher, United States Commissioner of Patents, told his audience in 1869, "The truth is we are an inventive people. Invention is by no means confined to our mechanics. Our merchants invent, our soldiers and our sailors invent, our schoolmasters invent, our professional men invent, aye, and our women and our children invent. Looking at the events of the last century, any one of these tinkerers, or inventors, is touched with enough genius to influence history."

Several of these creative tinkerers developed products that improved the standard of living, eliminated hard labor and provided more leisure. Aaron Montgomery Ward formed Montgomery and Ward in 1872 and created the first mail-order business with a printed catalog. After several years of struggle, Alexander Graham Bell received the patent for the telephone in 1876 and in the same year, Nicolaus August Otto received a patent for the first practical four-stroke internal combustion engine. Three years later, in 1879, Thomas Edison patented the first commercially viable incandescent electric light bulb. Five years later, in 1884, William H. Walker and George Eastman developed negative photo film to be used in standard glass plate cameras, and the

S. Short, *Flight Not Improbable*, Springer Biographies, https://doi.org/10.1007/978-3-031-24430-8_2

Kodak box camera with pre-installed negative film was patented in 1888 by George Eastman. In 1891, Whitcomb L. Judson, born and raised in Chicago, invented the zipper, still commonly used today.

The same year that the zipper was patented, the U.S. Patent Office celebrated its centennial, and Octave Chanute was one of the invited speakers. He selected the effect of invention upon the railroad as his main topic, but he also discussed inventions that pointed the way to a better life for regular people. "Improvement has followed upon improvement, because invention has been more active and successful than at any prior period in the world's history" [1].

A decade earlier, Octave Chanute had stated boldly at the American Society of Civil Engineers (ASCE) meeting, "If engineers desire to take a higher rank than they now occupy in this country, they must study new paths for themselves and be no longer men of routine, ordered hither and thither by promoters of schemes and the magnates of Wall Street." And he made a prescient prediction: "There are signs that a new motive power will be invented, which shall be safer, of greater energy and less wasteful than steam. And, with a new motive power, perhaps will come the solution of the last transportation problem which remains to be solved. There may be engineers in this room who will yet see men safely sailing through the air" [2].

2.1   Aëronautics in Illinois

Chanute moved with the railroad crew from the Hudson River Railroad from New York to Illinois in 1853, when something triggered his curiosity in manflight [3]. Did the reports of the circus performer Silas M. Brooks flying his balloon catch his fancy? Brooks' first performance took place in late May 1855 in Evansville, Indiana. An estimated 4,000 persons paid the 25 cents entrance fee to watch the inflation of the balloon, while many more folks stood outside the fenced area [4]. In the months to come, Brooks performed in many communities in Illinois and Iowa, and the newspapers eagerly reported the news, creating a certain balloon mania on the eve of the Civil War. The *Peoria Weekly Republican* promoted Brooks' arrival with a large display advertisement on July 25, 1856 [5]: "The citizens of Peoria and surrounding country are respectfully informed that Mr. S. M. Brooks, the great American Aeronaut, who has made more successful voyages throughout the heavens than any other man living, will have the honor of ascending in his Mammoth Balloon, the Hercules, from this city on Thursday, 31 July 1856." Due to weather the event was postponed (Fig. 2.1).

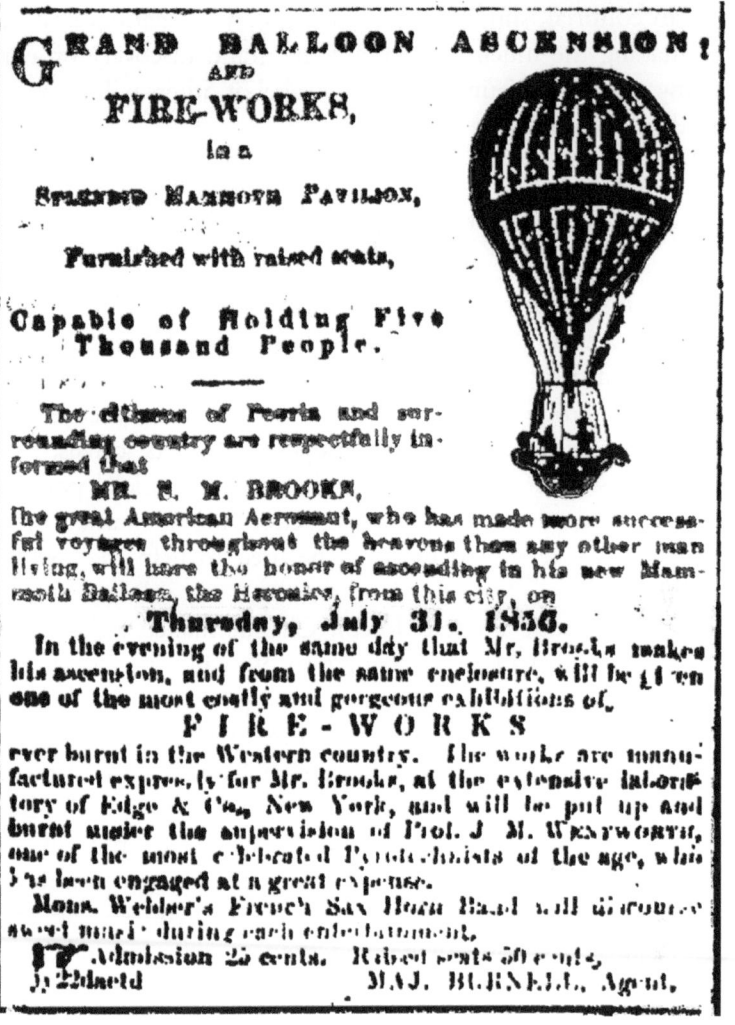

Fig. 2.1 Advertisement for the balloon ascension by Silas M. Brooks in Peoria, Illinois. *Peoria Weekly Republican*, July 25, 1856

One week later, Brooks set-up a large circus tent, lectured on ballooning prior to inflating his balloon and when all was ready, the ropes were released, and the Professor majestically ascended. The balloon landed about four miles east of town and was brought back to the city without much fanfare [6].

Watching Brooks take off was special, but to Chanute, the clear-thinking engineer, the balloon was simply drifting with the wind; the operator could not change the direction, speed or altitude. There had to be a better way of traveling through the air. A few months later the *Scientific American* published an editorial, simply stating "Who, or what nation, shall have the honor of making the much-needed and the grand discovery of a true and useful 'flying machine'? Young America, awake! or others will reap the glory" [7]. Should he become more involved?

2.2 The Power of Nature

The Royal Institution of Civil Engineers was formed in 1828 in England, and Chanute ordered copies of several publications. Here he read what Thomas Tredgold wrote in their charter, "Civil Engineering is the art of directing the Great Sources of Power in Nature for the use and convenience of man as applied in the construction of roads, bridges, aqueducts, canals, river navigation and docks, and in the art of navigation by artificial power for the purposes of commerce, and in the construction and adaptation of machinery" [8]. Thus, nature should or could be used to help the civil engineer accomplish his tasks.

Knowing of his son's interest, Joseph Chanut mailed a 32-page French pamphlet on the history of aerial locomotion [9], where the author described not only the many attempts to fly during the past centuries, but also how inventors envisioned using the power of nature to achieve manflight.

Having moved to Chicago in early 1864, Chanute received the *Chicago Tribune* regularly, as he needed to keep up with the latest news. Reading about Camp Douglas, the "80-acres of hell" prisoner-of-war camp on the southern edge of town, he agreed that improvements in the living conditions should be made. To make life a bit easier, the prisoners had requested among others pine sticks, paper, paste and twine, and soon a "half-score of six-cornered kites were ready to take a sail." The camp commander remembered his days of flying kites and saw no harm in this activity. The kites flew again the next day, but the following evening, the string of some kites happened to break, and the wind carried the kites far away from the camp. No suspicion yet. The next evening the kites floated again and the strings broke again. Now the commander ordered a guard to follow the kites, who found them in a field with several letters attached to the tail section, addressed to a Chicago "copperhead", a confederate sympathizer, who was to forward the letters to their final destinations. The newspaper summarized, that "… the rebs will have to devise a more ingenious postal scheme than this" [10].

To deliver mail through the air using a kite was a clever idea. Could a similar device be used to achieve manflight? Chanute submitted his thoughts on the "True Theory of Flying" to the *Scientific American*, signed only with

"C" [11]. "What is analogous to the driving power of the wings of a bird? What power is there in nature which we can lay hold on and turn to our uses, which need no cumbersome boiler, no heavy fuel, no complicated weighty machinery, each are death to the flying machine theory." Next, he discussed the available agents to propel machines. "There was steam, but it was inadequate to carry the necessary weight of machinery. Electricity was probably weaker than steam in its present mode of use. Compressed air may do for short flights, but there were other agents sufficiently powerful, which could drive a flying machine with abundant force, like gunpowder or dynamite. However, how this power could be applied or regulated needed to be ascertained by ingenious and educated engineers." He concluded, "there were a few brave spirits spending time and money in endeavoring to obtain the art of flying, which has so long bid defiance to human skill." Could he be the brave engineer to solve the problem?

A decade later, at the mid-December 1880 ASCE meeting, Chanute's friend Charles Shaler Smith presented a paper on "Wind pressure on bridges." The two engineers had travelled to St. Louis a few hours after a major storm had gone through the area in spring 1871. Participating in the discussion Chanute wondered: "It is somewhat curious that nobody has yet called attention to the influence which the form of bridge members has upon the resulting pressure. We all know that a convex surface presents less and a concave surface more resistance than a flat plane to the wind. I dare say that some of us have ascertained this fact experimentally with an umbrella, but I do not know that this knowledge has ever been applied to the calculation of the wind strains on bridges. The umbrella offers a resistance of 76%, if turned with its convex side to the wind. I believe that Sir George Cayley found that a sphere offered a resistance of but 42%, that of a flat plane taken through its mid-section, and as the surface of the semi-sphere is equal to two great circles, the average of the pressure in that case per square foot is but 21% of what it would be upon a flat surface of equal area. If we apply this knowledge to bridge members, we see at once that convex objects, like round ties and round closed columns, will offer less resistance than plane surfaces" [12] (Fig. 2.2).

At the second American Society of Mechanical Engineers (ASME) convention in November 1881, President Robert Thurston declared that the world was awaiting the appearance of three inventors, greater than any who have gone before. "The first is he who will show us how to produce the electric current; the second is the man who will teach us to reproduce the beautiful light of the glow-worm and the firefly, a light without heat, the production of which means the utilization of energy without a waste." And the third great

Fig. 2.2 We all know that a convex surface presents less and a concave surface more resistance than a flat plane to the wind, but I do not know that this knowledge has ever been applied to the calculation of the wind strains on bridges

genius of inventors would be the man who would fulfill Darwin's dream and build a great chariot to fly on wide-waving wings through the fields of air. "We have not yet learned to fly like Daedalus, and thus have escaped the fate of Icarus, but the navigation of the air is on the point of real advancement. We are apparently approaching this far goal, with progress being observed year by year, and there is no department of engineering in which the art of the mechanic has opportunity for greater achievement" [13].

Chanute could not have agreed more with his friend. Years later Chanute explained: "I gathered from time to time as was to be found on the subject, and added thereto such speculations as suggested themselves. After a while, this grew absorbing and interfered with regular duties." But any report on flying machines, published in newspapers or periodicals, was clipped and filed in unlabeled wooden storage boxes, placed neatly in his bookcases. No "house cleaning cyclone" could have guessed the contents in these boxes (Fig. 2.3).

Early in 1883, the now 51-year-old Chanute resigned his high-paying position as chief engineer and assistant general superintendent at the New York, Lake Erie and Western Railroad Company to start a life without being ordered hither and thither. He moved his family back to Kansas City and opened an engineering consulting business. Leaving the constraints of a full-time job should provide more time to do the things that he wanted to accomplish before stepping out. No, he did not experience a mental collapse, he simply considered doing the unthinkable: he wanted to help solve man's age-old dream to fly with mechanical wings. "This presented the attraction of an unsolved problem, which did not seem as visionary as that of perpetual motion. Birds gave daily proof that flying could be done" [14]. And if successful, this would be the greatest accomplishment in his life, and

Fig. 2.3 Storage box with Lawrence Hargrave's material sent to Chanute. University of Chicago Library (Cr 929.13)

he would achieve a goal that had captivated dreamers and experimenters throughout recorded history.

2.3   A New Branch of Scientific Inquiry Was Born—The 1886 Buffalo Meeting

Being altruistic, Chanute favored the free pooling of knowledge to solve problems. Believing that manflight was solvable, he wanted members of the engineering profession to accept its study as reputable. More people needed to become interested in aeronautics, as "the greater the number of minds that can be brought to bear upon a particular problem, the greater is the chance of early success."

The American Association for the Advancement of Science (AAAS) was resurrected after the Civil War in New York City, and Robert Thurston agreed to chair the first gathering of the "Mechanical Science and Engineering" or Section D, in Philadelphia in 1884. Knowing of Chanute's interests, he urged him to join and become acquainted with other members of the scientific community.

Two years later, Chanute agreed to chair Section D. As the previous meeting did not attract many attendees, he now puzzled what might be of interest to members. Having read a communication in the *Report of the Aeronautical Society of Great Britain* by an Israel Lancaster from Chicago on the flying habits of soaring birds [15], he wondered if knowing more about birds achieving soaring flight could give an insight into the potential for humans to fly. Traveling from New York back home to Kansas City, Chanute arranged to meet Lancaster in Chicago and listened to his explanation of how soaring birds use air currents to remain airborne. Following up a few weeks later, he wrote: "I would be glad to have you present a descriptive account of what you did and what you saw concerning the flight of birds—together with your reasons for believing that no muscular effort was expended in soaring. What you said to me interested me greatly and would doubtless produce the same effect on the gentlemen who are to meet in Buffalo" [15]. Lancaster agreed, and Chanute submitted his name for membership in the AAAS. He also mailed him guidelines on how to prepare a talk for an academic audience.

As chair of Section D, Chanute discussed progress made in various scientific areas and reminded his audience, "how dependent men are upon each other in developing abstract scientific knowledge or discoveries into useful appliances or inventions, and how attendance upon these meetings may result not only in enlarging knowledge, but in valuable developments in the practical affairs of life. Nothing is more remarkable in the field of invention, than the multitude of minds and facts, which are required to perfect even the simplest machine. At times generation after generation of inventors wear their lives out, before a needed machine becomes an accomplished success ... The history of almost all important inventions exhibits the same characteristics of gradual development through the efforts of many inventors, and impresses us with the vast amount of toil and study necessary for success." And, "inasmuch as I have noticed that whenever an imaginative writer pretends to give an account of future mechanical achievements, the first thing described is always a flying machine" [16].

The next afternoon, attendees crowded the lecture hall to hear Lancaster talk on "The Soaring of Birds." He described his models and his observations of birds gaining altitude by using the upward moving air. Next, Lancaster discussed his models, or "effigies," which acted like a soaring bird, floating steadily in an ordinary breeze. He also stated that his effigy was picked up by the wind and without any internal power could advance against the wind without falling [17]. Lancaster became the hero of the hour with reporters besieging him for interviews and college professors following him to learn more.

The *Science* editor reported somewhat sarcastically: "As a set-off to the papers of certain value, and perhaps for purposes of recreation, the section listened to a paper detailing observations and experiments, mixed up with some remarkable theories upon the flight of birds." He commented in closing that "we were scarcely prepared to hear that gravitating force as a continuous motive power could be accomplished by so simple a device as a bird's wing, rough in one direction and smooth in the other—but the section no doubt needed recreation" [18].

The report in the *Scientific American* was more favorable. Here only four papers (out of a total of 252 papers presented) were highlighted, with Lancaster's paper being discussed first. "A rather fanciful and highly wrought, yet interesting and suggestive paper was read by Mr. Lancaster of Chicago who has for many years made a special study of the flight of birds." The reporter then described the phenomenon of soaring flight and stated, "Considering the extraordinary claims that had been made, the general feeling was that the gentleman should have gratified the association by launching at least a single little model" [19].

And the reporter of the *Buffalo Courier* wrote with admiration, "Yet how gratifying would it be to believe that he had discovered the germ of a new principle of aerostatics destined to enable men to fly through the air with as much steering accuracy as they are able to command while navigating the waters" [20].

When asked, Lancaster agreed to demonstrate his effigy the next day. Seeing the "soaring effigy" was disappointing, it was a simple wooden frame with paper covered flat wings (Fig. 2.4).

The next day, instead of demonstrating his model of soaring birds or "effigy," the Illinois farmer tried to explain the lifting action of the wind mathematically and caused a row. Chanute was not present, and therefore could not help [21]. De Volson Wood from the University of Michigan made

Fig. 2.4 Israel Lancaster's "soaring effigy" as drawn by De Volson Wood. From *Scientific American Supplement*, October 16, 1886

some calculations on the spot to prove Lancaster's theories erroneous [22]. "It is impossible for the plane to float against the breeze, in accordance with Mr. Lancaster's theory. It also follows that a weight cannot be maintained statically on this principle, much less move horizontally, but must descend, like the boy's kite when the string is broken. We venture to assert that if Mr. Lancaster's effigies moved in a horizontal plane for hours, they were supported by the buoyancy of the air, on the principle of the balloon, and that they never moved against the breeze unaided." Thurston was not convinced that the presenter's data was erroneous, but he thought a mechanical engineer should do more research. Another attendee, Samuel Langley, was fascinated by the heated discussion and wondered what mechanical power would be required to sustain a given weight in the air and make it advance at a given speed. "This inquiry had to precede any attempt at mechanical flight, which was the very remote aim in his efforts" [23], Chanute recalled years later.

Lancaster's controversial talk stirred more discussion than any other talk at that AAAS meeting in Buffalo, with one member offering a prize of $50 for the best paper on the subject at the next meeting. Chanute had achieved his goal; this meeting is often considered the beginning of aeronautics as a new and challenging scientific development in the United States.

2.4   Air Machines to Come—The Beginning

Being in his late 50s, accomplished and professionally secure, and not fearing ridicule for advocating an absurd or questionable topic, Chanute was ready to devote his full attention to researching manflight, especially since members of his own family did not object. He was sure that he could not solve the problem alone, but he firmly believed that progress could be made if others collaborated.

Attending a monthly ASCE meeting in New York, a fellow member mentioned Jules Verne's latest book, *The Clipper of the Clouds* [24]. Chanute bought the science fiction novel and read it traveling back to Kansas City. Verne believed that mankind should not master the skies using a balloon; a successful flying machine should be heavier than the air surrounding it. The author, or his brilliant but mysterious engineer Robur, also thought that it would not be necessary to copy nature in detail, even though she never made mistakes. "Locomotives are not copied from the hare, nor are ships copied from the fish. To the first we have put wheels, which are not legs; to the second we have put screws, which are not fins. Besides, what is this mechanical movement in the flight of birds, whose action is so complex?"

To conceive aerial locomotion, Chanute had at least two hurdles to tackle: one was the negative attitude of the general public toward anyone showing an interest in manflight, while the other hurdle, to develop a flying machine capable of carrying a person, appeared to be rather complicated. To start, he took the heuristic approach and defined a series of questions to study: What are the basic aerodynamic requirements? What should the wing look like? Does the air flow over or under a wing? What are the forces acting on wing surfaces as they cut through the air? Using uplift from the slopes of the dunes or mountains would allow longer flights to test and gain experience in operating a flying machine. Creative thinking, imagination, careful observation and experimenting should yield answers. He then built a "dirigible parachute" with flat wings, even though the true experts in gliding, the soaring birds, perform their flying with wings, more or less convex on top and concave beneath. His model, carrying small bricks as "passenger," was dropped from the top floor of his three-story home at Ritchie Place, Chicago, in the early morning when only the milkman was about. They usually glided downward steadily in all sorts of breezes, but the angle of descent was much steeper than that of birds [25]. Experimenting with these models did not provide answers.

Always keeping his eyes open for aeronautical news, Chanute saw in the January 6, 1888, issue of the *Railroad Gazette* [26] a letter by Charles Latimer, who had received a large number of flying machine drawings. A member of the editorial staff commented, "The *Railroad Gazette* tries to keep a sharp eye on those agencies of transportation which move on the surface of the land, or go to but little depth below it, with an occasional glance at the Erie Canal and the great deep, but it has not yet tried to follow transportation into the heavens above. Mr. Latimer's question is therefore referred to those of its readers who have." Chanute responded to Latimer directly: "Noticed in last week's *Railroad Gazette* that you have fallen heir to some papers on aerial navigation, and the editor pokes mild fun at you. Now this is an old hobby of mine, but I say very little of it, partly because the scoffers would laugh, and partly because I came to the conclusion that aerial navigation was only practicable by the use of a motive power a good deal lighter in proportion to its energy than any motor known. I have thought of devoting some leisure time to inquire whether the invention was enough advanced to make success possible" [27].

In his response, Latimer explained that Eugene F. Falconnet's widow from Nashville, Tennessee, had hoped for some financial benefit from her husband's airship patents. After studying the patents, Chanute wrote to the widow [28]: "There are scores of shapes which might be driven through the air, if only we could have power without weight. My belief is that Mr. Falconnet did not solve the problem of aerial navigation, but he may have covered in his patents some devices which other inventors will be compelled to use to achieve success." He regretted that he could not help make her deceased husband's inventions financially profitable.

Later in December 1888, the *Chicago Tribune* published a report on "A Ship for the Skies" [29], describing Peter Campbell's recently patented airship. Chanute noticed several errors and explained to the editor that Campbell was fortunate to have worked on a perfectly calm day to display his airship. "While he has probably made a great step in advance in designing appliances to improve the steering of balloons, he will not accomplish a real success until he develops a motive power with much greater energy in proportion to its weight than any with which we are now acquainted." In his cover letter, Chanute asked not to include his name if this communication was printed; his comments, signed with "Engineer," were published on December 28, 1888 [30].

His long-time friend Matthias Forney had sold his share in the *Railroad Gazette* at about the same time that Chanute had resigned from the Erie Railway. But retirement in New York City was not as envisioned. The now 51-year-old bachelor decided to go back into the publishing business, purchased the struggling *American Railroad Journal* and the eclectic Van Nostrand's *Engineering Magazine*, combined the two and introduced on January 1, 1887, *The Railroad and Engineering Journal*, an illustrated 48-page monthly, having more the character of a magazine than a trade journal. To grow the readership, the journalist looked for articles on every aspect of engineering, and aeronautics surely fit into his scheme of "non-standard" articles. As Chanute had offered his help to make the new enterprise a financial success, Forney shared his thoughts and Chanute replied dryly: "Why have you put temptation in my way, by proposing such an excellent subject for a book, and such a good plan and propose a piece of work that I should like very much to do. I will not promise to undertake it, for I am a slow literary workman, and I might not be able to get it done as soon as you wish, but I will talk it over with you when next in New York" [31].

Knowing that he would be in France, Chanute submitted his first aeronautical contribution for the April 1889 issue. His article, "The Latest Rapid Transit Scheme," discussed Reuben Jasper Spalding's patented "Machine for Navigating the Air," a purportedly easy to operate apparatus for an imaginary way of travel by the most direct route. Chanute pointed out that the patent drawing showed the aeronaut wearing a leather jacket with feathered wings, suspended from a balloon. "The passenger just ready for flight with his broad pinions spread, and attention may be called to the magnificent and stately appearance presented by these pinions and by the steering apparatus, to speak politely." Spalding claimed that the balloon was not a necessity, as the wings were to operate with the same effect as the wings of an eagle. To this Chanute added his lighthearted comment, "which might be a dangerous admission were the eagle in a position to file objections in the Patent Office" [32].

2.5 The 1889 International Aeronautical Congress in Paris

Having completed his latest consulting job with the Atchison, Topeka & Santa Fe Railroad, designing and erecting all the major bridges on the 500-mile-long Kansas City—Chicago "Airline", Chanute anticipated no new assignment. Reading of an aeronautical congress to be held at the World's Fair in Paris convinced him to attend and meet like-minded enthusiasts. He was sure that the neighbors would never hear about him being part of such a congress, and no one would point fingers at members of his family and say: "There is the family of the fool who believes he can fly."

Serving as the U.S. delegate, Chanute offered to present a paper on the reactions of air flowing against various-shaped objects, a topic close to his heart as it had effected the speeding up of railroad transportation in the past. He ceremoniously removed the red tape from his clipping collection, bundled up fourteen years earlier, and started to study. And he contacted some friends, hoping that they would share their knowledge without making fun of him.

One of the first letters went to Charles Hastings from Kansas City asking about the best size of a flying machine, the weight of the sustaining surfaces, the resistance and power requirements, takeoff and landing, and whether sustained flight could be achieved by soaring [33]. He described in his nine-page detailed letter how he envisioned flight to be achieved and asked Hastings to sign an appended statement that he had read the letter. "This may prove useful in establishing a date, should I ever folly this subject further."

The letter to Thurston, now director of Sibley College of Mechanical Engineering, Cornell University, showed another approach: "I trust you will not think me a lunatic if I say that I have had in mind for years to devote part of my leisure time to the opening up of an inquiry, whether man can ever hope to fly through the air! A single person cannot carry out such an investigation, and I am looking for a number of other visionary students of science who will give thought to the subject and correspond with me." Continuing his letter, he inquired, "What are the pressure and the resolution of forces when moving air strikes a plane at an angle? Can you refer me to tables of experiments upon air pressures on inclined planes at various angles?" [34] Thurston had some ideas, but also suggested to contact Langley. With thoughtful comments from his friends and discovering data on moving bodies in fluids and velocities in sails, his paper made good progress.

Discussing his travel arrangements to Paris, Chanute's wife Annie was not interested, she preferred to spend time with her family in Peoria. Their youngest daughter, eighteen-year-old Nina, wanted to be with her mother, but his oldest daughter, thirty-year-old Alice, still emotional about her ugly divorce and then losing her young son, and his middle daughter, twenty-five-year-old Lizzie, were eager to join their father on their first trip to Europe.

The proud father with his two grown daughters left Chicago on March 23 on the Lakeshore Limited, traveling at fifty miles per hour through the night, and arriving in New York City the next morning just before ten o'clock. To make the trip across the Atlantic more sociable, the Chanutes joined a small group of civil engineers sailing on the "City of Chicago." This steamship, owned by the Inman Line, had four masts rigged for sails, and a single-screw steam engine that propelled her at a speed of fourteen knots. "Daughters got sea-sick, but not I," were his diary notes. They arrived in Liverpool on April 6 and spent one week with ICE members before continuing their trip to France.

Spending springtime in Paris was like a dream come true for the two Americans; Alice and Lizzie enjoyed strolling up and down the boulevards where the *Parisiens* met the *Américains*. Walking through the city, one could see the Eiffel tower from everywhere. Some Parisians had initially protested against building this "useless tower," but the *Tour de Eiffel* quickly became a focal point of the city, an epicenter of technological progress, modernity and national prestige. Chanute agreed with fellow engineers that it was a thoroughly well-designed piece of engineering.

The World's Fair opened its doors on May 6, and the Chanutes were part of the throng. Six weeks later, a group of 275 American engineers arrived at Calais. The greeting committee, which included Chanute and his daughters, escorted the group on a special train to Paris. To make the American engineers feel welcome, Gustave A. Eiffel, President of the French Civil Engineers Society, introduced the visitors to the Prefect of the Seine and then gave a guided tour of his tower. The view at Paris from almost 1,000 feet up high was spectacular.

Hearing of Chanute's talk, Eiffel was eager to discuss this topic with his American visitor. They both had gathered similar knowledge in their work with railroads; they both had designed spectacular iron railroad bridges and knew that gravity and air resistance played an important part in speeding up train travel (Fig. 2.5).

Eiffel then invited Chanute to his laboratory on the second level of the tower, or 377 feet above ground, and demonstrated how he measured air resistance of various shaped objects using different instruments [35]. Everything was compact in this mini-apartment and laboratory, outfitted for maximum comfort and efficiency, equipped with all the necessary tools for scientific experiments, but inaccessible to anyone who was not invited. Or as Verne stated, the laboratory was like "floating in perfect space, in the midst of perfect silence" [36], a place that could have been the cabin of a real airship. The two engineers then measured the effect of drag on objects dropped from the three different levels of the tower, with the stop watch in hand, an unforgettable experience for Chanute.

During the initial stages of organizing the aeronautical congress, President Jules Janssen and Secretary Abel Hureau de Villeneuve decided that all presenters should use the same wording to avoid confusion among the preparers, attendees and the readers. They clarified that the word "*aéroplane*" was to

Fig. 2.5 Gustave Eiffel's aerodynamic laboratory on the second level of the Eiffel tower. *Travaux Scientifiques executes a la Tour de trois cents metres, de 1889 a 1900*

describe all heavier-than-air craft flying horizontally like the birds, and the word "*aerostat*" should be used for lighter-than-air craft or balloons.

The *Congrès International d'Aéronautique* opened on July 31, with almost one hundred researchers in attendance; Samuel Langley was the only other American who had registered. The first speaker of the morning session on August 1 was Étienne-Jules Marey, discussing his twenty-year study of bird flight, which provided a good base of the state of knowledge at that time. To record birds in flight, Marey had designed a special camera that could take up to fifty photos in one second. Next, he developed a technique to separate the overlapping images to show the actual wing movement of the bird. Seeing the spectacular slides, Chanute wondered if the just introduced Kodak camera could take similar photos.

After a brief discussion it was Chanute's turn. At first, he considered reading his paper in the French language, but he did not want to stumble, so he asked to read his paper on the resistance of air to inclined planes in motion in English [37]. The request was granted; he was told that everyone knew English. To begin his talk Chanute explained that Pénaud had stated in 1876, that the laws of fluid pressures, as laid down by Newton, did not agree with experiments using inclined planes. He theorized that when air is compressed, the molecules press not only so that the cleavage of the air, so to speak, flares out at an angle from the plane. He believed that a further action takes place behind the plane in consequence of the partial vacuum produced by the air current rushing by. New and interesting information came up during the discussion that followed.

Stéphane Drzewiecki was the next speaker. He recommended building the flying machine as a simple aeroplane with outstretched wings, similar to a large soaring bird, placed on a framework with wheels on which it could roll until it acquired sufficient speed to raise itself from the ground. A light motor to drive a screw (or propeller) would then support the apparatus in the air. He also told his listeners that lift and drift had to be recalculated for each curved wing, as the wing shape was highly important. Success and failure of any proposed flying machine would depend upon the sustaining effect (assuming that a light motor could be found) of a properly curved wing surface to achieve sufficient lift [38].

The four-day conference was a good learning experience; Chanute felt privileged to meet so many experimenters, all willing to share their knowledge and experimental data. Listening to the various comments, however, he wondered if an intelligent engineer could solve aerial navigation. "There were numerous formulae, promoted by various physicists, but these gave such discordant results that arrangements were being proposed in France to try an entire set of new experiments with air currents to be produced by an enormous air blower" [39]. He believed that a government-sponsored facility should be initiated for scientific testing of flat, concave and convex surfaces to determine the most efficient wing shapes to help researchers comprehend what was needed to invent an aeroplane.

During the ocean voyage back home, Chanute incorporated his newly acquired knowledge into his next paper and submitted "Resistance of Air to Inclined Planes in Motion" as an abstract [40] to the AAAS meeting in Toronto.

While Chanute was in France, Chicago politicians proposed hosting the next World's Fair in their hometown. As it was to illustrate progress in arts, industries, inventions and transportation, Mayor DeWitt Cregier asked WSE members to join the planning committee [41]. Chanute arrived back in Chicago just in time for the next WSE monthly meeting, where he discussed the efficiency of French organizers in creating the World's Fair. Having only two years to create an event grander than the one made by the French in four years, the mayor asked Edward Jeffery, the former manager of the Illinois Central Railroad, and Chanute, the independent civil engineer, to go to Paris and gather details about their Fair, offering an all-expense-paid trip plus $1,000 a month for up to three months. Even though Chanute had just returned from France, he accepted, but added a note to the contract that the money should go to his wife if something unforeseen would happen to him [42].

Being keenly aware of traveling speed and luxury accommodation, Chanute suggested buying tickets on the "*City of Paris*," which had just earned the "Blue Riband" for the fastest westbound travel. Jeffery and Chanute, with Charles Dawley and Charles Schlacks as stenographer and secretary, arrived in Paris on October 15, two weeks prior to the Fair closing. During the teardown workers explained the various aspects of their highly successful event, allowing the four Chicagoans to study everything for their own planning.

Eiffel invited the Americans to join the next Civil Engineers monthly meeting where the good and bad of the just closed fair was to be discussed. Speaking French, Chanute explained that they were in Paris to study what was done, so that they could share the information with organizers in Chicago. The attending correspondent from *The World's* Paris office reported the next day. "Mr. Chanute of Chicago did not make a big hit with his lecture on 'Why Chicago is the Place for the next World's Fair.' Nobody knew that it was going to take place, and consequently, the lecture, in spite of Eiffel's patronage, was a complete frost. Chanute seems to take his mission very much *au serieux* and has got printed on his cards the words, 'General Manager of the World's Fair of 1892 at Chicago.' Nobody here, however, believes that there is any chance of Chicago getting the show" [43].

The four Americans left Liverpool on December 7 on the "*Umbria*" of the Cunard Line, but due to the weather, the crossing of the Atlantic took nine days instead of seven, which allowed more time to prepare their report for the Chicago mayor.

2.6   The Student of Bird Flight from Egypt

Étienne-Jules Marey had arranged an "Aviators Dinner" just prior to the opening of the 1889 Aeronautical Congress; this is where Chanute heard of Louis-Pierre Mouillard and his book *L'Empire de l'Air*, published in 1881. He purchased the book right away and started reading. Just like Verne, Mouillard suggested to use a fixed-wing craft with cambered bird-like wings to gain the skills needed to pilot an aeroplane. Chanute decided to translate the 280-page book so the information could be shared with English speaking enthusiasts.

In the preface, Mouillard explained, "There were two roads to possible success, the one broad, beautiful, smooth, and bordered with flowers, but after all leading to no result; it was that of the aerostation, or of balloons lighter than the air. The other way was contrary wise, a rough, narrow, rugged path, bristling with difficulties, but still leading to something; it was that of aviation, of rapid transit by machines heavier than the air. Most of the would-be inventors have taken the easy road, and from the height they have gained, pityingly look down upon the unfortunate aviators still floundering in the quagmire, with little thought that they might have to come down to this same quagmire in order to get somewhere. Oh Blind humanity! Open thy eyes and thou shalt see millions of birds and myriads of insects cleaving the atmosphere. All these creatures are whirling through the air without the slightest float. Many of them are gliding therein, without losing height, hour after hour, on pulseless wings without fatigue; and after beholding this demonstration given by the source of all knowledge, thou wilt acknowledge that aviation is the path to be followed. It is therefore the apparatus 'heavier than the air' which I propose to study. I expect to have as a guide and as a support that potent creator of all prodigies, Nature herself" [44]. Langley published a shortened translation in the *Annual Report of the Smithsonian Institution* for 1892, and John Brisben Walker published an extract in his *Cosmopolitan* Magazine [45].

The two-year younger **Louis-Pierre Mouillard** (1834–1897) had studied art and painting; when his father died, he took over the Algiers branch of the family business. He sold the farm in 1865 and moved to Cairo, Egypt, to teach at the military school. Chanute wrote his first letter to Mouillard in March 1890, and shared his theories on manflight half a year later [46]: "It is rather easy to get the transverse equilibrium, however it is much harder to get longitudinal equilibrium without an increase of resistance. The shape of the bird and the necessity to fold its wings give it an unstable equilibrium and for these reasons, the bird became an acrobat. I should not even dream of imitating the bird, but I believe that it is possible to apply the same principles in a different manner, and perhaps the theory will indicate the means with which to obtain a stable equilibrium. A bird is able to act against the motion of pressure simply by varying its center of gravity. But take a large machine with a 5-meter-wide wing. The center of pressure may vary by 1½ meters between 45° and 1° of inclination but there is no chance to vary the center of gravity quickly enough over this distance." In his response, Mouillard explained his theories: "When the bird is about to turn, the wing tip twists … The feathers obstruct the motion; they stop it, forming a long lever on this point. This causes a variation of motion and a rapid change of direction".

Mouillard mailed the manuscript of his second book, *Le Vol sans Battements* (or "Flight without Flapping") to Chanute, where he also described his earlier flying experiences. Chanute translated and published his story in the January 1893 *Railroad Journal*:

"It was on my farm in Algeria that I experimented with my apparatus … Nearby there was a wagon road, raised some 5 ft. above the plain. I thought that I might try it armed with my aeroplane; I took a good run across the road and jumped at the ditch as usual. But, oh horrors! Once across the ditch my feet did not come down to earth; I was gliding on the air and making vain efforts to land, for my aeroplane had set out on a cruise. I dangled only one foot from the soil, skimming along without the power to stop. At last, my feet touched the earth. I fell forward on my hands, broke one wing, and all was over; but goodness, how frightened I had been! I then measured the distance between my toe marks, and found it to be 138 ft. … I cannot say that on this occasion I appreciated the delights of traveling in the air, and yet never will I forget the strange sensations produced by this gliding" [48].

Continuing their correspondence, Mouillard wrote in early summer 1892 that he wanted to build an aeroplane to do maneuvers, not just simple glides. Wenham had discussed similar ideas in his 1866 paper: [49] "Two propellers were to be attached, turning on spindles just above the back. They are kept drawn up by a light spring, and pulled down by cords or chains ... By working this cross-piece with the feet, motion will be communicated to the propellers, and by giving a longer stroke with one foot than the other, a greater extent of motion will be given to the corresponding propeller, thus enabling the machine to turn, just as oars are worked in a rowing boat." Thus, to turn a flying machine in the air, more thrust needed to be generated on one side than on the other, and Chanute agreed that the flying machine would bank just like the locomotive going into a curve on a raised track.

As these ideas could lead to controlled flight, Chanute suggested that they compile the claims; he then submitted the claims with the application fee to his patent lawyer in September 1892. The steering of the machine was envisioned as: "In order to provide for the horizontal steering of the apparatus, that is, the guiding it to the right or left, I substitute for the ordinary rudder a novel and more effective arrangement. A portion J' of the fabric at the rear of each wing is free from the frame at its outer edge and at the sides. A pull upon one of these handles causes the portion J' to curve downward, and thus catch the air, increasing the resistance upon that side of the apparatus and causing it to turn in that direction ... For turning in the horizontal plane, the resistance on one wing is increased, so that it tends to travel at a less velocity than the other. This is done by warping the apparatus provided for the purpose." Creating drag worked as an effective brake in a two-dimensional system, but banking in two dimensions, as the locomotive on the track, is different from banking in three dimensions, like a flying machine in the air. Neither Chanute nor Mouillard considered the lifting effect, which in practice would bank the machine in the opposite direction from what they intended. It took a few more years to develop an efficient turning of an aircraft (Fig. 2.6).

Soon after receiving the claims, George Whittlesey, Chanute's patent attorney in Washington, reported that the examiner declared "Mouillard's device is incapable of ascension," because there was no gas field. Chanute responded quickly that everyone has seen a kite rise, it does not use a gas field and neither did their machine. "Such an objection indicates that we shall have more serious trouble when we come to the important part of getting the claims allowed" [50]. Chanute included $100 for expenses so that Whittlesey could discuss their claims with the examiner, or even the examiner-in-chief.

A year later, the United States Patent Office mailed a circular letter to every patent attorney who had previously submitted an application for flying machine patents. "On taking up this case for examination it is found that the invention disclosed thereby is, as a whole, incapable of practical operation, since without the assistance of a gas field or equivalent, the device will be incapable of ascension. In other words, it is not useful within the meaning of

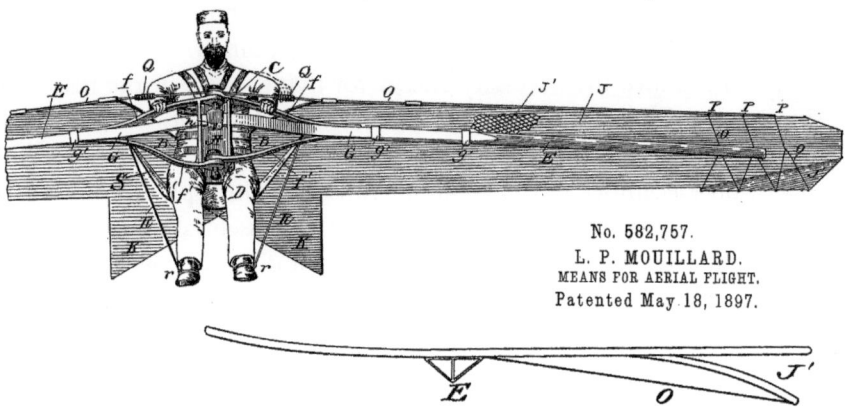

No. 582,757.
L. P. MOUILLARD.
MEANS FOR AERIAL FLIGHT.
Patented May 18, 1897.

Fig. 2.6 Patent by Louis-Pierre Mouillard, of Cairo, Egypt, assignor of one-half to Octave Chanute, of Chicago, Illinois, *Means for Aerial Flight*, showing the warping of the wings for turning an aeroplane

the law. No evidence will be considered sufficient to demonstrate the operativeness of the device and warrant the withdrawal of the objection made by the Office, except a working model—that is, one with which it is possible to actually ascend and direct the course of flight at will." Thus, it was almost useless for an inventor to apply for a patent until he had actually succeeded in flying. "It is barely possible that such a stand as this in regard to certain inventions would not be amiss as there are more impracticable ideas patented than the world at large has any idea of. While it may be considered a hardship by flying machine inventors to be debarred from the protection of the Patent Office, yet until they have actually produced something which will accomplish the results claimed, they certainly have no right to overlap the claim of protection on the Patent Office" [51]. Nevertheless, after much debate between the applicants, the patent lawyers and examiners, the Patent Office granted the Mouillard/Chanute patent "Means for Aerial Flight" [52] in May 1897.

As Mouillard needed financial help to build his aeroplane, Chanute mailed a draft for 2,500 francs, just asking for details of the machine and regular progress reports in exchange [53]. Mouillard then puzzled over the actual value of an aeroplane, and Chanute responded laconically, "In answer to your question about the lucrative uses of your machine, if it succeeds, it always seemed to me that profit would consist in selling the machine either to the Government or to the fans who want novelty. I even believe that all your time will be taken up in perfecting the machine and in fighting the imitators. For the present, these are all midsummer dreams" [54]. And Chanute mentioned that he too was working on a flying machine. "You are mistaken if you believe that I use a propulsive apparatus. I believe that human strength is amply sufficient to produce the necessary energy for flying, without the help of an outside energy. Gliding flight is accessible to man, but I also believe that it needs extraordinary skill, which might not be achieved for a long time" [55].

Finally, on December 7, 1895, Mouillard mailed three photos of his aeroplane, which looked clumsy to Chanute. "Still, the essential thing is to know if your invention will give you the two controls of direction and equilibrium. Do not get discouraged too soon, but be careful that you do not get hurt" [56]. A few weeks later Mouillard reported, "We took the machine to the desert and tested the action of the wind upon those surfaces, which at first sight appeared enormous. I released this great monster down a slope of 5% and even though we had a wind from the south with an estimated speed of at least 10 meter per second, no significant rise was obtained. I scoffed at the precautions I took so that the aeroplane might not be carried away at night by some sudden gust of wind. On 3 January there was a wind from the south, cold, with a speed of at least 21 meters. I decided to launch the machine in this fearful stream of air. It took six men to move it. At one moment all six of us were knocked over and caught under the machine. We righted it, joking about its flying whims, and set out carrying it to another slope" [57].

No additional news came from Egypt. One year later, Mouillard wrote that he was paralyzed after suffering a stroke; he died on 8 September 1897.

2.7   Looking, Learning and Communicating

A good friendship had developed between the civil engineer, Chanute, and the six-year younger mechanical engineer Robert Thurston. Both men enjoyed periodic excursions into the border-land of science, with flying being a favorite. "Should the time ever come when the practical difficulties of constructing a light-weight engine can be fully overcome, it is evident that success in aerial navigation will promptly follow. Today, however, man, with all his wonderful powers, is in this field beaten by every bird that flies, and even by so minute an insect as the gnat" [58].

Thurston joined Sibley College of Mechanical Engineering in 1885 and pioneered a balance between mechanical skills and engineering laboratory work on one side, and mathematics and physics on the other. To expand the knowledge base of his students, he invited non-resident lecturers to lecture on various engineering topics. For the 1889/1890 winter term, he invited the best-known aeronautical researchers, Samuel Langley, Alexander Graham

Bell, Walter LeConte Stevens and Chanute, to come to Cornell and lecture on their aeronautical work, stirring interest in this new technological concept.

The fifty-eight-year-old Chanute went to Ithaca on May 2, 1890, and Thurston announced in the college's paper *The Crank*, "Mr. Chanute's lecture will be especially valuable, treating as it does a subject which has usually been considered visionary in the extreme. He had investigated this subject from a theoretical point of view, and had developed the mathematical side of the theory to a practicable degree." Preparing and delivering this lecture was pivotal in Chanute's latest endeavor, but the civil engineer turning scientist thought that enticing inventive young engineering minds to supply fresh ideas should help solve the centuries old problem [59]. In his introduction, Chanute explained the two approaches to successful flight: (1) aeronauts believe that success will come by using an apparatus lighter than the air which it displaces, and (2) aviators point to the birds, believing that the apparatus must be heavier than air. And each group thought the other so wrong as to have no chance of ultimate success. It is generally believed that formal instruction in aeronautical engineering started with this lecture at Sibley College.

One of the attending students was the twenty-eight-year-old Albert Zahm who had transferred to Sibley after earning his master degree at Notre Dame. He recalled the lecture four years later [60], describing Chanute as "a silver-haired gentleman of fifty-eight, full of faith in the art, but apologetic enough for identifying himself with a pursuit so generally condemned. Mr. Chanute began his remarks with a hesitancy amounting almost to reluctance, seeming to entreat the young men not to believe that the study of such a subject was a more than probable indication of failing mental vigor. If he could appear before you this evening it would doubtless be very gratifying to him to realize that an audience of sober scientists and engineers were willing to listen for an hour to hear about the venturesome art of flying through the air."

The Sibley lecture was published as an abstract in the student's paper, *The Crank* [61], but Chanute wanted the information shared with a wider circle of engineers and scientists. He submitted his manuscript to the *Scientific American*; the editor published an extracted discussion on "Motors for Aerial Machines" in the *Supplement* [62] two years later. While waiting for his write-up to appear, Chanute looked for another prestigious journal; he then asked his friend Forney, if he would publish the complete lecture in his *Railroad Journal* with all the drawings, tables and images, which were costly to reproduce. Forney did not see a problem spending money on this topic and published the lecture in five installments starting in July 1890, signed with *Octave Chanute, CE of Chicago*. As a special treat, Forney also reprinted it as a 30-page pamphlet.

The reviewer in the *Railway World* was not sure what to make of this lecture. "Persons who are slow to believe in aerial transit may yet read with interest Mr. Chanute's summary of the ingenious attempts hitherto made at solving the problem." The title in the Chicago *Tribune* was clear: "Progress in Flying Machines. Inventors of such structures no longer thought to be insane" [63]. The editors of *Appleton's Cyclopedia of Applied Mechanics* were quick in inviting Chanute to contribute an entry for their next project, a "*Cyclopedia on Modern Mechanism*," dealing solely with the most useful engineering advances of the past ten years. They believed that readers would be inclined to consider aeronautics creatively and "to follow the poet where he saw the heavens fill with commerce and may enjoy reading an illustrated article on this subject" [64]. A short summary of the lecture appeared first under "Popular Miscellany" in *The Engineering Magazine*, and a full write-up was published in October [65] with several large engravings of flying machines, capturing the imagination of those readers who longed to imitate the birds (Fig. 2.7).

Thinking back to the stunning lantern slides that Marey showed in 1889, Chanute purchased a Kodak, hoping that every family member would use this camera to catch birds in flight. Everyone learned quickly that it was not easy to take a meaningful photo, the birds were either too far away or they moved too rapidly or there were no neighboring objects to judge dimensions! But everyone had fun trying.

At the annual WSE Meeting on January 7, 1891, President L. E. Cooley called on Chanute to say something on aerial navigation. Even though he usually felt comfortable talking about his eccentric hobby in the company of open-minded engineers, this time, in front of 120 members and guests, he hesitated a little. "The subject is not of my choosing, but I have laid myself open to being selected to speak on this topic by discussing it at the last meeting of the Society. I suppose the committee said to themselves, here is a man with a crotchet, we will have some fun with him, for I acknowledge that I do believe that it is not entirely impossible that the century which has seen the development of the steamer and the railway may yet see a successful attempt to navigate the air. But I have several other reasons for responding on this subject tonight, one of them being that it seems appropriate after supper to discuss a subject which proposes to abolish gravity, and another is that I feel sure that if any professional men are entitled to wear wings, they are the civil engineers. I can see that an engineer soaring over the country with a 'Kodak' would have real advantage over the present tedious method of

The Kodak Camera.

ANYBODY CAN USE THE KODAK.

Price, $25.00.

Fig. 2.7 So far as the wants of the engineer or surveyor are concerned, the art of photography can now be learned in a few days' practice, if reasonable attention is given to a few simple details of manipulation. *Proceedings of the Michigan Engineering Society for 1889*

plodding along on the ground, and I also see that for exploring inaccessible places, an engineer fluttering around the edges of a precipice with a barometer strapped to his back, as well as a pair of wings, would have great facilities over the present method of being let down with a rope. The principal difficulty hitherto has been the lack of adequate motive power, or a sufficiently light motor in proportion to its energy to accomplish what birds daily perform, but during the past two months announcements have come from three different parts of the world that lighter motors than any now known to exist are being developed. From France comes the statement that Commandant Renard, in charge of the aeronautical department of the French army, has developed a motor from which he obtains seventy horsepower, weighing but 946 pounds, or a weight of only 13½ pounds per horsepower. From England comes the news that Mr. Maxim, who is celebrated as an inventor of an electric light and the quick-firing gun which fires 100 shots a minute, has invented a motor of 100 horsepower which only weighs 600 pounds. From our own State comes a still more wonderful fairy tale; from Mount Carmel comes the information [E. J. Pennington's airship] that a gas motor has been invented which exerts 100 horsepower and only weighs 250 pounds. If one-half of the story of Mount Carmel be true, and we can obtain 50 horsepower with 250 pounds of weight, or even if a quarter be true, or to speak more accurately and professionally, if only the square root of it be true, and we can get a motor of ten horsepower weighing only 250 pounds, then an enormous step will have been made towards the solution of the problem. Perhaps the present generation will see the beginning of the solution, and our members will come to future annual meetings not only by rail or by water, but in flying machines through the air" [66]. Forney gave a brief report of the meeting in the next *Railroad Journal* with the title "Flight not Improbable" [67].

Late in 1890 Chanute received again the nomination for the ASCE presidency, this time he accepted. The election was uncontested, receiving 690 votes out of 694 cast. Members knew that their society had to emerge out of its organizational rut, and Chanute knew that he needed strong support for the upcoming engineering congress in Chicago during the World's Fair, now scheduled for 1893.

As president of the most prestigious engineering society, Chanute was invited to speak at the American Patent System Centennial Celebration, but he became ill and had to mail his paper to Washington; J. Howard Gore, from Columbia University, read it [1]. To explain how things had changed, Chanute mentioned that one mighty invention, steam, brought a multitude of other inventions along. "Invention has been more active and successful than at any period in the world's history. The sea, the land, and the air are experimented on to gain higher speeds or more economical modes of transit. I know personally of eight or ten perfectly sane men throughout the world who are experimenting with real flying machines."

One of these "perfectly sane men" was the next speaker, Samuel Langley, who ventured to predict, "that the air may probably be made to support engine driven flying machines before the expiration of the present century." After experimenting with flat wing surfaces on a sixty-foot whirling arm, he published his findings as "Experiment in Aerodynamics" [68].

To keep up with current events, Chanute used a subscription service to receive newspaper and magazine clippings on flying machines published around the country. Receiving sample clippings from the Durrant's Press Cutting Agency in London, he signed up to receive 125 cuttings per year on "flying machines," excluding balloons. Receiving material from the *Courrier de la Presse* in Paris, he also signed up. These clippings uncovered the names of researchers in this fledgling new science, and Chanute contacted just about every person mentioned.

Having received many complimentary comments on Chanute's lecture on "Aerial Navigation," Forney now wanted a monthly column on aeronautics. Actually, Chanute had also thought that progress in aeronautics was advancing at a snail's pace, as investigators generally knew nothing about the work done by others. There were three periodicals devoted to aeronautics in France, two in Germany, one in Italy, but none in the United States. Maybe an article series on the progress made with flying machines could provide such a communication link?

Seeing an article in the British *Nature* magazine on the "Soaring of Birds," Chanute abstracted it for the March 1891 *Railroad Journal* and explained that "The problem of soaring is one of the oldest problems in mathematics, but unfortunately the formulae given by mathematicians do not agree with the facts" [69], admitting that he too could not provide a mathematical solution or even give a good explanation of the necessary power to support the bird in the air.

A French magazine discussed Clément Ader who believed that the secret of aerial navigation lay in imitating the flight of large birds. After observing the vultures in flight, Ader designed and built a machine, driven by a light steam engine, the *Éole*, which he flew for about 150 feet on October 9, 1890. According to historian Charles Gibbs-Smith, this was the first tentative flight of a powered, man-carrying aeroplane in history [70], but Ader did not wish to share details with the public. Naturally, this reticence excited curiosity; almost eight months later, a French paper reported: "Nobody has seen anything, nobody knows anything, but *L'Illustration* has its friends everywhere. One of them was hunting lately in the environs of Paris, when he caught a glimpse through the leaves of a strange object resembling an enormous bird of bluish hue." The report ended with a simple statement typical of the times: "Therefore, for the present, and until we have witnessed a convincing experiment, at which we shall have seen with our own eyes the generator of the power employed, we shall remain skeptics." Chanute shared the opinion of the French reporter, translated the article and submitted it to Forney for the September 1891 issue.

The October 1891 issue of the *Railroad Journal* featured Chanute's first monthly article on the progress of flying machines [71]. Forney gave this new series special editorial mention: "The navigation of the air has heretofore been hardly considered a practical question by most men, chiefly from the reason that those who have attempted it have generally been persons without sufficient theoretical or practical knowledge to meet the conditions involved. It has now been undertaken, however, by another class of men, who are thoroughly equipped for dealing with the problems involved. Experiments conducted by such authorities as Professor Langley of the Smithsonian Institution and Mr. Chanute, President of the American Society of Civil Engineers, promise some valuable results" [67].

In his introduction Chanute explained that he would focus on the flying machine, or heavier-than-air apparatus, which derive support from and progress through the air like the birds. "The first inquiry in the reader's mind will probably be, whether we know just how birds fly and what power they consume. The answer must unfortunately be that we as yet know very little about it. Here is a phenomenon going on daily under our eyes, and it has not been reduced to the way of mathematical law … We know comparatively little of the laws and principles which govern air resistances and reactions, and the subject will be so novel to most readers, that it would be difficult to follow the more rational plan of first laying down the general principles, to serve as a basis for discussing past attempts to effect artificial flight. The course will therefore be adopted of first stating a few general considerations, and of postponing the statement of others until the discussion of some machines and past failures permits of showing at once the application of the principles." Next, Chanute explained why such an article series needed to be written: "Inventors, in their ignorance of the laws of air reactions and resistances, have proposed all sorts of devices for compassing artificial flight and experimented with quite a few." This information needed to be shared with the interested public to promote further progress.

To highlight his articles he used the engravings by M. Emmanuel Dieuaide that he had assembled in the *Tableau d'Aviation* [72], published in 1880 in Paris. There was also Gaston Tissandier's well researched book on aerial navigation [73], and the pamphlet [9] his father had mailed two decades earlier. Using these sources allowed Chanute to give an account of what had been proposed in the past, but also point out why the machines failed. "Men must learn from twenty failures how to succeed the twenty-first time in one thing. Failures, it is said, are more instructive than successes; and thus far in flying machines there have been nothing but failures."

As a practical engineer, Chanute always felt the need for proper wording. In this article series he formalized some of the nomenclature which would become standard: aeroplane (Chanute adopted the French word *aéroplane*, even though he felt that the word aerocurve might describe the flying machine better), lift (the force perpendicular or at right angles to the direction of the airflow), and drift (the force parallel to the direction of the airflow). He introduced the term drift to describe the drag of the wing necessary to calculate the resistance of the wing surfaces; this word was later replaced with drag or hull resistance.

Starting his section on aeroplanes [74] in the June 1892 issue, Chanute went back in history and discussed the story of the eleventh century English Benedictine monk Oliver of Malmesbury. The legend relates that "having manufactured some wings, modeled after the description that Ovid has given of those of Daedalus and having fastened them to his hands, he sprang from the top of a tower against the wind. He succeeded in sailing a distance of 125 paces; but either through his impetuosity or whirling of the wind, or through nervousness resulting from his audacious enterprise, he fell to the earth and broke his legs. Henceforth he dragged a miserable, languishing existence attributing his misfortune to his having failed to attach a tail to his feet." And Chanute explained, "Commentators have generally made merry over this last remark, but in point of fact it was probably pretty near the truth. To perform the maneuver described, of gliding downward against the breeze, utilizing both gravity and the wind, Oliver of Malmesbury must have employed an apparatus somewhat resembling the attitude of a gliding bird, but being unable to balance himself fore and aft, as does the bird by slight movements of his wings, head and legs, he would have needed even an ampler tail than the bird spreads on such occasions in order to maintain his equilibrium. He would have failed of true flight in any event, but he might have come down in safety."

Later in the same article, Chanute discussed the fourteenth century Italian mathematician Giovanni Battista Dante who reportedly sailed over a lake near Perugia on artificial wings; Chanute agreed that "the selection of a sheet of water to experiment over was very happy, as it would furnish a yielding bed to fall into if anything went wrong, as is pretty certain to happen upon the first trials," and it "cannot be too strongly urged upon any future inventor who desires to make similar experiments." Could Langley, who received the *Railroad Journal* regularly, have been inspired to launch his aerodromes from a house boat and fly over the Potomac River after reading this article?

To make each article accurate, Chanute looked for the true experts. He had read more than once Wenham's paper "*On Aerial Locomotion*," presented at the first meeting of the Aeronautical Society in London, England, in 1866. "For it may be remarked that the swiftest flying birds possess extremely long and narrow wings and the slow, heavy flyers short and wide ones," recognizing the advantage of a high aspect ratio wing for a flying machine. Wenham also thought that the same lifting power of one larger area wing could be obtained by dividing the same wing area into several smaller planes arranged in tiers. In his first letter to Wenham, Chanute explained: "I am writing a series of articles on 'Progress in Flying Machines,' and for more than a year, I have been endeavoring to get your address, so that I might communicate with the pioneer who first advanced a rational engineering view of the subject ... My general idea is to pass in review what has hitherto been experimented, with a view to accounting for the failure, clearing away the rubbish, and pointing out some of the elements of success" [75]. Chanute discussed Wenham's work in the September 1892 issue.

Another Englishman, Horatio F. Phillips, recognized the importance of curved wing surfaces. He exposed wings of different curvature to artificial currents of air of varying strength, induced from a steam jet in a wooden trunk. The wings were loaded with a weight, applied one third of the width back from the forward edge, believing this to be the center of pressure, and were swung by two wires attached to their leading edges. When the wings assumed a lifting attitude in the velocity of the current, the thrust or drift was measured and recorded. Phillips patented his most efficient wing shapes in October 1884 and published the data in *Engineering* (London) [76].

Reading Phillips' report, Chanute thought that the angles of incidence should be given, as Phillips assumed the center of pressure to be uniformly one-third of the distance back from the front edge and did not give the actual lifting attitude of the wing in regard to the air current. Chanute drew up his own tunnel design with a lift balance and asked some correspondents for comments [77], but no wind tunnel was built, he gave the calculated angle as 15° (Fig. 2.8).

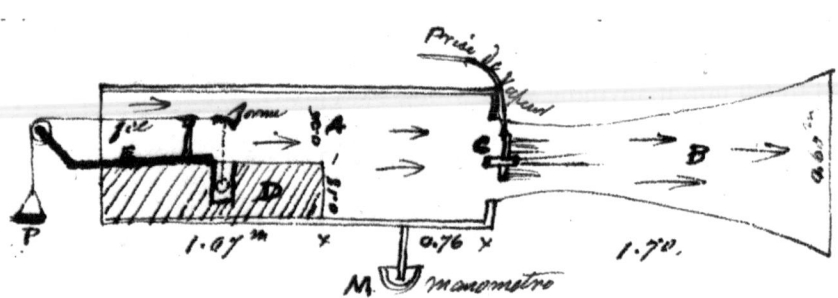

Fig. 2.8 Chanute's proposed wind tunnel design, letter to A. Goupil, December 17, 1892

In his March 1893 article, Chanute concluded, "In any case it seems very desirable that further scientific experiments be made on concavo-convex surfaces of varying shapes, for it is not impossible that the difference between success and failure of a proposed flying machine will depend upon the sustaining effect (with a given motor) between a plane surface and one properly curved to get a maximum of "lift".

Late in 1892 the Smithsonian Institution received several issues of the *Prometheus* magazine with articles by a German experimenter, Otto Lilienthal. George Curtis translated the text for Langley and then mailed the material to Chanute. Lilienthal reported soberly what he had done in practical flying and invited others to not only repeat his experiments, but also to improve on them. For him the most important part was to learn how to handle the glider in the air and practice ... practice ... practice. The Smithsonian published a translation in their 1893 report, and Chanute discussed Lilienthal's work in the July 1893 *Railroad Journal*. "Among the most systematic and carefully conducted series of experiments that have ever been made in the direction of artificial flight are those of Herr Otto Lilienthal, of Berlin, Germany, a mechanical engineer and a prominent member of the German Society for the Advancement of Aerial Navigation" [79]. These were the first reports of the "Flying Man," published in the United States.

The "Progress in Flying Machines" article series was concluded in January 1894. To define a flying machine, Chanute explained, "To the possible inquiry as to the probable character of a successful flying machine, the writer would answer that in his judgment, two types of such machines may eventually be evolved: one, which may be termed the soaring type, and which will carry but a single operator, and another, which may be termed the journeying type, to carry several passengers and be provided with a motor.

"The soaring type may or may not be provided with a motor of its own. This must be a very simple machine, as this type will have to rely upon the power of the wind, just as the soaring birds do, and whoever has observed such birds will appreciate how continuously they can remain in the air. Such an apparatus for one man need not weigh more than 40 or 50 lbs., nor cost more than twice as much as a first-class bicycle. Such machines are likely to serve for sport.

"The other, or journeying type of flying machine, must invariably be provided with a powerful and light motor, but they will also utilize the wind at times ... It seems quite certain to the writer that flying machines can never carry even light and valuable freights at anything like the present rates of water or land transportation, so that those who may apprehend that such machines will abolish frontiers and tariffs are probably mistaken".

Summarizing in the conclusion, Chanute stated that stability and control of the aeroplane had to be worked out, but equilibrium was the most important and most difficult problem to be solved. As the failures of past experiments resulted from many different causes, he believed that each problem could be solved in more than one way. "The general problem having been thus decomposed into its several elements, and each element considered as a separate problem, it will be seen that the mechanical difficulties are very great; but it will be discerned also that none can now be said to be insuperable, and that progress has recently been achieved toward their solution." He listed what he considered the ten most critical problems:

1. The resistance and supporting power of air.
2. The motor, its character and its energy.
3. The instrument for obtaining propulsion (=propeller).
4. The form and kind of the apparatus (=airframe).
5. The extent of the sustaining surfaces (=wings and tail).
6. The material and texture of the apparatus.
7. The maintenance of the equilibrium.
8. The guidance in any desired direction.
9. The starting up under all conditions.
10. The alighting (=landing) safely anywhere.

Clearly the motor and propulsion were just two obstacles, but Chanute also theorized that "it is possible to utilize a still lighter power, for we have seen that the wind may be availed of under favorable circumstances, and that it will furnish an extraneous motor which costs nothing and imposes no weight upon the apparatus. Just how much power can be thus utilized cannot well be told in advance of experiment; but we have calculated that under certain supposed conditions it may be as much as some 6 h.p. for an aeroplane with 1,000 sq. ft. of sustaining surface; and we have also seen that while but few experimenters have resorted to the wind as a motor, those few have accomplished remarkable results."

As part of his fact gathering, Chanute had identified flight enthusiasts of all levels of knowledge and education. Becoming the "guiding light," he distributed information, shared the latest news of what was going on, offered encouragement, sound advice and criticism and occasionally financial support. But in the real world, people usually still rolled their eyes and quietly left the room when the conversation turned to flying machines.

2.8   An Undergraduate Student Discovers Aeronautics

The University of Notre Dame at South Bend, Indiana, was the first Catholic University to introduce formal civil engineering classes in 1873. John A. Zahm, after graduating in 1875, joined the university as Professor of Physics; he wanted Notre Dame to become a research university, dedicated to scholarship and may have convinced the university to purchase books on aeronautics in 1877.

John's ten-year younger brother, **Albert Francis Zahm** (1862–1954) arrived at Notre Dame in 1879. The fresh undergraduate student enrolled in the classical science but also attended engineering classes, where aeronautical topics were discussed. He recalled many years later, "One day in a Greek class, the professor told the story of Daedalus and Icarus, the two mythological characters who fashioned wings for themselves so that they could fly. I decided then and there to find a method of flight."

On November 11, 1882, Zahm read his first paper on "Flying Machines" before the Notre Dame Scientific Association, published in the *Notre Dame Scholastic.* He wondered, "Shall man ever be able to fly is a question frequently asked, and a question deserving of more attention than it usually receives. For the most part a hasty and decisive answer is immediately given. 'Tis either a knowing YES, from the whimsical and ignorant, or an arrogant, confident NO, emphatically pronounced by the still more ignorant and superficial." Puzzling on how to achieve flight, one could either use a motor or utilize the forces of nature as the soaring birds do. "What could be more amusing than to see a number of our students sailing to and fro above the College grounds in beautiful, well formed, controllable machines of this kind?" He then gave details on how to build a machine and utilize the energy of the March winds. "Fasten together the large ends of four fish poles at right angles, such as form the 'sticks' of a kite. The whole could be strengthened by cording and bracing. Now, to the center attach a handle perpendicularly, like an umbrella handle; also, from the center, suspend four strong cords, to which can be fastened a platform for the manager to stand upon, just beneath the handle. This done, the machine is ready for use" [80].

Zahm mentioned years later that his brother John pointed out that soaring in a uniform horizontal current is equivalent to soaring in a calm, which means no less than perpetual motion or the creation of energy. "I saw then that flight on rigid wings must require a wind having either an upward trend, or a variable velocity, or a variable direction, as it appeared evident that such an upward trend could not prevail so generally as the flight of vultures and gulls would demand. I concluded that these must derive their motive power largely from the pulsations of the wind or from its variations of velocity and direction. Doubtless many of the strange phenomena which have hitherto puzzled experimenters with sailing and flying apparatus, the sudden pitchings and the mysterious soarings, can be attributed directly to the fickleness of the wind" [81, 82].

After earning his Bachelor in 1883, Zahm worked for his Master Degree and became more involved with aeronautics. Several students built and experimented with man-carrying gliders and a primitive wind tunnel. Some reports were published in the *Notre Dame Scholastic*, while others were recalled fifty some years later. Here is just one of the more colorful stories. "From the ceiling of the museum he suspended a flying machine operated by foot-power. On this, his shop assistant, George Archambault, made flights about the museum to test the merits of various kinds of propellers. As a diversion the rider was mounted on a huge bird with two long black wings which, beating up and down, drove the suspended weight as if in real flight. Brother Bernard, coming suddenly on the scene was greatly impressed and reported the 'flight' as a marvelous demonstration. Reverting to Brother Bernard, one noted next day some concern on his kindly face. During his morning patrol as the janitor, he had found on the museum wall a black foot print. Only the devil could walk there, of course, by current demon lore. Hence what he had seen last night, flapping through the dark with long black wings, must have been Satan seeking someone to devour" [83].

Another full-size glider was built in the late 1880s. "It was a rectangular sail with strong parallel bars joining the front and rear spars. To the rear extensions various types of tail could be attached for trial. The operator, with the bars under his armpits, would run forward, leap from a long bench and glide to the ground. Another launching device was a long horizontal beam jutting from the roof of Science Hall. A rope attached to the glider passed over a pulley in the beam end and down to a power windlass in the basement. Now one man worked the windlass, while another man dropped with the glider" [83] (Fig. 2.9).

These were apparently the first experiments with what one could describe as a full-size man-carrying glider in America; little is known about these early efforts, even though the *Notre Dame Scholastic* publication was widely shared with other catholic schools in the United States. Zahm wrote his first letter to Chanute in March 1891, but did not feel comfortable discussing his undergraduate work with the eminent civil engineer.

Fig. 2.9 Painting by Robert F. Brian at the Hessert Center for Aerospace Research, Notre Dame University, showing Albert F. Zahm's glider experiments in the 1880s. Author's collection

2.9 The International Conference on Aerial Navigation in Chicago, 1893

As part of the initial struggle to stage a World's Fair, Chicago politicians had to convince Congress that their hometown was not a "cow-town," and that its citizens could stage a prestigious fair that would draw a crowd from around the world. Chicago, the railroad center in the middle of the country, could illustrate progress in transportation within all its branches, on land, on water, or through the air and further the image of the United States as a modern, industrialized nation [84]. Most assuredly it was Chanute's influence to include aerial transportation. At that time no one really knew whether human ingenuity would ever lead to a successful aeronautical undertaking, but the subject was fascinating and worth studying.

The 1893 World's Fair expected to show Chicago in its best light. Civic boosters were motivated to create an engineering triumph that dwarfed all its predecessors. Several WSE members had ideas, but then came a proposal from George W. Ferris, President of a Pittsburgh engineering firm specializing in bridge designs. He suggested to erect a huge rotating wheel from which visitors would be able to view the entire fair ground and the city in the distance. According to Ferris, a tower and a wheel are more or less adapted from the principles of the cantilever bridge with the former being a bridge set on end and the latter being a bridge where the extremes are united to form a circle. The Ferris project was so grandiose that the committee dismissed it as unrealistic, but WSE members were intrigued; fair organizers finally granted the concession to Ferris in December 1892. Early in 1893 workers started to erect the 264-foot-diameter wheel in the Midway Plaisance. Between the two rims of the wheel, thirty-six cars, with a seating capacity of forty passengers each, lifted riders high above the fairgrounds. Fifty cents paid for a two-revolution ride of the wheel with a bird's-eye view of the fair, the city of Chicago to the north, and Lake Michigan and Northwest Indiana to the east and southeast.

Several WSE members subscribed to the Fair's capital stock by "investing" $1,000, but their goal differed from those of Chicago politicians. They wanted to stage an engineering congress to highlight their profession, the first such congress to be held in the United States or possibly in the world [85]. Chanute was convinced that the civil engineer would emerge from the lower plane of a trade into a true profession, if this congress was a success. To help make it happen, he accepted the presidency of the General Committee of Engineering Societies and became vice-chairman of the World Congress Auxiliary, coordinating the various congresses under the motto "*Not things, but men*" [86].

Hearing of the proposed engineering congress, Albert Zahm thought that a conference on aerial navigation should be included, just as the French had done. He made an appointment to discuss it with the President of the World's Congress Auxiliary and then the vice-chairman, Chanute, who was in favor of the basic idea but refused to chair such an event, simply stating, "Not on your life; I have a family to support, and must retain some reputation for sanity" [87]. Hearing of this, several WSE members, including Bion Arnold, Ira Baker, Elmer Corthell and William Karner, stepped forward and offered their full support to help decipher the mystery of flight; they did not think that crackpots would disrupt this academic gathering. Chanute then reluctantly accepted the chairmanship with Zahm serving as secretary. With the help of WSE members, the first circular letter was mailed in December 1892 to the foremost experimenters and aeronautical societies around the world [86], requesting papers on aeronautics and its allied sciences. Writing to Carl Myers, in Frankfort, New York, Chanute explained, "We shall desire to elicit facts, positive knowledge and records of experiments, rather than theories or novel proposals" [88]. He later introduced Myers as an "aeronautical engineer," the same title the military had assigned to balloonists during the Civil War.

A truly international assembly of speakers agreed to prepare papers on topics ranging from meteorology to bird flight, and balloons to flying machines. As a certain safeguard, the WSE appointed a committee to read and approve every paper before presenting it to the public. And to prevent uninvited "cranks" from disrupting the conference, every person who wished to attend had to purchase, in advance, a "personal card of admission" for $3, which entitled the bearer to a reserved seat at the conference and to participate in the discussion.

On Monday, July 31, several hundred engineers from around the world attended the opening ceremony of the Engineering Congress. Chanute, as president of the General Committee of Engineering Societies, opened the Congress in the large Hall of Washington in the recently erected World's Congress Art Palace, today the Art Institute of Chicago (Fig. 2.10).

The next day, Tuesday afternoon, August 1, the aeronautical conference opened in the adjacent Hall 7. In his address, Chanute summarized the current state of affair. "The truth of these assertions seems to indicate that it is not unreasonable for us as engineers, as mechanics and as investigators to meet here to discuss some of the scientific principles involved and to interchange our knowledge and ideas" [89]. The first session, "*Scientific Principles*," covered the properties of the air, propellers, motors, and materials of construction. Langley who had originally agreed to chair this session, had mailed his paper "*The Internal Works of the Wind*" [90]. Chanute read it first, as it promised to be an important breakthrough. "This internal work of the wind might conceivably be so utilized as to furnish a power which should not only keep an inert body from falling, but cause it to rise. They are paradoxical at first sight, since they imply that, under certain specified conditions, heavy bodies, entirely detached from the earth can be sustained indefinitely without any expenditure of energy from within. This is, without misuse of language, to be called a physical miracle." DeVolson Wood agreed with Langley, but the extent and the method remained to be explained. "Upon the whole, there seems to be some principle not yet fully understood, involved in the soaring of birds, and the possibility of reproducing it artificially does not yet appear

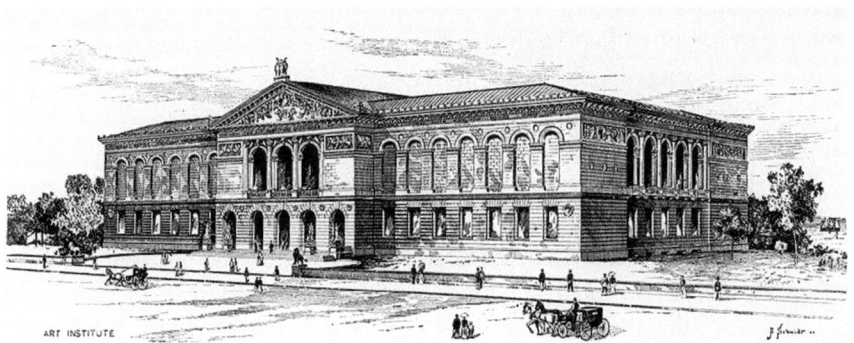

Fig. 2.10 The World's Congress Art Palace, today the Art Institute of Chicago. *Twenty-second annual report*, 1901

to be within the grasp of man" [91]. It took a few more years to understand how a motorless craft can fly against the wind, using the lift provided by the wind.

The recently deceased Charles Hastings had prepared two papers on the theoretical aspect of aeronautics; William Breithaupt, who had worked with Chanute on building the bridges on the Santa Fe Airline, read them both. Carl Vogt from Denmark discussed in his paper the differences of a screw pushing a ship in the water and a propeller driving a balloon or aeroplane through the air, describing the peculiarities of both supporting media. Thurston read his paper "*Materials of Aeronautic Engineering*" last, stating in the introduction that it would be bold to say that aerial navigation was impossible to achieve. He presented facts on the properties and qualities of lighter weight metals, where strength and weight were combined. His metal of choice for the future aircraft was steel, but his research had shown that adding small quantities of magnesium to aluminum strengthened it considerably, thus he recommended this metal as his second choice. Thurston's paper was republished in *Cassier's Magazine* of September 1894, giving it a wider audience.

On Wednesday, Thurston chaired the "*Aviation*" session. Twelve papers discussed heavier-than-air flying machines that had to propel themselves forward to overcome gravity. About half the papers detailed the sailing flight of birds, while others discussed aeroplane designs and learning to fly. Zahm read a paper on "*Soaring Flight*" by an experimenter from Tennessee, Edward Huffaker, and then his own paper on "*Atmospheric Gusts and their Relation to Flight.*" In his second paper, "*The Equipoise of Flying Machines,*" he intended to draw further discussion by attendees. In closing, Zahm stated: "I would suggest three methods of learning to rise in the air: (1) by direct lift, using a vertical screw; (2) Lilienthal's method, by learning to ride the aeroplane before the motor is added; (3) by gliding rapidly over a field of ice or other smooth surface, not allowing the machine to rise more than a few inches until the pilot has acquired skill and confidence."

The "*Ballooning*" session on Thursday had the advantage that balloons had provided the means for humans to fly in the past. Looking for someone to chair the session just prior to the opening of the conference, Thurston suggested to ask Lieutenant Colonel W. R. King of the Army Corps of Engineers, who had just read a paper to Sibley students on the "*Military Engineer and his Work.*" Even though King had very limited knowledge of ballooning, he agreed. As the first speaker, Carl Myers described a flight by his wife Carlotta, in which she had reached an altitude of 21,000 feet and landed safely after traveling 90 miles. As the "gas-bag" had no precedent in the ornithological world, De Volson Wood suggested in his paper "*Flotation vs. Aviation*" a departure from the bird method in flying machines. "Discarding the flapping wings as a motor, the continuously rotating propeller is the only form of driver that commends itself. This may be used either to drive an aeroplane or a floating vessel." In closing he stated: "A speed of 30 miles per hour produces relatively a brisk breeze, and to talk of carrying a man 100 miles per hour in an airship we consider rather 'flighty'."

There was a great deal of interest in the various aeronautical topics, so organizers arranged a supplementary session on Saturday, August 4. Several speakers discussed topics that had not been previously approved by the committee. One of the speakers was John J. Montgomery from Santa Clara, California, who discussed flying his glider and the benefits of using a parabolic wing curvature, as he had studied a decade earlier. Another presenter was Charles E. Duryea from Peoria, who proposed "to suspend a flying machine from a captive balloon, anchored by several ropes. By means of a rope passing through a pulley block attached to the balloon, and thence to a windlass on the ground, the machine to be experimented with may be drawn into the air to a sufficient height to clear the gusty air conditions found near the ground; there the sky-cycler might gradually gain skill and confidence preparatory to trusting himself to actual flight" [92].

In closing the congress, Chanute theorized, "Success with aeroplanes, if it comes at all, is likely to be promoted by the navigable balloon. It now seems not improbable that the course of development will consist, first in improvements of the balloon, so as to enable it to stem the winds most usually prevailing, and then in using it to obtain the initial velocities required to float aeroplanes. Once the stability of the latter is well demonstrated, perhaps the gas-bag can be dispensed with altogether."

With the engineering congress going on in the adjacent larger hall, several creative giants in the engineering profession joined the aeronautical congress at times. One such engineer was Francis A. Pratt, co-founder of the Pratt and Whitney company with whom Chanute had worked more than ten years earlier, trying to establish uniformity of hardware used by the Erie Railway. Other attendees included John P. Holland, who considered building a flapping wing machine; Alfred Mayer, professor at Stevens Institute, who had experimented with flying models in the 1870s; Thomas Edison, known for inventing the light bulb, was interested in helicopter designs; Nikola Tesla, an electrical engineer, had long puzzled over a flying machine; and Samuel Cabot, an amateur bird watcher from Boston, who owned a wood preservation business. The enthusiastic twenty-four-year-old Charles Steinmetz shared what he had learned with fellow engineers back home, which resulted in the formation of the "Mohawk Aerial Navigation Company, Ltd." in 1894, the first flying club in America or possibly even the world [93]. The group built two not very successful Lilienthal-type gliders but members had fun.

About one hundred persons attended each session of the aeronautical congress, and Zahm noted in his diary: "Large attendance and papers pronounced very valuable and interesting. Three hours all too short for reading of papers. Reporters and stenographer in constant attendance" [87].

As expected, Chanute suffered some raillery from fellow WSE members for being interested in an inchoate art, but he firmly believed that aerial naviga-tion was a serious study and that responsible men were willing to work toward an eventual success. To share the presented research with others, he wanted the papers with their discussions published as a *Proceedings*; a Chicago printer showed some interest, but Forney had a better idea. He suggested to start a new monthly, *AeronauticS*, charging a subscription price of $1 for twelve issues for new subscribers and $0.50 for current subscribers of the *Railroad Journal*, and publish one paper after the other every month. With the pages typeset, the *Proceedings* could be published in book format without added expenses.

The highly successful 1893 aeronautical conference attracted not only a new generation of enthusiasts, but also legislative attention. Senator Cockrell introduced Senate Bill S.1344 in December 1893 to secure aerial navigation. The bill was transferred to the Interstate Commerce Committee and Mr. Calvin S. Brice, Senator of Ohio, reported to the Senate in February 1895: "The trackless ether presents such limitless highways that the regulations for travel would be comparatively simple and travel in a given direction would follow laws both as to altitude and latitude. Rapidity of flight would be by no means the chief advantage. As the railroad across the plains, in comparison with the ox cart, has made travel delightful, so the little aerial car fitted up for the family party would excel the costliest parlor car of Mr. Pullman" [94]. The Committee on Interstate Commerce acknowledged the importance of aerial navigation, but did not recommend the passage of the bill due to the "con-tinuing deficiency in revenues to meet ordinary appropriations." Two years later James Means and Chanute redrafted the bill and asked Senator Henry Cabot Lodge from Massachusetts to present it. Senate Bill S.302 suggested the "Secretary of the Treasury to award the sum of $100,000 to any person, who shall at any time prior to January 1, 1901, construct an apparatus that will on a verified report of a committee of three members, appointed by the Secretary of War, demonstrate within or near the city of Washington the prac-ticability of safely navigating the air at a speed of not less than thirty miles an hour, and capable of carrying passengers, weighing a total of not less than 400 pounds" [95]. As before, the bill was referred to the Committee on Interstate Commerce and then tabled.

The conference surely brought aeronautical investigation into the lime light of public acceptance. This was a risky stance in an environment that still ridiculed the idea of manflight, but Chicago politicians believed they should take the lead. Over the next decades, vast sums of money were spent on promoting aeronautics in Chicago, partially due to Chanute's work [96]. His contribution in advance planning of the World's Fair helped the engineering profession, the Columbian Exposition and Chicago to achieve its ultimate goal and success. He had used his leadership talents to bring together his vast acquaintances, and his experience synthesized success in a new endeavor, which in turn extended his personal network, a resource that continued to aid him throughout his life's next accomplishments.

2.10 A Landmark Publication—*Progress in Flying Machines*

Only five months after the aeronautical conference concluded and more than seven years after Forney had first suggested this project, Chanute's articles on aeronautical progress were ready for publication in book format.

Compiling the preface last, Chanute gave his personal reasons for writing the articles in the first place. He wanted to satisfy himself that men might eventually fly, but he also wanted to provide an understanding of the principles and the known experiments to prevent wasted effort by experimenters in trying devices, which had failed in the past. He wanted to present the results thus far accomplished, to enable an investigator to distinguish between an inadequate proposal, sure to fail, and a reasonable design that might succeed. He also hoped that the information would help avoid duplication of failed experiments, wasting energy and funding, and would result in eventually accomplishing manflight [97]. His academic acumen was clearly the result of scientific curiosity without any thought of immediate application or even financial benefit.

Forney shipped the typeset pages to Chanute for proofreading and to create an index, which turned out to be a "devil of a project." Besides correcting typographical errors in the text and figure captions, he also wanted his name on the title page to appear as "O. Chanute," instead of showing his full first name. Questioning why there was no publication date on the title page, Forney explained that the text was reprinted from his copyrighted magazine, thus no date was needed.

Discussing the color of the cloth cover binding over dinner, one daughter suggested gray-green, gray for the author and green for the purchaser. Forney knew better and bound the books in a light-blue cloth. The artwork on the cover brought more discussion; Forney suggested a balloon engraving, but Chanute did not think this appropriate for a book on flying machines; he suggested showing an angel or a goddess. But then he received a copy of *La Science Illustrée* [98] with an article on Lilienthal which included a full-page engraving of the glider in flight. This was perfect! (Fig. 2.11). He forwarded the page to Forney and suggested to use this image, embossed in gold, for the cover. And the angel drawing became the final graphic as the end piece in the book.

Forney had everything ready to go to the binder when Chanute received Lilienthal's latest article on the benefits of curved wing surfaces. This article "so fully sustains the views set forth in this book, and holds out such promise of success in the near future" [99], that including this article would surely help sell the book. George Curtis from the Smithsonian translated it quickly, and Forney included it as an appendix.

In late April the 308-page book *Progress in Flying Machines* arrived at Chanute's home. Forney had printed about five hundred sets of pages and

LA SCIENCE ILLUSTRÉE.   153

L'HOMME VOLANT. — Expérience de M. Otto Lilienthal.

Fig. 2.11 *"L'Homme Volant."* Illustration from *La Science Illustrée*, February 5, 1894, used on the cover of the book *Progress in Flying Machines*, published two months later

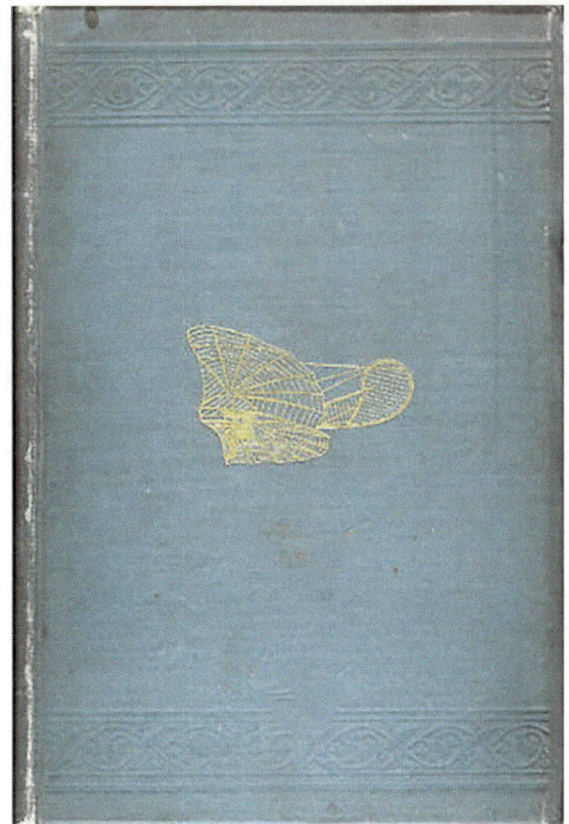

Fig. 2.11 (continued)

had two hundred books bound. Chanute received twenty-five unbound page sets, and he purchased forty books for $2.50 each, minus 45% discount, to send to his friends.

Forney shipped review copies to influential people, publishers and several university libraries. The public was probably surprised to learn how far man's knowledge of a possible future success had accumulated within the last few years. "As an engineer Mr. Chanute is able to prick many a bubble and fallacy, while his genuine enthusiasm for the subject leads him to examine carefully the most unlikely project," was the reviewer's comment in the *Electrical Engineer* [100].

Another laudatory review was published in a rather unlikely magazine, *The Internal Revenue Record and Customs Journal*. "It gives us a realizing sense of our progress toward conquering the realm of the air to receive from the American Engineer and Railroad Journal an octavo volume by O. Chanute, C. E., on 'Progress in Flying Machines.' When such eminent authorities as

Mr. Chanute and Prof. Langley deem the subject of aerial navigation worthy of serious discussion, we may be sure that 'there is something in it,' as Abraham Lincoln said of the lady's stocking. Mr. Chanute gives an historical review of the effects and experiments of inventors to accomplish flight as birds do. He has gathered all the records of such experiments and has endeavored to show the reasons for their failure, and to explain the principles, which govern flight, and to satisfy himself and his readers whether we may reasonably hope eventually to fly through the air. His conclusion is that this question may now be answered in the affirmative" [101].

Effective January 1, 1896, Forney sold his *Railroad Journal* but agreed to stay a little longer, inserting book ads periodically to fill space. Up to the end of 1898, Forney sold ninety-nine books and Chanute purchased a similar amount, receiving a royalty payment of $0.14 for each book sold. Needing more books, Chanute inquired about a softcover binding, but Forney was not in favor. The title page of the new binding shows M. N. Forney as the publisher and the 1899 date. In April 1902 Forney had the remaining page sets bound without making any changes.

Compiled by the world's most knowledgeable authority on the history and current (at that time) state of aeronautics, *Progress in Flying Machines* became an essential reference for the next generation of experimenters who achieved the dream of powered flight in a heavier-than-air flying machine. The book was and still is an aviation classic, even though it never became a bestseller. Its editing by an engineer of Chanute's status gave the subject a prestige not afforded to most aeronautical literature of the day. There is little question that the book heavily influenced the work of the next generation of aeronautical experimenters.

2.11   An Experimenter from Eastern Tennessee

Showing an early talent for mathematics, **Edward Chalmers Huffaker** (1856–1937) from Knoxville, Tennessee, attended Emory and Henry College and then earned a Master Degree from the University of Virginia in 1885 [102]. Having watched the birds for many years, he began experimenting with flying models and wrote a letter on the value of curved wing surfaces in early January 1893 to Langley. Next he compiled a paper on "*Soaring Flight*" [103] and forwarded the manuscript to Zahm, who read it at the Conference on Aerial Navigation. As a follow-up, Huffaker mailed another essay with a drawing of his "soaring machine," and mentioned his interest to build a model, if funds could be furnished. Zahm gave the letter to Chanute who responded to Huffaker: "I am not disinclined to advance a small sum, but I should want to be sure that it was not to repeat one of the many experiments which have already been tried by others and failed. I also note that you allude to a device for securing automatic stability, which you have discovered" [104].

This led to a working relationship between Huffaker and Chanute, that lasted for almost ten years. Many questions pertaining to soaring flight were discussed, including the ability of birds to soar in apparent calm. Not believing this to be possible, Chanute visited Huffaker in Chuckey City, Tennessee, in September 1894, taking along his anemometer. With amazement he watched one buzzard soaring apparently in a dead calm. "The bird was not simply gliding, utilizing gravity or acquired momentum, he was actually circling horizontally in defiance of physics and mathematics. It took two years and a whole series of further observations to bring those two findings into accord with the facts. Curiously enough the key to the performance of circling in a light wind or a dead calm was not found through the usual way of gathering human knowledge. The mystery was, in fact, solved by an eclectic process of conjecture and computation, but once these computations indicated what observations should be made, the results gave at once the reasons for the circling of the birds, for their observed attitude, and for the necessity of an independent initial sustaining speed before soaring began" [105]. Until that time Chanute did not quite grasp that a column of hot air, or what we call today a "thermal", could make this happen, he was still thinking of "aspiration".

Having heard of the work with aerodromes at the Smithsonian, Huffaker asked Chanute for help to obtain an interview. He joined Langley's team as an assistant in the aeronautical department on January 1, 1895, on a one-year contract with the task of designing efficient wings. On May 6, 1896, Langley and his team journeyed out to Quantico, just outside of Washington. It was understood that if the models No. 5 and No. 6 would not fly, the long series of experiments would come to an end. Model No. 6 was mounted first on the track on top of the houseboat; it dived at once into the water and came to grief within twenty feet of where it started.

Huffaker, who was responsible for building the wings of aerodrome No. 5, recalled several years later what happened next. "The secretary then gave orders for No. 5 to be placed upon the way, which was done with every precaution. Everyone waited in silence for the signal to release the machine, knowing that the next few seconds would tell the tale of final success or failure. The secretary gave the order as he stood disconsolately waiting on the shore with Graham Bell at his side. For the smallest fraction of a second there was hesitancy in the machinery, immediately after it did launch into the air. This time there was no hitch. The propellers were whirring, the wings working nicely. After a fall of two feet on leaving the houseboat, the machine rose and began flying like a thing of life. The hour of success had come, not partial, but full and complete, for the machine continued to fly, and to mount upward in a sure and sustained flight, turning to the right as it rose and flying in circles at a height of 150 feet. In its flight it covered a distance of more than a half-mile and came down only after the supply of fuel was exhausted, it descended slowly and gracefully and settled upon the water without any mishap.

"I have witnessed the flight of other machines since then, but have seen none that thrilled me as did that first sustained flight by mechanical power in the history of the world. The two old men on the shore shook hands again and again in sheer delight of realizing that at last the problem of the ages had been solved and that a man-made machine had demonstrated its mastery over the air" [106].

Having grown up in the mountains of Tennessee, Huffaker did not like working for Langley or being in Washington. He resigned on December 16, 1898, and moved back to Tennessee.

2.12   The College Inventor from California

Wanting to attend the Electrician's Congress in the middle of August 1893, **John J. Montgomery** (1858–1911) from Santa Clara, California, arrived in late July in Chicago. Stopping at the Art Palace, he was surprised to hear of "learned men" to discuss flight, something he had tried a decade earlier [107]. He introduced himself to Zahm and Chanute and told them that he had built and flown gliders. Even though he had not purchased the required admission ticket, he attended most of the lectures. On the make-up day, Saturday, he discussed his research on wing shapes and how birds master soaring flight [108].

Being curious about Montgomery's flying activity, Chanute invited him to his home, and then published his story in the December 1893 *Railroad Journal*, providing the first written description of Montgomery's glider flying trials.

"Mr. J. J. Montgomery, of California, had, some years previously, constructed a soaring apparatus, consisting of two wings, each 10 ft. long by an average width of 4½ ft., united together by a framework to which a seat was suspended, and provided with a horizontal tail which could be elevated or depressed by pulleys. The wings were arched beneath, like those of a gull, and afforded a sustaining area of about 90 sq. ft. The weight of the apparatus was 40 lbs., and that of the experimenter some 130 lbs. more. Mr. Montgomery took this apparatus to the top of a hill nearly a mile long, which gradually sloped at an angle of about 10°, and placing himself within the central framework, the rods of which he grasped with each hand, ready to sit down, he faced a sea breeze steadily blowing from 8 to 12 miles an hour, and gave a jump into the air without previous running. He found himself at once launched upon the wind, and glided gently forward, almost horizontally at first, and then descended to the ground, finding that he could meanwhile direct his course by leaning to one side or the other. The total distance glided was about 100 ft., and the sensation was that of firm yet yielding and soft support, being quite similar to the experience of M. Mouillard, as already described, except that there was no apprehension of disaster. Mr. Montgomery carried his machine back to the top of the hill and prepared to repeat the experiment, but as soon as he got into position the apparatus began to sway and to twist about in the wind; one side dipped downward, caught on a small shrub, and, as quick as a flash, the operator was tossed some 8 or 10 ft. into the air, overturned, and thrown down headlong. He fortunately fell without serious injury, and found, as soon as he recovered himself, that one side of his machine was smashed past mending. Montgomery built two additional gliders; his last apparatus proved an entire failure, as no lifting effect could be obtained from the wind, sufficient to carry the 180 lbs. it was designed to bear. Mr. Montgomery then turned his attention to other matters, but he has since made a more careful and complete study of the principles involved, and he expects to resume his experiments."

After submitting his write-up to Forney, Chanute wondered if he could or should collaborate with Montgomery and work with him on a real flying machine. With winter approaching, Chanute's wife Annie joined her brother's family in San Diego in mid-January. She and her daughters then took a ride to Otay Mesa hoping to meet Montgomery at his farm. Seeing her visitors arrive, Montgomery's mother told the visitors that her son was out-of-town, and Chanute received a letter from Montgomery a few days later, that he could not resume experiments [109].

2.13 The Pioneer from Down Under

Aerial navigation excited not only enthusiasts in Europe or America, but also in Australia. Born in London, England, **Lawrence Hargrave** (1850–1915) joined his father in 1865 in Australia and became interested in flight about

1872. Looking for an area with favorable winds, Hargrave moved to Stanwell Park, south of Sydney, and started to experiment with cellular kites. His first models were not much larger than toys and were powered by stretched rubber bands to drive the propellers, but as the models grew in size, he became interested in applying more power.

Reading of Hargrave's aeronautical work in *Engineering* (London), Chanute contacted him in 1891. Hargrave responded and shared his philosophy: "Workers must root out the idea that by keeping the results of their labors to themselves a fortune will be assured to them … The flying machine of the future will not be born fully-fledged and capable of a flight for 1,000 miles or so. Like everything else it must be evolved gradually. The first difficulty is to get a thing that will fly. When this is made, a full description should be published as an aid to others. Excellence of design and workmanship will always defy competition." Writing to Langley a few months later, Hargrave explained: "Will you impress on your coworkers the fallacy of secrecy? Cooperation and a free interchange will hasten success in which all will share. There are so many forms of flying machines possible that it is hopeless to think any inventor will be able to monopolize the profits by a corner" [110].

Hargrave's research was published in the *Proceedings of the Royal Society of New South Wales* and shared with the scientific community. With a certain admiration, Chanute wrote, "If there be one man, more than another, who deserves to succeed in flying through the air, that man is Mr. Laurence Hargrave, of Sydney, New South Wales. He has now constructed with his own hands no less than 18 flying machines of increasing size, all of which fly." And adding, "Mr. Hargrave is probably correct, for the history of all new methods of transportation teaches that the original inventor seldom receives pecuniary reward for the contrivance which is the first to succeed, but nevertheless he is certainly broadly liberal in giving to the world gratuitously the results of his constant studies and labors."

Looking for papers that showed actual progress in aeronautics, Chanute invited Hargrave to submit one for the 1893 Aeronautical Congress and Zahm read his paper on "*Flying Machine Motors and Cellular Kites.*" Hargrave also shipped his No. 14, the best of six models driven by compressed air, which arrived in Chicago in late 1894. Not knowing what to do with this model, Hargrave responded to place it where people could see it. It was given to the Field Museum in Chicago. As this model really did not fit into their holding, they decommissioned the model a few years later, and, with Hargrave's permission, Chanute shipped it in December 1905 to New York to be displayed at the first aeronautical exhibition, arranged by the Aero Club of America. After the show closed in late January 1906, the artifact was transferred to the National Museum; it still is part of the National Air and Space Museum in Washington [111].

2.14   The Aeronaut from Scotland

Reading of Lilienthal's flying activity, the Scottish marine engineer **Percy Sinclair Pilcher** (1866–1899) became intrigued with manflight; he built and flew his first glider in early 1895. Receiving Forney's *Railroad Journal* on a regular basis and reading its monthly articles about aeronautics, Pilcher decided to share the news about his experiments with the readers of this American journal. His report, "*Pilcher's Soaring Machine*," was published with a photo in August 1895 [112], possibly the first report by or about Pilcher and his glider, the "Bat." Here Pilcher wrote, "The machine was made to try to repeat the very successful experiments made by Herr Lilienthal of Berlin. It consists of five parts, i.e., a body piece, a triangle, wings, and a rudder." On the first two days of flying in late June 1895 the wind was calm, so Pilcher just ran downhill. On the third day there was a rather puffy 15 miles/hour wind blowing up the hill. "Having rigged at the foot of the hill, he cautiously and with some difficulty proceeded up the hill backward, the wind taking all the weight of the machine; then, slightly elevating the front edge of his wings, he was taken up 4 ft. into the air, and remained there poised for ten seconds, when, throwing his weight slightly forward, he came down on exactly the same spot as he went up from. Afterward he ran down the hill several times, taking, without any effort, leaps of up to 60 ft. in length at about 2–8 ft. from the ground. Being caught by a side puff, the machine was blown over, and the front starboard spar was too much broken to mend on the field. The following week Mr. Pilcher repeated these experiments with much the same success, but again broke one of the spars." Thus for about ten seconds the Scottish pioneer experienced his first flight in a glider of his own design and construction, and the readership of the *Railroad Journal* in the United States was the first to know about it [113]. Forney was pleased that his publication enticed enthusiasts from around the world to share their work and Chanute enjoyed reading the report.

Following in Lilienthal's footsteps, Pilcher built the "Gull" and "Hawk" gliders and learned to control them in flight. Although his sights were set on powered flight, he knew the importance of learning to fly first before attaching an engine.

Chanute wrote his first letter to Pilcher late in November 1897, mailing photos of his 1896 and 1897 flying camps and a reprint of his WSE lecture "*Gliding Experiments*." Reading this paper fired Pilcher's enthusiasm to build a Chanute-type biplane or triplane. Writing from San Diego, Chanute responded with his typical supportive attitude on February 10, 1898: "You are quite right: I am much more desirous of having the general problem of flight solved, than that I shall have all the honor; and for a long time, I fancy, there will be nothing but honor to be expected." He continued explaining the details of his multiplane and wrote, "You are very welcome to make use of my multiple wing machine, and I hope that you will hit upon improvements, which shall be all your own. The full plans are in the Aeronautical Annual of 1897, but you will find further plans and explanations in a British Patent, which will probably issue to me about the time that you receive this letter, being Application No. 13,372 of May 31, 1897. I have applied for this patent, 1st to establish priority, and 2nd to prevent premature exhibition and accidents of wandering acrobats, for as you seem to have well discovered, one has to select the best conditions for success in the present stage of development, and this cannot be done if exhibitions have to be given at set times. I believe the plans on the patent are full enough to give you what you want, but if not, I shall be glad to send you such explanations as you may desire." But he regretted, "that I cannot extend to you the same license to reproduce a 'two-surface' machine. This has a regulator, designed by Mr. Herring, who believes that money is to be made out of it, and who has left me to work on his own account" [114].

Receiving this information, Pilcher designed a triplane in 1899, showing a strong resemblance to Chanute's triplane, in which he planned to install a small 4 h.p. two-cylinder engine. Noteworthy features were the forward-mounted engine with a long driveshaft passing over the pilot's head to a two-bladed pusher propeller behind the wing cellule; the single-spar wings, each comprising four panels, the twin fins; and the small wheels on the bottoms of the inner wing struts. It was estimated to weigh about 266 pounds, including the engine and pilot, and would have needed to attain a flying speed of 25–35 mph. On that fatal day, Pilcher flew his monoplane *Hawk* glider, and crashed in front of the assembled spectators [113] (Fig. 2.12).

Fig. 2.12 The late Frank Munger created this artist's impression of Percy Pilcher's triplane. Courtesy Philip Jarrett

2.15   Hiring a Helping Hand

Dreaming of inventing the aeroplane, **Augustus Moore Herring** (1867–1922) had experimented with rubber-powered models in the late 1880s. Seeing the German patent for Lilienthal's flying machine, Herring built two Lilienthal-type gliders in the summer of 1894. The first machine had a wing area of 129 square feet, was 20 feet wide and weighed 14¾ lbs. "It was my lack of knowledge in sailing it that caused me to break it. The longest flight was ~80 ft., with a descent of 17 ft. The second machine was built stronger to withstand any heavy shocks, which might occur from my unskillful handling. I was able on one occasion to fly 150 ft. or more." He was proud that he could experience "the sensation of sailing through the air so near the ground without the least suspicion of jar" [115]. Herring mailed his report with two photos to the editorial office of the *Railroad Journal* who forwarded everything to Chanute. The write-up sounded interesting, but Chanute questioned why Herring's Lilienthal-type glider was so light weight, nevertheless, he published the report in January 1895; this was the beginning of an at times fractious relationship.

With Mouillard progressing slowly and Montgomery not interested, Chanute wondered if Herring might like to work with him. But then he became concerned if he could trust Herring with his ideas. Prior to mailing his sketches, he recorded, in the French language, what he knew and his thoughts on Herring's Lilienthal-type machine. Having documented the facts, he then mailed his sketches and wrote: "This present design is for a soaring apparatus for one man, but it is intended to add eventually a motor and two screws, revolving in opposite directions … The primary object is to provide for automatic stability, by causing the center of gravity to move, so as to coincide with the center of pressure in horizontal flight, so that the operator need only intervene when he wants to change his direction, either up or down or sideways" [116].

Combining a Rapid Transit Commission meeting in New York with getting to know Herring was the next step. The twenty-seven-year-old Herring liked the Chicagoan, who was about the same age as his father who had died a few years earlier, and the almost sixty-three-year-old Chanute liked this young engineer, who seemed to have the qualifications he wanted to foster. Without hesitation Herring accepted Chanute's job offer for $5 a day of actual work. At that time a carpenter made about $2.80 and a brick mason about $4 a day, working for eight hours.

Chanute's first assignment for Herring was to build three different wings to determine the most efficient wing curvature and to test each design by mounting it on a bicycle, driven at a steady speed to collect reproducible data (Fig. 2.13).

When the wings were ready for testing, Herring wired Chanute to come to New York, but "Business must go before pleasure, I cannot come right now to join in the flying." Herring reported back that the wings performed better

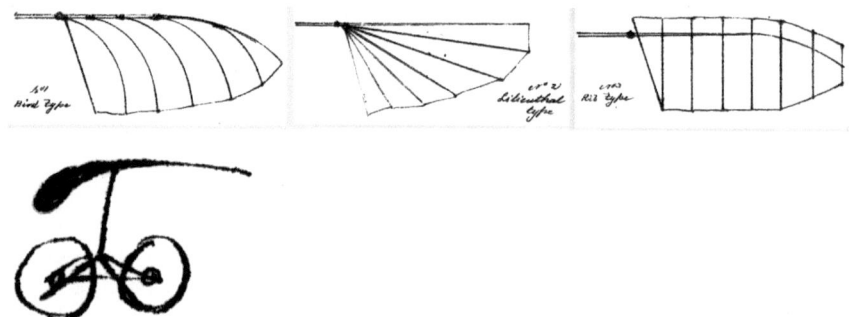

Fig. 2.13 Chanute suggested that Herring build three different wing shapes and experiment with them using a bicycle. Letter to Augustus Herring, April 28, 1895

than the ones he had built previously; he then proceeded to build several larger flying models to prove both of their theories.

Late in April 1895, Samuel Langley paid Herring an unexpected visit. Assuming that Langley looked for a new staff member, Herring showed his latest Lilienthal-type glider and demonstrated the various models, built for Chanute. Langley then wrote to Chanute about his intent to hire Herring, whereupon Chanute laconically told Langley that he wanted Herring to build him a full-size soaring apparatus, "but you are so much better equipped than myself, that I now believe that greater progress will be made toward a solution of the problem by my foregoing my purpose, and Mr. Herring joining you. If you can make an arrangement, I propose to let you avail of whatever novelty and value there may be in my own plans, models or ideas. I should expect in return a like frank access to your results, when the proper time arrives" [117]. Herring accepted Langley's job offer, even though the pay was not significantly higher than what Chanute had paid, but he, just like Huffaker, hoped that the Smithsonian would be a more prestigious workplace and provide a steadier income. Two weeks later Herring left for Washington, leaving his wife and young child behind in New York.

Only one week later, Herring reported back to Chanute: "Started work for Langley on Monday, 20 May, for $150/month. Even in a day's work under him I could foresee a possibility that he and I might not be able to agree" [118]. Just like his co-worker Huffaker, Herring applied his knowledge to Langley's models. "The success of Langley's model in May 1896 depended upon the curved surfaces and upon the elastic hinging of the tail. Until I came to Langley he had not employed curved surfaces, but I induced him to make the change" [119]. The 28-year-old Herring believed that he was solely responsible for the success of the aerodrome, either No. 5 or No. 6, but Huffaker, hired a few months earlier, also believed that his application of curved wings contributed to the aerodrome's success.

When Langley returned from his summer sojourn in Europe, the unhappy Herring wrote to Chanute: "One of the disagreeable features is Mr. Langley's inability to distinguish between the ideas of other people and his own, whether this is intentional or not I can't say—the effect is the same." A couple of weeks later, he wired: "Mr. Chanute: Expect resign December first. Have you any work for me?" Even though Langley had confided that Herring's reliability was questionable, Chanute ignored the perhaps ominous statement about ownership of ideas and wired back, "I want you here to build a soaring machine."

2.16   Shall We Ever Fly? Why? Yes, of Course!

After concluding his article series on the progress in flying machines, Chanute needed a break. Spending the late winter month in San Diego, he wrote to Langley, "Breezes here in San Diego impress me, have my Richard anemometer with me and make measurements of fluctuations. I think this would be a capital place to make experiments. Climate here is most charming, I am now writing in my shirtsleeves with the thermometer at 72. Unbroken sunshine and gentle breezes reign, and the vultures, pelicans and sea gulls are soaring" [120].

Wondering how nature could produce such constant, steady and moderate wind velocities around San Diego, with a land breeze in the morning and a sea breeze in the afternoon, Chanute stopped at the Weather Bureau office and introduced himself to Ford Carpenter and Willis L. Moore, who explained the phenomenon of unequal heating of land and water. A short time later Moore transferred first to Chicago and then to Washington to become chief of the U.S. Weather Bureau. To provide a better weather forecast, Moore introduced in 1895 Chanute's ingenious idea of raising kites, instead of balloons, to higher altitudes with instrumentation to gain better knowledge about the weather pattern and provide a more meaningful forecast. Kites, reinvented by Lawrence Hargrave, soon became serious scientific apparatus in the hands of the meteorologist. Moore proudly informed Chanute and the press, that their department had succeeded in flying a kite, carrying its self-registering equipment, to a height of 15,800 feet.

With the kites becoming more useful, Charles Marvin, who was working for Moore at the Weather Bureau, explained: "In this age of progress even the boy's kite is made to serve a useful purpose and investigations are now being made under the special direction of Prof. Willis L. Moore, Chief of the Weather Bureau, with a view to employing kites for the purpose of sending meteorological instruments to high elevations, so as to gain better information respecting the nature and causes of atmospheric phenomena than can be done from observations at the surface of the earth. We are sometimes told by naturalists, that man is a descendant of the monkey and that by processes of evolution the tail has been entirely dispensed with. Exactly this same sort of evolution is going on before our eyes today. Kites are rapidly losing their tails, and those of the future are sure to be made altogether without tails. Among the great variety of sizes and shapes tried by the Weather Bureau, none have tails" [121].

To promote research and increase knowledge in aeronautics, Chanute mailed a check for $100 to the Boston Aeronautical Society to be used as a prize for the best monograph on kites. Hearing of the award, Marvin showed his just finished paper on the theory of the mechanics and stability of kites to his supervisor Willis Moore, who forwarded a copy to Chanute, wondering if this was what he had in mind when offering the prize. Yes, it sure was, and the Boston Aeronautical Society awarded the "Chanute Prize" to Charles Marvin for his kite paper [122].

The well-known American inventor Hiram S. Maxim, who had moved to England in the early 1880s, had experimented with a flying machine for some time. Spending about $100,000, he developed, built and then tested his steam-powered, heavier-than-air machine on July 31, 1894, at Baldwyn's Park outside of London. On his third launching attempt, Maxim's huge 8,000-pound white bird with four wings and two 17-foot two-bladed propellers rose up for a few seconds and was damaged. Chanute described the launch objectively in the September 1894 *Railroad Journal*, while the editor of *Scientific American* reported sarcastically, "A moment later it lay stretched on the ground, like a wounded bird with torn plumage and broken wings." Maxim's craft, a marvel of engineering ingenuity, did not fly at ninety miles an hour for a thousand miles as predicted by its inventor, but, according to the noted historian Charles Gibbs-Smith, this was the second successful attempt in powered flight carrying a man, with Clément Ader being first [70] (Fig. 2.14).

Spending a few late winter weeks in Thomasville, Georgia, provided Chanute more opportunities for bird watching. He described his observations to Wilhelm Kress in Austria: "The soaring vultures (C. Aura) are abundant here and they are aloft all day long, whenever there is wind. When the wind is light, they sweep in circles, but when the breeze is brisk, they progress in straight lines in every direction, even dead against the wind, without losing

THE MAXIM FLYING MACHINE

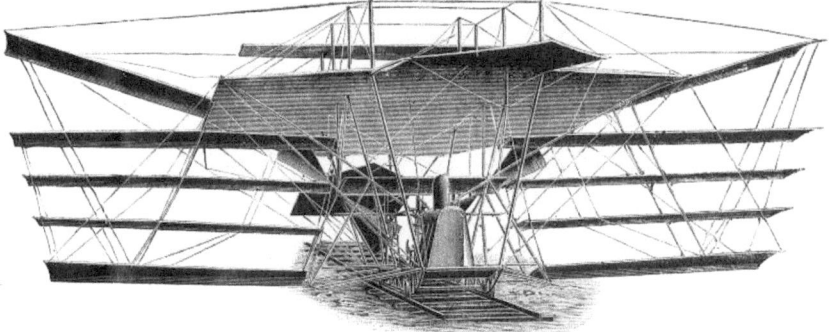

Fig. 2.14 Maxim's flying machine. *Engineering* (London), August 10, 1894

height, giving a practical demonstration of what the French call 'aspiration'. I am more and more convinced that there is a phenomenon which we do not yet understand, and which, when thoroughly mastered, may enable us to join the birds in the sky" [123].

After spending almost ten years researching, lecturing and writing, even building and flying models, Chanute was highly aware of the many unknowns blocking the way toward a practical flying machine. In 1892 he concluded that the technology for powered flight was almost in hand, but "in such complicated matters theory cannot progress in advance of experiment," and "Science has been awaiting the great physicist, who, like Galileo or Newton, should bring order out of chaos in aerodynamics and reduce its many anomalies to the rule of harmonious law. It is not impossible that when that law is formulated all the discrepancies and apparent anomalies which now appear, will be found easily explained and accounted for by one simple general cause, which has been hitherto overlooked" [74].

During this crucial time of development, the self-taught civil engineer Chanute continued his personal learning by maintaining a steady correspondence with the foremost aeronautical researchers of the times, including Hiram Maxim, Thomas Moy, Horace Phillips, Percy Pilcher and Francis Wenham in England, Lawrence Hargrave in Australia, Otto Lilienthal and Herman Moedebeck in Germany, Wilhelm Kress and Karl Milla in Austria, Stéphane Drzewiecki, Gustave Eiffel, Étienne-Jules Marey and Victor Tatin in France, and Samuel Cabot, William Eddy, Samuel Langley, Albert Merrill and James Means in the United States. But he also communicated with lesser-known enthusiasts who had either contacted him or whom he had contacted after reading about them in the newspapers.

But in the public's opinion the flying machine and perpetual motion were mostly still regarded in the same questionable light. This all changed in the course of one year. 1896 was a turning point in the development of aeronautics.

3

Theory—Investigating—Practical Flying

Interest in aerial navigation is still chiefly speculative, but a possible invention in the inchoate art has begun. It is a good deal like an old-fashioned garden maze about a summer hotel. It is easy to enter, but once in its midst, so many paths offer choices that it is hard to find the way out, until somebody points out the right way. We are not certain whether we shall ever be able to navigate the air with the certainty as demonstrated by the birds.

Octave Chanute to Science Club, Washington City. January 7, 1896

Almost one-hundred years earlier, Sir George Cayley prophesied, "I feel perfectly confident, that this noble art [of flying] will soon be brought home to man's general convenience, and that we shall be able to transport ourselves and families, and their goods and chattels, more securely by air than by water, and with a velocity of from 20 to 100 miles per hour" [1]. In the early 1890s, this noble art of flying was still a dream, but an editorial in the *Evening Star* of October 18, 1890, announced boldly that "the era of aerial volitation is approaching rapidly."

Aerial volitation, or flying by mechanical means, was the goal of Carl E. Myers and his wife Carlotta, owners of a balloon farm near Frankfort, New York. They had developed a "Sky-Cycle" in the early 1890s and Carl explained to a reporter of *The World*: "What I have invented is an intelligent balloon. It is a machine which will rise in the air, stay there all day long, and can be moved more or less at pleasure. This is the nearest approach that has yet been made to navigating the atmosphere" [2]. A *The World* reporter

© The Author(s), under exclusive license to Springer Nature
Switzerland AG 2023
S. Short, *Flight Not Improbable*, Springer Biographies,
https://doi.org/10.1007/978-3-031-24430-8_3

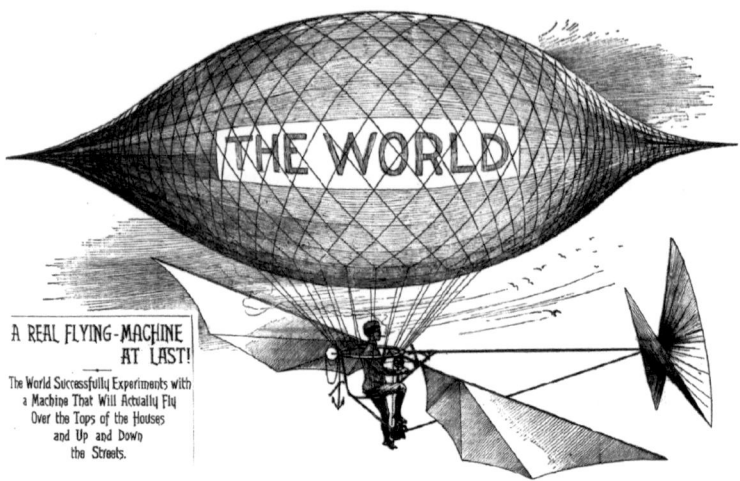

A REAL FLYING-MACHINE AT LAST!

The World Successfully Experiments with
a Machine That Will Actually Fly
Over the Tops of the Houses
and Up and Down
the Streets.

Fig. 3.1 The reporter described this craft, designed and built by Carl Myers, as a real flying machine with wings and a big propeller, and not a balloon. *The World*, August 4, 1895

flew the huge egg-shaped balloon, driven by two propellers, successfully over Brooklyn, much to the delight of the public and the newspaper (Fig. 3.1).

But was this mode of moving through the air the solution of the manflight problem? Alexander Graham Bell did not think so and explained to a reporter, "What is wanted for rising in the air is not the cumbrous balloon nor an unmanageable imitation of the wing of a bird or insect, but a mechanical contrivance, that can be readily operated by ordinary motors."

Even the casually interested person could sense that there was "something in the air," and soon the "Come Fly with me" slogan stopped to be a mere figure of speech. The efforts of three long-time researchers pointed the way in 1896 for the next generation of experimenters to achieve success and bring the flying machine one giant step closer to reality; they left their mark on all who followed. These were the mechanical engineer and industrialist Otto Lilienthal from Berlin, Germany; Samuel P. Langley, the self-educated physicist and Secretary of the Smithsonian Institution in Washington, DC; and Octave Chanute, one of the country's leading civil engineers, self-educated and self-employed, from Chicago.

3.1   The "Flying Man" from Berlin

The study of bird flight had always been of interest to **Otto Lilienthal** (1848–1896) and his one-year younger brother Gustav. The teenagers began by experimenting with bird-like wings in the 1860s, but the human muscles were not strong enough to reproduce the bird's motion of flapping their wings.

To better understand the requirements to fly, the mechanical engineer Lilienthal began by analyzing the flight problem systematically. He concluded that flat surfaces present undue resistance, and that success in manflight can only be expected from concavo-convex surfaces. He compiled his theories and conclusions into a manuscript with tables, graphs and drawings, and offered it to several publishing houses in Berlin; there was no interest in such a literary project. He finally found a publisher who agreed to print 1,000 books, but only if the printing and distribution expenses were shared. Lilienthal's 200-page book *Der Vogelflug als Grundlage der Fliegekunst* [3] was ready for distribution late in 1889.

To reach a broader international audience, review copies were also mailed to several universities in the United States, including to Harvard and the University of Chicago. One of the first book reviews appeared in early February 1890. "In a book on the flight of the bird as a basis of the art of flying, a German author, Herr Otto Lilienthal, describes the result of 23 years of experimenting by himself and brother on the form of wings best adapted for carrying heavy bodies. He concludes—as have others—that the real secret of a bird's flight lies in the arching of its wings, which accounts for the small expenditure of strength, and he believes that close imitation of the birds is the only method of solving the problem of human flight. The artificial wings of the investigators have been so effective as to carry half the weight of the operator, and the apparatus, worked by foot-levers, being made to rise with a person weighing 160 pounds when a counterweight of 80 pounds is helping to lift by means of suitable pulleys" [4]. Lilienthal's description of his whirling arm experiments in 1868 must have fascinated the reviewer. Seeing the write-up on the wire service impressed the editor of the *Daily Inter Ocean* from Chicago; he republished the last sentence as a news item in late June. More than thirty local papers picked it up, including *The Evening Item*, produced for the residents of the West Side of Dayton, Ohio, by Orville and Wilbur Wright [5]. Was curiosity or sensationalism the reason for republishing it in their paper?

Having developed his theories, Lilienthal built his first full-size glider and began practical tests in the summer of 1891; he taught himself to fly by jumping from a three-meter (about ten feet) high diving board to the lawn below in his backyard, while trying to hold the wings steady. Slowly raising the board to thirty feet he discovered that he could change the direction of flight by changing the position of his body (or the center of gravity) with corresponding leg movements, and he told his co-workers that sailing flight was not the exclusive prerogative of the birds anymore.

In the early 1890s the economic conditions were as bad in Germany as they were in the U.S., and Lilienthal hoped that the sport of flying could bring in money. He had submitted the claims of his *Flugapparat* [or flying machine] to the German Patent Office in November 1891, and the patent was issued in September 1893. One year later, he submitted revisions with new drawings and then mailed the claims to Austrian, British and American Patent Offices [6]. Writing to his sister Marie who had recently emigrated to New Zealand, "I have been pushing the flying thing a bit lately. I think the invention can make money, and I am now looking for an investor who can build a sports ground for flying exercises near Berlin. When I now launch such a flying sport, assuming that my patented flying machines were to be used, a good source of income would open up for me. The machine shop is now very miserable, it is not possible to think of earnings this year" [7].

Writing to Hermann Moedebeck, Lilienthal explained: "My recent publications concerning this invention have aroused such public interest that I feel you will soon hear more about attempts at glider flight. I even had to construct a special kind of factory for the machine and have trained a 'flight technician' [Paul Beylich] who runs it for me. My customers are so anxious to order my machine that they send the 300 Mark down-payment right away, which I require, to assure prompt delivery. If this continues, it could become quite nice for me. Next year should bring me a great deal of progress, as the flight of my glider has occasionally shown a glide ratio of 1:10 without the aid of powerful beating wings" [8].

To improve his flying skills and make longer flights, Lilienthal built an almost 50-foot-high "Fliegeberg" (aviator hill) with a glider "hangar" at the hill top to store his equipment near his factory in Lichterfelde. He estimated that his best glides were about 0.25 km (about 820 ft) in length, but he never took the time to actually measure the distances flown. He also guessed that he made about 1,100 flights or glides in the summer of 1894, being airborne for almost three hours.

By demonstrating his glider regularly, Lilienthal's fame spread quickly. Stories and photographs of the "Flying Man" appeared in newspapers and magazines, as the recently introduced half-tone printing process allowed photos to be reproduced at a relatively low cost. Seeing a write-up in the Boston Sunday *Herald* about the "*Man who soars like a bird*," Chanute wondered if the tale was really true. His previous two letters to Lilienthal did not receive a response, so he wrote another one and mailed it to Lilienthal's company [9]. "I need hardly say that I will be pleased to correspond with you, especially if you write in English, as I do not read German." On the very bottom of the letter Chanute added that he was Ex-President of the ASCE and still President of the Associated Engineering Societies of the United States [9]. This time Lilienthal responded, writing in German, and mailed his latest article from the January 1894 *Prometheus*. A fellow Western Society of Engineers (WSE) member translated it and Chanute published "*Why is artificial flight so difficult an invention*" in December 1894 [10]. Here Lilienthal wrote, "The mercilessness of the wind toward all flying machines is a new difficulty in imitating bird flight. With me the wind has often played catch-ball, when, in my sailing [or soaring] practice, I was surprised by gusts of wind in the middle of my course. I was suddenly lifted house high and thrown hither and thither in such a way that, before I became accustomed to it, I lost my breath. In this, one may become an air-gymnast in the boldest meaning of the word, and I believe I am entitled to an opinion on the effect of the wind on free flying machines and the reason of its destructive action. Many tricks are necessary, like moving the legs to shift the center of gravity in order to bring the machine unharmed back to *terra firma*. If we did not actually see birds flying with ease and battling with the wind we should despair of flight." By translating and republishing the articles, Chanute created the scientific link between the "Flying Man" from Germany and the aeronautical community in the United States.

Lilienthal used the forces supplied by nature to obtain support and propulsion to achieve flight [11], a philosophy shared by Chanute. "The method of carrying on these adventures is for the operator to place himself within and under the apparatus, which should be light enough to be easily carried on the shoulders or by the hands, and to face the wind on a hillside" [12]. The operator should glide parallel to the sloping ground, "Go dead against the wind and practice alighting until you gain confidence. The key to success is practice, practice, practice" (Fig. 3.2).

Several American engineers visited Berlin and watched Lilienthal perform, "Dr. Lilienthal began by putting on his fair-weather wings, and, ascending to the top of the hill, he jumped off. Something however was wrong about the apparatus, and in a moment, he landed prostrate on the ground. Fortunately, he was not injured, and, taking off the unfaithful wings, he put on a second pair, which is similar in construction, but of smaller spread, being destined for windy days. With this apparatus he essayed a second flight, with success, passing over the heads of the spectators, and landing safely a little beyond the foot of the hill" [13]. A few months later Samuel Langley also stopped in Berlin and reported to Augustus Herring in Washington: "Although I was only part of two days in Berlin, I managed to go out to Lilienthal's place

LILIENTHAL'S LATEST FLYING-MACHINE

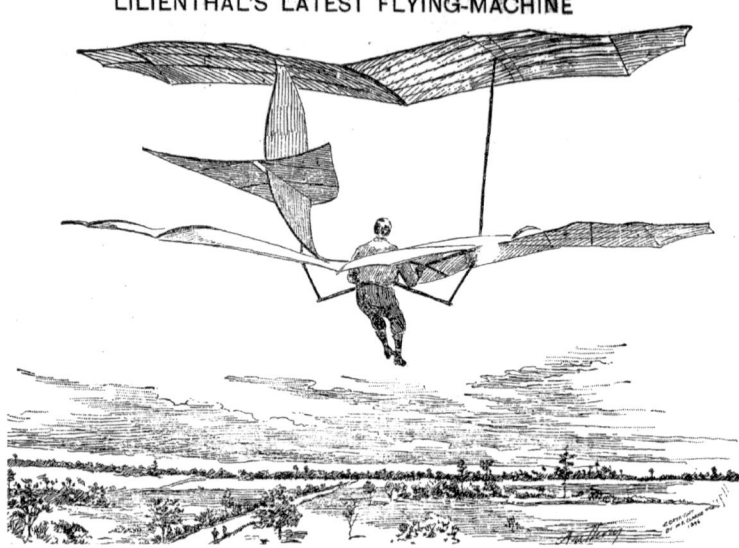

Fig. 3.2 The reporter explained that Otto Lilienthal from Berlin, Germany, had done much towards solving the interesting problem of manflight. *The World*, February 23, 1896

and see him take several flights. They were interesting, but I did not feel that I learned much. His flying machine looks unnecessarily heavy. Could not describe the curves, as I had no opportunity of measuring, but the aspect of the whole was heavy and clumsy—however, handsome is as handsome does" [14]. Indeed, Lilienthal's glider structure was very different to Langley's aerodrome.

Between 1891 and 1896, Lilienthal designed and built at least twenty gliders with the care and expertise of a trained mechanical engineer. Making gradual improvements with each new design, Lilienthal solved some of the initial problems of manflight and picked up unknown bits of knowledge regarding the air surrounding him. With great prudence, he was the only person in the world to fly reproducible and reliable flights in his gliders. Yes, these flights were all unpowered and controlled by shifting the weight of his body, but it was true flying, witnessed by thousands.

Chanute shared the lessons he learned from Lilienthal with his readers: "It seemed reasonably possible for designers of soaring machines to experiment with their apparatus without further search for some hidden secret, for Herr Lilienthal says that his experiments have taught him that there is no mystery about sailing flight, and that the wind is sufficient to account for it. Inventors need not look for some new mysterious force, nor need they be afraid that they will be considered lunatics."

Lilienthal crashed on August 9, 1896, after an unsuccessful attempt to steer his favorite monoplane glider through a strong thermal (called a heat eddy at the time) [15]. He died the following day in a Berlin hospital, ending a promising career. Newspapers around the world published the sad news in the next few days, stating: "The apparatus worked well for a few minutes, when it suddenly got out of order, and man and machine fell to the ground."

Several European correspondents shared the news with Chanute. Wilhelm Kress, from Vienna, wrote a five-page detailed letter in the cursive script handwriting, which one of the older German WSE members needed to read and translate. Here Kress described his last visit with Lilienthal a few weeks earlier and assumed that he had crashed flying this biplane that he saw. Chanute recalled the letter many years later [16]: "Two weeks before that distressing loss to science, Herr Wilhelm Kress, the distinguished and veteran aviator of Vienna, witnessed a number of glides by Lilienthal with his double-decked apparatus. He noticed that it was much wracked and wobbly and wrote to me after the accident: The connection of the wings and the steering arrangement were very bad and unreliable. I warned Herr Lilienthal very seriously. He promised me that he would soon put it in order, but I fear that he did not attend to it immediately. In fact, Lilienthal had built a new machine, upon a different principle, from which he expected great results, and intended to make but very few more flights with the old apparatus. He unwisely made one too many and was the victim of a distorted apparatus. Probably one of the joints of the struts gave way, the upper surface blew back and Lilienthal, who was well forward on the lower surface, was pitched headlong to destruction."

Contemporary sources however confirm that Lilienthal did not make his last flight in an "out of perfect order" biplane glider. He flew a monoplane, built the previous year, which was in good working condition, but the misinformation from Kress, repeated by Chanute, still propagates through the literature today.

"This deplorable accident removed the man who has hitherto done most to show that human flight is probably possible, who was the first in modern times to endeavor to imitate the soaring birds with full-sized apparatus, and who was so well equipped in every way that he probably would have accomplished final success if he had lived" [12]. As sad as the accident was, Lilienthal's successes, followed by his abrupt death, were an attention-getting catalyst. By sharing what he had learned, Lilienthal laid the foundation for further research and motivated individuals to take aeronautics to the next level. Some pessimists took his death as proof that humankind was not meant to fly, however other individuals reacted to the news with even greater determination to get into the air.

3.2 Journalism off its Feet!

Being a bit of a showman and a salesman, Lilienthal stated that his wings were sufficiently improved to provide a new sport. "Who can doubt for a moment that this motion through the air affords greater joy than any other kind of sport? Who can doubt the indescribable pleasure of this novel power? Whoever has launched himself upon the flying trapeze knows the matchless thrill of the moment when the hands leave the bar and stretch forward easily for the one hanging in the air beyond. The suggestion of flying as a sport is worthy of serious attention by all our athletic associations. What athletic club will be the first to see that a sufficient number of its members practice with wings, and then to put up proper prizes for the men who sustain themselves for the longest distance in the air? Here we have a new kind of sport, bewildering in its attractiveness, and promising to give us all but perfect victory over the air which has teased and baffled us with increasing dissatisfaction since the youthful Icarus sank into the Aegean Sea. Bring on your wings! Let us fly!" [17].

In late April 1896, only one month after this editorial appeared, William Randolph Hearst, owner and publisher of the *New York Journal*, imported a Lilienthal standard glider, hoping that stories of inspired reporters trying to fly it would boost the circulation of his newspapers. The *Journal* published its first report on May 3 with the title "*A Flying Machine at last that really flies*" [18], and the *New York Sun* published its report the next day [19], describing the first attempts of flying the glider at the residence of J. Harper Bonnell, near Garrettson, Staten Island.

"In and around the city there are a dozen men or more who have at one time or another tried to fly, and the majority of them are now walking lame or have not recovered the full use of their arms or have repaired ribs or some other disability consequent upon their efforts to invade the realm of air. This doesn't prevent other enthusiastic amateurs from trying to play bird, and as surely as comes the spring, so surely does some aspiring genius bump himself violently against Mother Earth through the medium of wings that fail to fly".

To fly the glider was not the easiest thing to do, especially without good instructions. "Attempted remonstrances by the artist were cut short by the enthusiastic shouts of the entire staff, and a rope was soon tied around the experimenter's waist. All concerned attached themselves to the other end of the rope, and started a run. Anyone who had ever seen a paper kite bunt and batter along the ground can appreciate the unpleasant features of the artist's progress for a few yards. All the breath was knocked out him, and he didn't recover it until he found himself twenty-five feet above the earth, hovering. He had shot up like an arrow from a bow (Fig. 3.3).

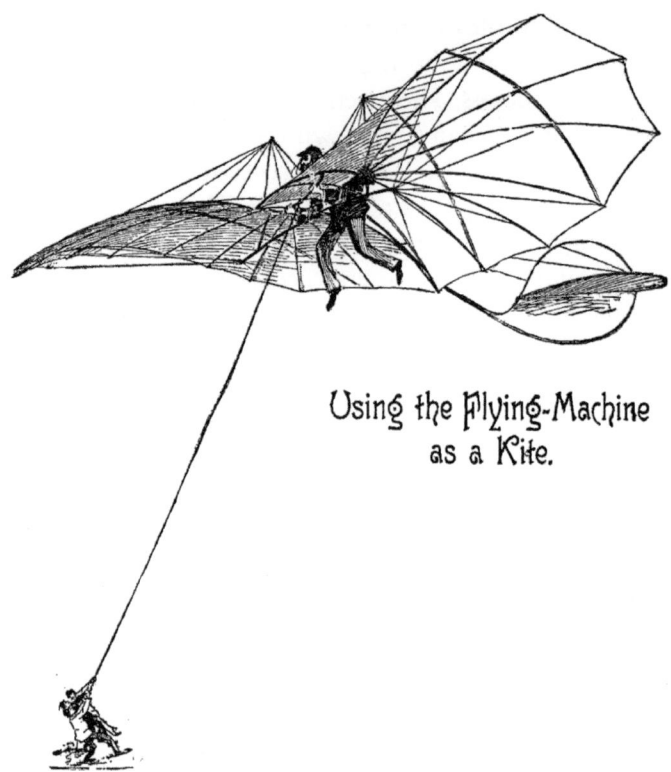

Using the Flying-Machine
as a Kite.

Fig. 3.3 Launching the flying machine as a kite was just another way to get the Lilienthal glider with its pilot airborne. *The Journal*, New York, May 3, 1896

"A peculiar thing then happened. The flying machine turned slowly backward, and with a dive slanted down to earth, landing on the tip of one wing with a tremendous crash. They extricated the amateur bird from the wreck and sent him to a doctor to be patched up. He had ploughed a hole in the ground with his head to the great detriment of one ear and part of his scalp, and several of his ribs were the worse for wear. As for the machine, that was in no condition for further experiments. The aerial editor who had his pictures of a mere speck in the sky, representing the flying machine, all ready for the next Sunday issue, had to give up all hopes of getting the article prepared in time. To find a man who could mend the thing was a hard job. One was found, and the athlete who had made the original attempt was persuaded to make another try. After a number of attempts he finally got himself launched, skimmed along the earth, his feet dragging, and finally, with the exhortations of the aerial editor and his staff to 'curl up those damned long legs' ringing in his ears, he succeeded in detaching himself from *terra firma* for a distance of about forty feet, after which he trailed some distance further before standing

on his head as a finish. Measurements of the space from start to this undignified finish were made, and the 'flight' was declared to be seventy-five yards.
It was voted a grand success. The meeting then adjourned after unanimously
resolving, 'Man can fly.' In comparison with this forty-foot-in-the-clear flight,
it is interesting to note that Herr Lilienthal succeeded two years ago in soaring
300 yards from a low hill against a stiff breeze. There seems to be no danger
of his record being broken at present in this vicinity" [19].

In the next couple months, several people tried to fly the great white
artificial bird with mixed success, but hearing of Lilienthal's accident in
early August convinced Hearst to put the glider in storage. In 1905, Hearst
purchased the *Cosmopolitan Magazine* from John Brisben Walker, who had
promoted aeronautics in the 1890s, and the glider was part of the financial
transaction. Walker agreed to display the fragile glider at the first Aero-Club
exhibit in New York in January 1906 and then donated the craft to the
National Museum in Washington in February [20].

3.3   The Secretary of the Smithsonian Institution

Being interested in astronomy, **Samuel Pierpont Langley** (1834–1906)
joined the Allegheny Observatory in 1867 and taught classes at the Western
University of Pennsylvania in Pittsburgh. Flight had always interested him,
but it was Israel Lancaster's talk at the 1886 AAAS meeting in Buffalo that
aroused his dormant attention. The discussion following the lecture was so
ludicrous that Langley decided to explore the basic laws of aerodynamics.
The only publications available at that time were from the French and British
aeronautical societies, but in these, as in everything then accessible, fact had
not yet been discriminated from fancy or fiction.

Shortly after the meeting in Buffalo, Langley was invited to the post
of Assistant Secretary of the Smithsonian Institution and was elected the
third Secretary one year later. Continuing his research work at the Allegheny
Observatory a little longer, he built a 60-foot diameter whirling arm, driven
by a 10 hp engine, and experimented with rubber-powered models using
flat wing surfaces, as he believed that the laws of aerodynamics could be
easier understood [21]. The Smithsonian Institution published his research in
1891 as "*Experiments in Aerodynamics.*" In a letter to the Assistant Secretary
of the Smithsonian, and probably much to the delight of Langley, Chanute
commented in April 1896: "It is significant that prior to the publication of
Langley's work, it was the rare exception to find engineers and scientists who

would fully admit the possibility of man being able to solve the centuries old problem of aviation. Since the publication of '*Experiments in Aerodynamics*' however, it is the exception to find an intelligent engineer who disputes the probability of the eventual solution to the problem of manflight" [22].

Moving to Washington, Langley focused his research on the action of the wind. The story goes that the almost sixty-year-old life-long bachelor stood on a bridge across the Potomac in a bitter cold December gale, watching a buzzard teetering and balancing in the cold blast. Responding to a reporter, "Did you ever think what a physical miracle it is for such a bird, like our common turkey buzzard, to fly the way it does? I am not speaking of birds, which fly by flapping their wings, but of those which fly without flapping their wings. You may see them any day along the Potomac floating in the air with hardly the movement of a feather" [23]. Langley concluded that it was theoretically possible for a heavy body to move to the east while the wind was blowing to the west, using only the lifting power furnished by the wind. "No satisfactory mechanical explanation would be offered in this connection by the writer were he not satisfied that it involves much more than an ornithological problem, and that it points to novel conclusions of mechanical and utilitarian importance. They are paradoxical at first sight, since they imply that, under certain specified conditions, very heavy bodies, entirely detached from the earth can be sustained indefinitely without any expenditure of energy from within. This is, without misuse of language, to be called a physical miracle" [24] (Fig. 3.4).

A few days after Hearst's Lilienthal glider was flown at Staten Island, Langley and his team launched their steam-powered aerodromes over the Potomac River. A few days later Alexander Graham Bell told a reporter from *The World*, "Last Wednesday, May 6th, I witnessed a very remarkable experiment with Prof. Langley's aerodrome on the Potomac River. Indeed, it seemed to me that the experiment was of such historical importance that it should be made public. The aerodrome, or 'flying machine,' in question was of steel, driven by a steam engine. It resembled an enormous bird, soaring in the air with extreme regularity in large curves, sweeping steadily upward in a spiral path, with a diameter of perhaps 100 yards, until it reached a height of about 100 feet in the air. When the steam gave out, the propellers stopped, and it settled as slowly and gracefully as it is possible for any bird to do, touched the water without any damage and was immediately picked out, ready to be tried again. A second trial was like the first. No one could have witnessed these experiments without being convinced that the practicability of mechanical flight had been demonstrated" [25].

PROF. LANGLEY'S FLYING-MACHINE REALLY FLIES.

Secretary of the Smithsonian Institution Has Invented a Steam-Propelled, Winged Aerodrome That Last Week Flew a Half-Mile, Raising Itself 200 Feet in the Air----Scientists Say He Has Solved the Problem of Human Flight.

DRAWN FOR THE SUNDAY WORLD FROM PROF. LANGLEY'S ORIGINAL MODEL.

Fig. 3.4 Aerodrome No. 5 in flight, *Langley Memoir on Mechanical Flight*. Smithsonian Institution, 1911, and the newspaper's more sensational interpretation of this model in the air, shown on the front page of The *World's Sunday Magazine*, May 17, 1896

The flight of aerodrome No. 5 was quite a success and an accomplishment that even today's modelers should admire. Langley believed that his machine had solved all problems of manflight, and he ventured to predict, "that the air may probably be made to support engine driven flying machines before the expiration of the present century." Reading in the papers about these flights, Chanute did not think that the problem of flight was solved. And then hearing from Herring that Langley did not want to discuss his work with anyone, Chanute simply stated: "I am rather sorry that he did not accept my offer to communicate my knowledge to him as frankly as it was made; and yet it is perhaps best that each of us should remain in his portion of the field: Professor Langley working out dynamic flight, and I experimenting on soaring flight" [26].

Wanting to now construct a larger machine, Langley puzzled over the efficiency of wings and decided to consult the best informed of all men on this subject [27]. He wrote a confidential letter late in 1897 to Chanute, who responded by sending a table showing the Duchemin data on flat plates, the Lilienthal data on curved surfaces, and some of his own calculations on drift vs. lift. The thankful Langley responded right away: "You and I have been working at substantially the same problem from opposite ends, and I should express the hope that we might meet in the middle, were I sure of going on; but the next step in my experiments would be an expensive one, the construction of an aerodrome capable of carrying a man, or men, for a long flight ... I consider it a perfectly legitimate subject for business enterprise. If you hear of anyone who is disposed to give the means to such an unselfish end, I should be very glad to meet him" [28]. Agreeing with some of Langley's statements, the two-year older Chanute responded that he would "truly like to meet in the middle before either of them flew off into the hereafter," but, he knew nobody willing to provide funding for a purely scientific experiment, "nor do I see what promise of financial profit, or of fame, could be made to a rich man furnishing the funds, while I do see a decided danger of an experimenter being injured, if not killed. It is for this reason that I have confined myself to the question of the best method of promoting safety, and have experimented wholly at my own expense. I am profoundly grateful that I have had no accidents."

The beginning of the unpleasantness with Spain, and a possible war, awakened Uncle Sam to the fact that if a successful aerial warship could be constructed, the United States should be the first nation to own one. Charles D. Walcott, Langley's assistant secretary, met on March 25, 1898 with Theodore Roosevelt and described Langley's aerodrome that had flown for a minute and a half two years earlier. The result of the meeting was a memo from Roosevelt to the secretary of the navy that "the [Langley] machine has worked. It seems to me worthwhile for this government to try whether it will not work on a large enough scale to be of use in the event of war." Roosevelt proposed forming a committee, composed of Army and Naval officers, and "outside experts like R. H. Thurston, President Sibley College, and O. Chanute, President of the American Society of Civil Engineers, at Chicago." This new government committee was to make recommendations as to the practical approach of constructing a flying machine and to submit estimates of the cost. It was, as Roosevelt noted, "well worth doing" [29]. As hoped, President McKinley became impressed with the possible value of an aeroplane and appointed a Board of Army and Navy officers who recommended that a full-size man-carrying flying machine should be developed for war purposes. In November 1898, the Board of Ordnance made an allotment of $25,000 to develop, construct and test an airship.

The optimistic Langley shared the latest news with Robert Thurston. "I mention in special confidence that official inquiries from the Army and Navy lead me to think it possible (I hardly can say probable) that an aerodrome, capable of a speed of 30 miles an hour maintained for three hours, carrying an aeronaut and possibly some missiles may be attempted." In closing, Langley wrote, "Have you any young man who is morally trustworthy (a good fellow) with some gumption and a professional training? Please let me hear from you by a letter or are you coming to Washington?" [30]. Thurston discussed this job with several students and a few weeks later, Charles M. Manly, a senior majoring in mechanical engineering, joined Langley's team, in charge of the construction of the manned aerodrome.

3.4   The Civil Engineer Takes the Next Step

To develop a flying machine, Chanute had to draw freely on his imagination, just as Albert Einstein stated a quarter century later. "I am enough of the artist to draw freely upon my imagination. Imagination is more important than knowledge; knowledge is limited; imagination encircles the world" [31].

Having received the 198-page "*Taschenbuch für Flugtechniker und Luftschiffer,*" edited by Captain Hermann Moedebeck of the German Artillery, Chanute appreciated seeing the ads for his *Progress* book but also for Otto Lilienthal's "*Sailing Apparatus.*" Lilienthal's article, "*Kunstflug*" (or translated "artificial flight"), with its graphs and tables, really piqued his curiosity, as he had created a table of normal and tangential pressures by tabulating values for the coefficients η (labeled as "normal") and ∂ (labeled as

116   THE AERONAUTICAL ANNUAL. (1897)
                         EXTRACTED FROM "SAILING FLIGHT"

TABLE OF NORMAL AND TANGENTIAL PRESSURES

Deduced by Lilienthal from the diagrams on Plate VI., in his
book "Bird-flight as the Basis of the Flying Art."

a Angle.	η Normal.	ϑ Tangential.	a Angle.	η Normal.	ϑ Tangential.
$-9°$	0.000	$+0.070$	$16°$	0.909	-0.075
$-8°$	0.040	$+0.067$	$17°$	0.915	-0.073
$-7°$	0.080	$+0.064$	$18°$	0.919	-0.070
$-6°$	0.120	$+0.060$	$19°$	0.921	-0.065
$-5°$	0.160	$+0.055$	$20°$	0.922	-0.059
$-4°$	0.200	$+0.049$	$21°$	0.923	-0.053
$-3°$	0.242	$+0.043$	$22°$	0.924	-0.047
$-2°$	0.286	$+0.037$	$23°$	0.924	-0.041
$-1°$	0.332	$+0.031$	$24°$	0.923	-0.036
$0°$	0.381	$+0.024$	$25°$	0.922	-0.031
$+1°$	0.434	$+0.016$	$26°$	0.920	-0.026
$+2°$	0.489	$+0.008$	$27°$	0.918	-0.021
$+3°$	0.546	0.000	$28°$	0.915	-0.016
$+4°$	0.600	-0.007	$29°$	0.912	-0.012
$+5°$	0.650	-0.014	$30°$	0.910	-0.008
$+6°$	0.696	-0.021	$32°$	0.906	0.000
$+7°$	0.737	-0.028	$35°$	0.896	$+0.010$
$+8°$	0.771	-0.035	$40°$	0.890	$+0.016$
$+9°$	0.800	-0.042	$45°$	0.888	$+0.020$
$10°$	0.825	-0.050	$50°$	0.888	$+0.023$
$11°$	0.846	-0.058	$55°$	0.890	$+0.026$
$12°$	0.864	-0.064	$60°$	0.900	$+0.028$
$13°$	0.879	-0.070	$70°$	0.930	$+0.030$
$14°$	0.891	-0.074	$80°$	0.960	$+0.015$
$15°$	0.901	-0.076	$90°$	1.000	0.000

Fig. 3.5 This table with normal and tangential data, assembled by Otto Lilienthal
and published in H. W. Moedebeck's *Taschenbuch* in 1895, was translated by Octave
Chanute and published in the *Aeronautical Annual* of 1897 as part of his article on
"*Sailing Flight*"

"tangential") vs. the angle of incidence. Chanute succeeded to calculate "lift" and "drift" on wing surfaces of a 1/12th curvature (Fig. 3.5).

Hearing from James Means that he needed material for his second *Aeronautical Annual*, Chanute agreed to write up what he knew about "*Sailing Flight*." To stress that the topic of sailing or soaring flight was thoroughly studied, even though no one could explain this phenomenon mathematically, he listed the names of more than 30 authors, "Many theories have heretofore been advanced to account for the paradox of sailing flight," and he explained, "It may be added that the simplest and most satisfactory explanation of soaring thus far is that which assumes ascending columns or trends of wind to exist at opportune times and places, but that it does not account for the cases where the wind is horizontal." As the Lilienthal table had helped him, Chanute included this table in his follow-up article, published the following year. This article with the table contributed significantly to the evolution of the aeroplane in the next decade [32].

Having studied Lilienthal's patent and photos, Chanute submitted his claims for an improved "*Soaring Machine*" to the United States Patent Office in December 1895 [33] (Fig. 3.6). His claims were to take the glider one step closer to imitating the bird's soaring flight, but also make it safer for enthusiasts to fly. The operator should only move the body to displace the center of gravity of the machine, "which is accomplished while the wings are held rigidly still, so far as any flapping movement is concerned. It is necessary, however, to provide for a fore-and-aft movement of the wings in order to preserve the equilibrium." Continuing his explanation, "The present invention is an improvement upon the flying machine recently patented to Otto Lilienthal, in which he provides a pair of wings fastened rigidly to a hoop and composed of fabric stretched over hinged ribs, so as to allow the wings to be folded for transportation … The aviator must keep his balance by moving his body so as to displace the center of gravity of the machine. In my invention I provide a stationary seat and pivot each wing on an upright pintle, so that it can be moved bodily forward and backward, as maybe required to preserve the balance of the machine and the aviator. I also provide a strong spring or springs exerting a constant tendency to pull the wings forward." Wire braces would keep the wings in position, "but permit them to be turned forward and backward. Strong springs between the hoop and the leading edge of the wing were to exert a constant forward pull on the wings. In operation, the springs are arranged to yield sufficiently to allow the machine to be properly balanced when soaring in wind of ten or twelve miles per hour. If the wind strengthens, the wings are forced farther backward. If the wind lessens, the springs pull the wings farther apart. In each case the shifting of the wings compensates for the change in the center of the wind-pressure due to the change in the velocity and keeps the center of pressure coincident with the center of gravity, thereby automatically preserving the equilibrium. By positively moving the wings the aviator can cause the machine to tilt up or down by the action of the wind, and so make it glide upward or downward at will." Chanute, the veteran engineer, envisioned a flying machine to stabilize itself automatically under any wind conditions, and he explained, "without assured equilibrium, safety is uncertain; and without a reasonable degree of safety, flight whether for pleasure or for business, is out of the question."

To maintain his leadership in all things aeronautical, Chanute had to move from theory to practice and test his own ideas about flying. He knew that to advance into the unknown, any researcher should start with experimentation. The first kind of experiment is needed to verify the validity of the work done by others, exchanging second-hand knowledge for first-hand knowledge, and separating fact from error. In the second stage, the researcher needed to proceed into the unknown, using the knowledge gained from the previous set of experiments. Both phases are essential to develop expertise and are part of the evolution process.

Shortly after moving to the Palmer Subdivision in Streeterville in northwest Chicago, Chanute had introduced himself to William "Bill" Avery,

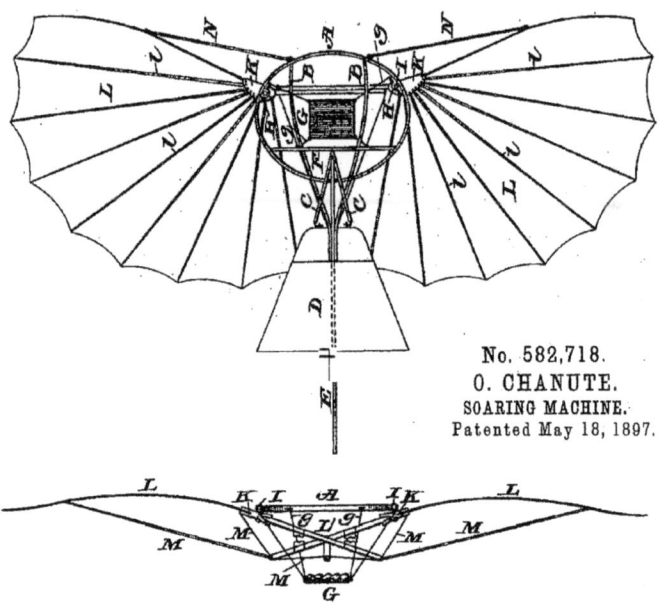

Fig. 3.6 Octave Chanute's patented ideas to make Lilienthal's monoplane safer to fly. No. 582,718

whose father owned a jobbing shop across the alley at 404 Superior Street. The 22-year-old Avery was an avid sailor with good knowledge of wind and weather, knew how to fix everything and was eager to learn a different aspect of the woodworking trade; he agreed to construct Chanute's proposed soaring machine. Being optimistic, Chanute told a friend, "Have made contact with an intelligent carpenter to build my own machine and we will see how much of a failure we can make" [34]. But, then, Avery had second thoughts about building the craft in his shop. Hearing of Herring being ready to quit his job with Langley, Chanute decided to rehire him.

Annie Chanute and her daughters had left Chicago in late December for a warmer climate, and he needed to attend an out-of-town meeting in New York, so he wrote Herring that he and his family could live at first in their house; the servants would show them around. And he enclosed his work instructions, "You will find all the plans, models etc. on my library table. You had better go to work at once. You can use my library and Avery's shop. The main arms are to be of spruce, the front edge of wings of ½ bamboo (c), and the ribs of willow, of which samples are in my room. I would make one wing first, and then proceed with the other" [35]. The Herring family with their two rat terriers, called Rags and Tatters (or Mitus and Titus), arrived in early January 1896. Herring also brought along his latest Lilienthal-type glider, which was to be rebuilt, as Chanute wanted to increase the wing area, incorporate some of his patented ideas and reinforce the over-all construction.

3.5   Steal the Bird's Art

Initially, Chanute thought of experimenting with his flying machines in southern California, but Willis Moore, who had recently transferred to the Weather Bureau's Chicago office as their new chief, had informed him that the southern Lake Michigan shoreline had a higher and steadier wind velocity than any other place in America. This sounded very good, as the Indiana dunes were closer to home.

In late spring, Chanute and Herring took the Lakeshore & Michigan Southern train from Chicago to Michigan City, traveling around the great bend of Lake Michigan to inspect the sixty-mile stretch of "wild waste regions of sand hills and marshes" (Fig. 3.7).

Few people were living there, which Chanute found ideal, as he simply did not want curious onlookers. Interestingly, this "wasteland" disappeared quickly a decade later, when the U.S. Steel Corporation established a manufacturing complex along the lakeshore. Situated between the iron ore beds of the northern Great Lakes and the southern Indiana coalfields, with the Lake Michigan sand dunes in the middle, U.S. Steel arguably took the most

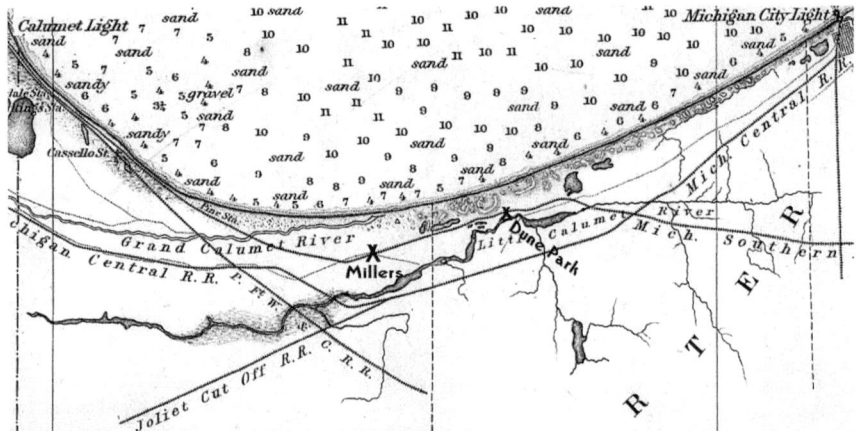

Fig. 3.7 The south end of Lake Michigan, extracted from *Lake Michigan Coast Chart.*
No. 4, Chicago to Kenosha, 1876. University of Chicago Library

dramatic step of any manufacturer in turning the lakeshore into one of the greatest concentrations of industry of the time.

The area around the former mouth of the Calumet River, about 30 miles southeast of Chicago, appeared to be an ideal location. The almost treeless sloping 60–80-foot-high dunes north of Millers looked perfect. Here the men could launch their gliders into several wind directions and land safely on soft sand.

On Monday morning, June 22, 1896, six weeks after Langley's launching of his models, Chanute and his team met at the Chicago Tie Preserving Company in Englewood, Chicago, and carried gear and supplies for the next two weeks to the Englewood Train Station. The Lakeshore & Michigan Southern Express left at 7:24 a.m., arriving one hour later at Millers. At that time, the railroad community consisted of the station, about twenty houses, a small school building, a general store and two saloons.

The colorful group stepping off the morning train caused a hum of excitement among the locals: Chanute, a distinguished-looking well-dressed older man with a gray mustache and imperial, his son Charley, Avery, Paul Butusov and Herring with his two dogs. Each person carried odd-looking luggage, including a Kodak camera, over the mile-long road through town and over the recently built bridge across the Calumet River to the dunes (Fig. 3.8).

The group pitched their tent and assembled the rebuilt Lilienthal-type glider first to begin their flying experiments with a known design before venturing into the unknown. Everybody was sure they could repeat Lilienthal's flying experiences.

Herring, the most experienced aeronaut in the team, took the first jump into the wind, looking like a huge butterfly with concave wings, slowly settling to the ground. Butusov tried next but had an upset. Avery then

OCTAVE CHANUTE, AERONAUT AND INVENTOR.

Fig. 3.8 Octave Chanute, aeronaut and inventor, *The Chicago Record,* June 29, 1896

asked if he could try. Writing a decade later, "My heart was in my throat, but in spite of my fears, I succeeded in making a very creditable flight" [36]. The Lilienthal-type glider, however, was "cranky." Watching the glider in the air confirmed Chanute's concerns about stability, as the pilot had to be an acrobat to keep the craft under control, just as Lilienthal had reported. After making about fifteen jumps, the glider turned over on landing without hurting its pilot and was broken past mending. "Glad to be rid of it," Chanute recorded in his notebook [37] (Fig. 3.9).

Much to everyone's surprise the first report of "fishermen seeing a huge contrivance flitting through space" appeared in the *Chicago Chronicle* newspaper [38] the day after the group had arrived in Millers. The *Chicago Tribune* sent their reporter next, which resulted in a front-page headline: "*Men Fly in Midair. Chicago Experts Make Experiments on Indiana Soil*" [39].

MR. CHANUTE'S 2½-FOOT MODEL, WHICH CAN LIFT FIFTY POUNDS.

MR. HERRING'S AEROCURVE KITE, WHICH PERFORMS "ASPIRATIONS."

MR. HERRING'S FLIGHT WITH THE IMPROVED LILLIENTHAL SOARING MACHINE.

Fig. 3.9 Mr. Chanute's 2½-foot model, which can lift (Chanute corrected this to "pull"). Mr. Herring's aerocurve kite, which performs "aspirations." And Mr. Herring's flight with the improved Lilienthal soaring machine. The *Chicago Record*, June 29, 1896

A reporter from nearby Westchester, Indiana, came on June 27 [40]. And the *Kansas City Journal* was just one national newspapers that picked up the story from the wire service, but they added three laudatory paragraphs about their former citizen, who was so serious about aerial navigation [41]. The last visitor to arrive in camp with his sketchbook was Frank Manley from the *Chicago Record*, who had interviewed Chanute a few months earlier on a bridge project in Chicago. His report, published on June 29, entitled "*Steal the Bird's Art*" [42], shows drawings of the Lilienthal machine and the two kite designs, all in-flight. Chanute thought this report was "less inaccurate than the others," so he mailed the two-column clipping, with his penciled-in corrections, to his correspondents worldwide.

3.6   From Researching to Flying—The Katydid Multiplane Glider

While the eagles and gulls flew effortlessly overhead, Chanute's team assembled the multiplane next. This innovative "soaring machine" consisted of twelve equally-sized wings, each six feet long and three feet wide, similar to the "ladder-kite," which Herring had built as a model and tested the previous year. To reproduce the curvature of the bird's wings, Chanute had consulted with John B. Johnson of Washington University about the characteristics of rattan. Using this wood allowed shaping the ribs in a circular-arc curvature with a height of about 1/12th of the width, trying to reproduce the wing shape of the soaring bird. Avery and Herring stretched the closely woven silk material tightly over the ribs and spars and varnished the cloth with pyroxylene to make it airtight; upon drying, the cloth surface then rang like a drum when tapped with the knuckles [43].

To achieve automatic stability Chanute thought that the wings should pivot to allow adjustment of the center of lift with the wind, but in this first attempt they were fixed in the front and flexible in the rear. He explained the arrangements to a fellow engineer: "The man was hanging down by his arm pits and guided the apparatus by swinging his legs about in the preliminary apparatus, which I made to test the safety and equilibrium of the machine. This will be modified in the next trials, but in no case will the operator recline, as some of the newspaper reports described. I should expect the full-sized machine first to glide downward in light breezes, à la Lilienthal, and

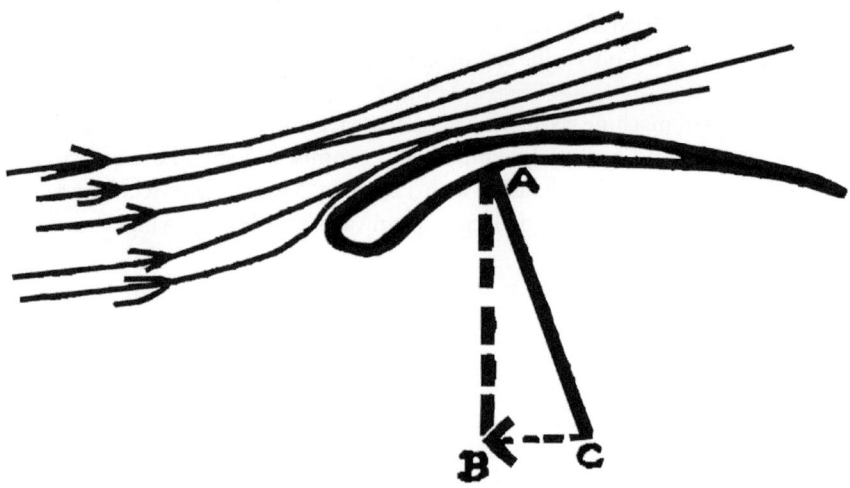

Fig. 3.10 Shape of airflow over a curved wing. Chanute's letter to A. M. Herring, April 26, 1895. The solid line A–C indicates lift; the dotted line A–B supports the weight (or gravity); the horizontal line B–C pulls the machine forward, thus producing "aspiration" (thrust). Chanute Papers, Manuscript Division, Library of Congress

eventually to soar in stronger winds like a bird, either in spirals in a ten-mile breeze, or in direct 'aspiration' in a wind of twenty miles an hour." In Chanute's vocabulary, "ascension" meant gaining altitude after takeoff, while "aspiration" was the performance of a bird using air currents, so it could glide forward in rising air currents to achieve soaring flight.

It did not take long to figure out that the wings of his multiplane greatly interfered with each other. In typical engineering fashion, Chanute played with primary lifting surfaces to achieve the best performance. To obtain maximum lift and steadiness in flight, he had to alter the glider's architecture to find not only the most efficient spacing between the wings, but also the placement of each set of wings (Fig. 3.10).

To see the movement of the air above and below the wings, Chanute had brought along several bags of down feathers. "The paths of the wind currents were indicated by liberating bits of down in front of the machine, and, under their guidance, six permutations were made, each of which was found to produce an improvement in actual gliding flight over its predecessors." After each flight, the men carried the machine back up the dune, while the boss remained near the landing site, ready to record all pertinent data of the next flight [37].

June 24, 1896 Tried the 12-winged plane. Found it steady but lifting too much in front. Tied on 12-lb. bag of sand in front.

June 25 Reset machine as 8-winged. Located c. of pressure just under 1st pair. It proves steadier than Lilienthal's and promises to soar. Found it better balanced and easier to handle than with 12 wings. Made tests of wind currents with down and found it was deflected downward after leaving wings.

June 27 Rigged up machine with 4 wings in front & 8 behind, the latter superposed to 0.4 their breadth. Found this more stable than any arrangement yet tried.

June 29 Found 4 wings front & 8 behind better balanced when standing still in wind, but the front was insufficiently sustained in flight. Next tested with 8 wings front & 4 behind. Found machine nearly as easy to control when standing still and sustaining much better in flight. Both Herring & Avery made glides of 76 ft. against a north wind and pronounced machine more stable in its present shape than the Lilienthal.

June 30 Raised top wings to test whether the increased leverage of wind would make machine more difficult to control, and whether the increased distance between wings would make the lift greater. Added a 5-lb. bag of shot in front.

July 1 Got flights of 30 ft. and found it as easy to handle as with the top wings lower down. After waiting till 5 P.M. for more wind, took off the upper back wing and inserted it in the wide space under front top surface thus having 5 sets wings in front (148 sq. ft.) and having 1 set at bottom in rear (29 sq. ft.). Called this the "Katydid." Found machine steady and manageable.

July 4 Made a number of excellent jumps. Best one: Avery, 78 ft.; Herring, 82' 6."

The team returned to Chicago on July 4. Each team member had learned how to launch and control the aircraft in flight and land safely.

In a step-by-step process, the multiplane had evolved into an ordinary glider. The resulting *Katydid*, named after the common grasshopper, could be handled in a twenty-mile wind. To control the machine in flight, the operator needed to move only two or three inches and not the acrobatic fifteen to eighteen inches that the Lilienthal-type had required. In a thirteen-mile breeze, the *Katydid* with its pilot could glide in an angle a little better than one in four.

The *Katydid* was rebuilt in Avery's shop in late July/early August. Writing to Wenham, "Now, I will try the next set of experiments, that of endeavoring to glide at right angles to a wind blowing up along the hillside" [44] or what Zahm described as "naval soaring" (or ridge soaring in today's language). And he indicated, "I may have something interesting to report in a week or two— or I may have a failure." Changes were mainly made to the wing roots, which were now pivoting on ball bearings, placed at the top and bottom of wooden uprights and trussed together. The pivoting process was envisioned as: "When the forward motion of the machine in flight is slackened, the springs pull the wings forward, thus advancing the center of pressure. The center of gravity, which has remained stationary, causes the supporting surfaces to oscillate towards the rear and to increase the angle of incidence, thus restoring the needed support. When the speed increases, the reverse effect takes place. In practice we have yet been unable to adjust the springs so that this action is entirely automatic, and the man has to move about one inch, as against two inches in the pivoted tail machine [the biplane glider]" [45] (Fig. 3.11).

Fig. 3.11 The multiplane *Katydid* glider and Chanute's envisioned improvements to achieve automatic stability. Chanute Photo Album, Manuscript Division, Library of Congress, Washington, DC (wb0039p-004-11)

Back in the dunes in late August, the rebuilt *Katydid* did not perform as hoped, but Chanute considered this craft the first step toward a stable flying machine. His patent attorney did not think that a patent could be obtained, so he submitted the claims to the British Patent Office as a communication. [46] The improvements over the Lilienthal patent were mainly the automatic preservation of equilibrium. "The rear portion of each wing, back of the longitudinal stretcher, is left free to flex upward by reason of the elasticity of the transverse ribs. This action compensates for abnormal changes in wind pressure when the machine undulates in flight, and the same effect can be produced by mounting the tail wings on a flexible rod. A vertical rudder is hinged to the stern post between the tail wings and may be governed by tiller ropes [or control cables in today's language]." Chanute envisioned a seat for the operator, suspended below the framework for better stability. "Foot-operated levers linked the wings to allow the pilot to operate the wing-swinging system. A cord with a spring 'interposed' ran from the lever around a pulley on the frame, to the front edge of the set of multiple wings. Another cord, also incorporating a spring, ran forward to the frame to act as a return device." The claim in the patent reads: "By shoving on both pedals, the wings will be pulled back, and the machine will tilt over slightly to the front and glide at a flatter angle. On letting both the levers swing back, the springs pull the wings forward, increasing the angle of incidence, tilting the machine up in front, and slowing its headway. This latter affect will be produced automatically in case the operator accidentally loses his hold on the pedals. By operating one pedal alone, the machine may be steered to the right or left." Unfortunately, this claim did not work in reality. If Chanute had thought of crossing the control lines, then a banking motion, utilizing lift instead of drag, might have been achieved by pushing the appropriate lever forward.

During the next decade, Chanute shared drawings of the multiplane glider freely with other experimenters, only asking that they acknowledge his invention.

3.7   From Triplane to Biplane Glider

Appreciating Wenham's glider configuration, Chanute sketched a triplane and handed it to Herring to have it built in Avery's shop. Knowing that the Pratt truss design gave the most rigid, yet flexible and lightweight structure, his new "fixed-wing" glider mirrored his bridges. The craft had three superposed wing panels with a sixteen-foot wingspan and a four-foot three-inch chord, rigidly trussed with "uprights" of spruce and steel wire. The bottom wing had a cut-out for the operator, and the curved wings were covered with Japanese silk and treated with pyroxylene. The resulting triplane with a 191-square-foot wing surface and Herring's regulating tail (an elastic cord with a spring and a hinge assembly between the main frame and the Pénaud tail) weighed 31 pounds [12]. Writing optimistically to his friend James Means, "I shall now pass from the toboggan stage of air jumping to some attempt at soaring. There is a vast difference between experimenting with models and with full-size machines, as the wind is constantly changing in trend, in direction and in form, and gliding becomes an acrobatic exercise" [47].

Several Chicago papers, including the *Chicago Tribune* [48], reported on August 12 that Otto Lilienthal had been killed in a lamentable flying accident. Now Chanute had to promise his family that he would take every precaution to avoid accidents. He made an appointment with his doctor, Dr. Walter H. Allport, and inquired, "somewhat quizzically, but in his usual serious and methodical manner, as to a young doctor who might not be too busy or too dignified to accompany him, for a consideration, on an expedition where he, the doctor, would be expected to do cooking during the few spare moments when he was not engaged in the professional duty of setting broken limbs" [49]. Dr. Allport recommended hiring Dr. Howard T. "Jim" Ricketts for $2 a day plus expenses. Fortunately, his skills in the medical field were not needed and opinions differ on his cooking abilities. Ricketts wrote that the daily breakfast consisted of coffee, roasted potatoes and eggs scrambled with small chips of fried bacon, and everyone, including the newspaper reporters, ate ravenously. Later Chanute told his doctor that "Ricketts may have been a good surgeon, but as a cook he was certainly the rottenest I ever knew."

To avoid the world of meddling curiosity, Chanute decided to take a different route for their second trip to the dunes [37]. On August 20, 1896, the team transported camping equipment, the *Katydid* and the new triplane from Avery's shop at East Superior Street to the Peshtigo dock (now part of Navy Pier). Everything was loaded onto the sloop yacht *Scorpion*, which also had an auxiliary gasoline motor; no one noticed the activity among the many other boats being loaded and unloaded. The *Scorpion* then travelled ten miles south on Lake Michigan to the 71st Street pier to load the Butusov machine, the third glider built since July 4. As it was getting late, the boat anchored at South Chicago overnight.

The next day, the *Scorpion* headed toward Dune Park, but ran aground on a sand bar in front of Windsor Beach. After two hours of hard work, the sailing vessel was freed and the group continued its journey. The *Scorpion* arrived off "experiment hill" at 12:30 p.m. about ten miles east of Millers, and just north of the Dune Park railroad station in an absolute wilderness. Seeing the sailing vessel going by from her two-bedroom homestead in Millers, Mrs.

Fig. 3.12 Pitching the tent in the dunes just north of the Dune Park Station. Chanute Photo Album, Manuscript Division, Library of Congress, Washington, DC (wb0039p-025-94)

Drusilla Carr took her flatboat over to Dune Park to help transfer everything to the beach. In the next few weeks, Mrs. Carr, and her handy-man Davy Crockett, an ex-fugitive slave, periodically checked on the group, and watched the flying activity with some amazement. She also brought lunch, which was better than what Ricketts prepared.

The group pitched the tent and got settled. "A fearful storm came up from S.W. at 3 a.m., blew the tent down and rained very hard. All the party got wet and provisions were damaged. Herring's wings blown away and smashed. Camp equipage scattered & damaged. Sent for another tent and got it that night." Its arrival at the train station made the groups existence known, and several reporters arrived in the next few weeks at their camp (Fig. 3.12).

The triplane, also called Herring's wings, was quickly repaired and flown on August 29. When first tried, Herring's elastic attachment to the Pénaud tail proved to be a failure, and the bottom wing always got caught in the sand during takeoff and landing. So, the experiment team went to work: Avery suggested removing the bottom wing and Herring repositioned the tail, resulting in a perfectly trussed biplane. "This machine proved a success, it being safe and manageable" [50] (Fig. 3.13).

Chanute had high hopes that no one would find them, but this was the age of yellow journalism and enterprising reporters. The first write-up appeared on August 25 in the Chicago *Inter Ocean*. "About thirty miles south of Chicago in the very heart of· the Indiana swamps, there has lately been seen a strange phenomenon. Hunters who started out with a full whisky flask

Fig. 3.13 Chanute holding biplane, with side panels as first envisioned (late August 1896). Chanute Photo Album, Manuscript Division, Library of Congress, Washington, DC (wb0039p-014-52)

and a record-breaking determination, in the dim gray dawn of the morning, have caught distant glimpses of a white-looking something that sailed slowly from the summit of a tall hill, landed a couple of hundred feet away, and crawled slowly back to repeat its performance. Sometimes there were three of them, looking like baby specimen of the great primeval lizard-bird taking first lessons in the art of navigating the upper regions o! the air. Nine times out of ten the hunters. who had unwittingly stumbled on the strange spectacle, withdrew with more haste than dignity, others brought back marvelous tales of a ghastly looking affair; the few who investigated found that they had only run across a new flying machine station" [51].

On September 4, 1896, many beautiful glides were made in calm weather and almost no wind. Avery proved to be a good pupil and quickly learned to manage the glider nearly as well as Herring. One week later, on September 11, a steady northerly wind of twenty-five to thirty miles per hour blew against the dunes, allowing several long flights. Chanute recorded in his diary, that the angle of descent was now much more shallow than in the glides with no wind [12]. Just for fun, Chanute suggested staging a mini-gliding competition, with Avery flying the *Katydid* and Herring the biplane. There was no clear winner. Watching the two men, Charley Chanute asked to give it a try; he tore his pants when landing awkwardly, the only reported incident. As Ricketts had developed a taste for flying, he too was allowed to make a few glides, and some visiting reporters also took turns in "doing some cruises" in the biplane.

Herring decided the next day to take the 23-pound biplane to a nearby dune, where no one could watch him, and reported his best flight to be 14 seconds, covering a distance of 359 feet. He was now ready to install an engine, but Chanute was not. Maybe he was irritated by the older engineer's caution or frustrated about not being the best performer, but he packed his belongings and took the next train back to Chicago.

Also going home for the weekend, Chanute took a later train.

Shortly after lunch on Monday, Herring stopped at Chanute's home and said that he did not believe Butusov's story and considered his machine dangerous; he wanted to withdraw. Knowing of the conflicts between his team members, Chanute paid him off and let him go.

Arriving back in camp a few hours later, Chanute explained why Herring did not come back, and Avery and Ricketts recalled various incidents and remarks showing unfairness on Herring's part while testing the machines. A few days later, Chanute wrote to Huffaker that Herring was violently opposed to Butusov's *Albatross* and was so elated at the success of "his" machine. Now he left "to go on his own hook and endeavor to build a power machine. I am afraid that Professor Langley was quite right about him" [52].

Just before breaking camp, several reporters from Chicago, New York and Boston arrived. The *Chicago Record* sent Frank Manley again with his sketch-book, who made a drawing of Avery flying high along the dunes in the "aerocurve." Chanute mailed this report to his many correspondents [53].

3.8   William Paul Butusov and His Albatross

Early in June 1896, a visitor came to Chanute's home and introduced himself as William Paul Butusov, but wanted to be called Mr. Paul [54]. He had joined the merchant marine as a sailor when 13 years old and settled in Chicago in 1882. Having watched the Albatross on his sea-faring travel, he reportedly built a soaring apparatus similar to Le Bris' craft and took it to an area near Mammoth Cave, Kentucky. He claimed to have soared for 45 minutes in 1889, "imitating the maneuvers of the soaring birds, gliding about in zigzags and spirals, raising upon the wind and going in any direction." Chanute had some doubts about Butusov's story, but after quickly checking his references, he hired him to first help in camp and then build a duplicate of his Kentucky flyer, the *Albatross*, which was built near Butusov's home at Wentworth Avenue on the south side of Chicago and re-assembled

Fig. 3.14 Paul Butusov's *Albatross* on the ramp, ready to be launched in September 1896, Chanute standing on the far right. Chanute Photo Album, Manuscript Division, Library of Congress, Washington, DC (wb0039p-025-95)

in camp. The 190-pound craft with its 40-foot wingspan and boat-like fuselage, could not be launched like the other gliders; it needed a ramp to assist the takeoff. Chanute's son Charley and his partner's son Joe Card delivered the timber for the trestle ramp and several locals were hired to erect it [55] (Fig. 3.14).

The ramp's fixed position, a precursor of today's aircraft catapults, required a north wind to launch the *Albatross*, but during this September, the winds mostly came from the south. Henry Bunting from the *Chicago Tribune* reported on one attempt to launch the ship, highlighting it with a sketch of the "toboggan slide for flying machine" [56] Chanute described the final launch on September 26 in his notebook: "The wind from the north finally set in at 8 a.m. at 3 miles per hour. By 3 p.m. wind 18 miles an hour. Paul in machine and all was ready. Cut the rope at 3 p.m. Wind turned to N.E. Machine slid down slowly and stopped on level portion. Fastened a rope at once to haul machine to top of slide. Substituted 90 lbs. ballast. Fastened ropes to front and rear of machine & manned them so as to give initial velocity. Sent machine off again at 3:30 p.m. As soon as the front of machine had fairly left the chute, the side wind blew the head around. The left wing struck the trees west of chute and was broken" [37]. No one was hurt and the Westchester *Tribune* simply reported, "*Ship fails to fly.*" Even though the

Albatross was never airborne, it did catch the imagination and attention of the various reporters.

The flock of mechanical birds was packed up, but Butusov's machine was left behind. "The white gull, circling high in the air over the sand dunes along the lonely shore of the lake, looked with pitying contempt at the wreck of the flying machine far below. 'In my opinion,' said the bird, 'no inventor will ever hatch a real flying machine out of his head. The human skull is too thick,'" read the editorial comment in the *Chicago Tribune* [57].

A few days later the *Kansas City Journal* reported that a replica of Chanute's *Albatross* would be part of Kansas City's annual parade [58]; they explained: "High hopes were entertained throughout the scientific world for the machine of Octave Chanute, but only the other day it failed as ignominious as did that of Darius Green." Kansas Citizens were sure that their former citizen, who had bridged the unbridgeable river, would surely succeed — sooner or later.

Sometime during 1896, Butusov had compiled a "*Memorandum of Agreement*" [59]. Chanute was to pay $500 for the construction of the *Albatross*, $20/week for living expense while building the machine and $200 for experimenting with the *Albatross*. If the *Albatross* was successful, Chanute was to pay $15,000 to develop a flyer with motor and propeller. They would submit the claims to the Patent Office, with one-half of the patent assigned to Chanute, who would pay all expenses for securing it. The patent claims, with its five engravings, was submitted to a Chicago patent attorney and granted two years later [60], listing only Butusov's name; seeing this, Chanute requested a correction in the patent book but not a full reissue.

After the unsuccessful showing of the *Albatross*, Butusov decided to enter the exhibition scene with a biplane glider. Hearing of Carl Myers opening an airship farm outside of Chicago to teach and demonstrate glider flying [61] Butusov joined this enterprise and agreed to perform. His glider was attached to Myers' balloon, carried up to about 1,000 feet before cutting loose. "Then he will proceed to give an imitation of a large and lively turkey buzzard and soar and soar and soar until he gets tired or the wind gives out and his machine settles gentle down to earth" [62]. After making several flights, Butusov had a bad fall and was paralyzed.

A few months after Chanute's death, Butusov read in William Jackman's book *Flying Machines, Construction & Operation* [16] about the events in 1896, and was angered as his name was not even mentioned. He went to the Chanute family, explained that their father was his partner and wanted their father's half of the patent signed over to him. Charley, always good hearted, paid him $100 [63].

3.9   Newspaper Coverage in 1896—Blending Fact with Fiction

The Democratic National Convention was held in Chicago in July 1896, indicating that political leaders at least regarded the midwestern part of this vast country as a battle-ground and a center of political influence. The 36-year-old William Jennings Bryan, who favored "Free Silver", campaigned hard for the presidency of the United States. Cartoons, poking fun at political events, were at that time as important as bold headlines on the front pages of newspapers. With Chanute's glider flying still in the news, several Chicago papers created cartoons of Bryan using adaptations of Chanute's flying machines to hopefully fly to the presidency (Fig. 3.15).

Chicago had come to be known as the cradle of startling stories of airships, and "yellow journalism" was in full swing. To distribute the latest news through the wire service, W. J. Lloyd of the Western Union Telegraph Company in Chicago asked Chanute very courteously for information on his gliding experiments.

In general, Chanute was not in favor of supplying information to correspondents, but this time he hoped to avoid sensationalism in any article which would be distributed through the wire service. And in a way he wanted to share the news and what his team did, but he surely did not want to be part of distributing yellow journalistic stories. He wrote two long letters to Lloyd, reporting unsensational facts on what had been done in the past few weeks,

DANGEROUS EXPERIMENT ON A PERILOUS VOYAGE.

Fig. 3.15 Cartoon highlighting William Jennings Bryan to became president of the United States. *Chicago Daily Tribune,* September 14, 1896

and he tried to make it clear that he did not appreciate reports of sensational news until he had accomplished what he hoped to do. "These flights have all been of rather short lengths, varying from about 50 feet to a trifle over 100 feet in length. They were made on the gentler slopes of the sand hills, where there was no grass or trees. Jumping from higher and steeper slopes might give longer flights, but this was not deemed prudent to attempt until I am certain that the apparatus is perfectly manageable, automatically stable and strong enough in every part" [64]. As could be expected the news was reported in several newspapers around the country. Chanute also supplied one photo that he took with his Kodak camera which was used as part of the story [65].

In general, reporters only came for a few hours asking many questions, but the *Chicago Tribune* reporter, 27-year-old Henry S. Bunting, who had arrived on September 7, 1896, stayed in camp for almost three weeks. The first of his many unsensational reports was published on the front page of the *Tribune* the day after his arrival. Forty some years later Bunting recalled his time in camp for the "*I Personally Award*" [66] writer's contest and closed his write-up with, "I did not know then, of course, that my newspaper reports of Chanute's experiments in the Indiana dunes reflected anything more than current news to be looked upon of epochal importance. Seven years later, Orville and Wilbur Wright flew the first glider that was powered with a motor in the Kitty Hawk sand dunes and they freely acknowledged that the scientific data obtained from Octave Chanute's experiments with motorless gliders had been basic and indispensable to their air triumph" (Fig. 3.16).

Another reporter from the *Chicago Chronicle*, Frank Hemingway, quoted Chanute, "For many years I have been studying whether it is possible to construct a machine, which shall be able to support itself, or to be supported by the atmospheric currents and pressure, without the use of any artificial means. Such a device, if one is ever constructed, would be a 'soaring machine,' to which, after the motor was added, the more pretentious name of airship might be applied. The term 'flying machine' is both inappropriate and misleading. I do not now know whether it will ever be possible to construct a device such as that of which I speak, and I should not be at all surprised to learn that it could not be done. Of one thing I am certain, no one man can contrive such device ab initio, and for that reason the basis of my several devices has been taken from the more or less successful plans upon which other men have constructed them." The reporter wondered, "Have you attained any degree of success with your outdoor experiments?" to which Chanute explained, "Yes, sir, with every machine. With a machine modeled after the one upon which Lilienthal lost his life last month, I glided 106 feet,

This Is the Latest
Flying-Machine.

Octave Chanute, a Famous Civil Engineer, Is the Inventor, and He Is Now Testing the Contrivance Among the Sand Hills of Indiana.

PROF. CHANUTE'S NEW FLYING MACHINE.
(From a Photograph taken for the Sunday Journal at Miller, Ind.)

Fig. 3.16 The multiwing *Katydid* in flight as reported in various newspapers and the original photo, supplied by Chanute to the wire service. *The New York Journal*, July 26, 1896, and Chanute Papers, Library of Congress

and have gone even farther with devices more original with myself. These semi-successes prove nothing. I hope that one of the three machines which we now have may be found to be a long step in advance, but it is by no means certain that such will be the case" [67] (Fig. 3.17).

3.10 Summing Up the Flying Activity of 1896

Members of Chanute's team and several newcomers made more than 400 flights of various length and duration in the summer and fall of 1896. Some flights were very short and Chanute called them "jumps", others were longer or "runs", and finally the longest were "glides" or "flights." These flights unlocked some of the vexing mysteries of mechanical flight with a reasonably good gliding descent of almost 6 to 1, or six feet forward for every foot in vertical descent. And Chanute explained, "The experiments were conducted

Fig. 3.17 Chanute and Herring Aerial Navigators in Action. The Herring Flyer on Saturday broke all records for long distance flights, gliding a distance of three hundred and eighty-three feet, from a height of only forty feet. *The New York Journal*, September 14, 1896

with two gliding or soaring machines, and not flying machines provided with a motor and propeller, but skimming machines, intended to study the problem of safely gliding through the wind. They are operated by running down a hill and jumping into the air, sliding down thereon and governing the apparatus in the wind gusts. The landing is performed by the legs of the operator in the way inaugurated by Herr Lilienthal" [64].

Breakdown of all flights made in 1896

22–26 June	About 15 jumps in *Lilienthal-type* glider
23 Jun–4 Jul	About 150 jumps, runs or flights in *Katydid*
4–23 Sep	About 30 flights in redesigned *Katydid*
29 Aug	About 5 jumps in triplane
31 Aug	Triplane was cut down to a biplane glider
31 Aug–23 Sep	About 220 glides and flights in biplane
26 Sep	Unsuccessful launch of the *Albatross*

The best glides in the biplane were as follows:

Operator	Length in ft.	Time in seconds	Angle of descent	Height fallen, ft.	Speed	Descent of ft. per sec
11 September 1896 (with everyone watching)						
Avery	199	8.0	10°	34.6	24.9	1 in 5.75
Herring	234	8.7	7 $1/2$°	30.4	26.9	1 in 7.69
Avery	253	...	10 $1/2$°	46.	...	1 in 5.50
Herring	239	...	11°	46.3	...	1 in 5.24
Herring	220	9.0	...	...	24.4	
Herring	235	10.3	...	...	22.8	
Avery	256	10.2	8°	25.5	25.1	1 in 7.18
12 September 1896 (reported by Herring, without observers)						
Herring	359	14.	10°	62.1	25.6	1 in 5.75

In one way, Chanute was pleased with the overall progress but a little disappointed that sustained flight over a longer distance and duration, or to soar like a bird and steadily rise upward with the wind, had not been achieved. "Not even the birds could have operated more safely than we, but they would have made longer and flatter glides and would have soared up into the blue" [68]. Chanute explained to James Means, "The bird easily glides [in an angle of] 1 in 10, and I have not been able to get an artificial machine to do better than 1 in 6" [69]. And he concluded, "None of the three machines were able to perform soaring flight."

3.11   The Mystery Winged Ship of 1896/7

A mysterious airship was sighted in late 1896 over Sacramento and San Francisco, California. It then reportedly traveled east and reports of mostly nighttime sightings appeared in many mid-western towns. News about the airship's appearance made big headlines around the country.

On April 2, 1897, the day after April fool's day, the airship was reportedly sighted over Chicago. The *Chicago Tribune* gave their report the eye-catching title "*See Airship or a Star*" and stated, "Either the long-expected airship from the Pacific coast reached Chicago at 8:30 last evening or the fixed star Alpha Orionis, shone with unusual brilliance, was augmented by the tricks of refraction. Hundreds of folks in Evanston, Niles Center, South Chicago and even in downtown Chicago saw the strange phenomenon off in the eastern sky and were convinced it was the storied vehicle of the heavens that has been worrying the inhabitants of Podunk and Squab Corners for a fortnight. At

2 a.m. the ship, or star, or meteor, or what-not, appeared over the western horizon. It hung in the sky for a time, and bore more the appearance of the full moon than a star." The write-up continued, "In Chicago there was a general disposition to laugh at the airship theory, but attorney Max L. Harmar, Secretary of the Chicago Aeronautical Association, does not smile at it. Word was received here several weeks ago that the ship would stop here for the purpose of registration. The end of the trip is Washington, where the ship will be brought to earth and given up to inspection. President Octave Chanute of the Chicago Society has full information concerning the ship. He, with a number of other wealthy men, has furnished the money for the venture. Mr. Chanute is in California at the present time. I would not care to furnish full details as to the experiment, as it would be unfair with the inventors and would take off the edge of public interest" [70] (Fig. 3.18).

Chanute read the write-up with amazement; he had never heard of such an association, he surely did not give an interview, he surely was not in California, and he surely did not give money for such an enterprise. When asked

THE AIR SHIP AS PHOTOGRAPHED OVER ROGERS PARK SUNDAY MORNING BY WALTER M'CANN.
[Reproduced From Photograph Furnished by Mr. McCann and Pronounced by Experts to Be Genuine.]

Fig. 3.18 The mysterious airship arrived over Chicago. The *Chicago Times-Herald*, April 12, 1896

by a reporter, he just responded that "he could not command patience to read the full account in the newspaper because of its absurdities."

The Chicago *Record* offered a different explanation of the heavenly soarer: "The airship disease with which we are at present struggling came suddenly and without warning. Chicago started it nearly a year ago with Mr. Chanute's exhibition of mechanical birds down in northern Indiana. But Mr. Chanute's large pine and canvas backed ducks and flamingoes were simply a sort of measles or croup in comparison with the terrible scourge of flying machines which is now ravaging up and down the country, attacking the very flower of the land and driving truth back to the bottom of her well. Starting from the small and humble beginning of the Indiana sand dunes and the toboggan chute down which Mr. Chanute used to slide his wooden birds, the plague of airships has virtually covered this country. According to the dispatches, one may flee from Grand Forks, N.D., because of the apparition of a large, brunette aeroplane with green eyes and a long, flowing mane and tail, and, lo, as he gets off the train at Dallas, along toward the middle of the night, his uneasy eyes, soaring heavenward, are paralyzed by the spectacle of the red, white and blue aircraft which burns danger signals like an apothecary's shop and chops around here and there in the upper ether, to the vast danger of the stars and church steeples. Or he may go to Oklahoma, and the very first thing he knows after a delegation has taken him out to show him proudly where the cyclone was, he is pounced upon and held entranced by a double-barreled sky ship, with large sprockets and detachable name plate. It's unfortunate, of course, but what can be done about it? Nothing. Absolutely nothing" [71]. The airship then headed east and disappeared from the news as mysteriously as it had appeared half a year earlier.

Reading the many reports by reliable observers, Chanute wondered if this incredulous airship story may have originated with John W. Keely, who had invited several railroad men, including Chanute, in the 1880s to see and then recommend adopting his motor, if it was true [72]. A decade later Keely thought of displaying his airship, propelled by jets of air, at the World's Fair in Chicago, but "he could not make his journey to Paris as a mere afternoon outing." Almost coinciding with the sightings of the mysterious airship, Keely had told the press that his new flying machine, weighing more than two tons, was able to go anywhere and that its "engine permits the craft to descend as slowly, as gracefully and as safely as a bird alighting." So far there is no solid explanation of the airship that thousands reportedly had seen.

3.12   The United States Signal Corps Shows Interest in Aeronautics

As the twentieth century approached, the United States Army found itself caught between its role as a frontier constabulary and the demands of an industrialized nation. A handful of astute soldiers recognized that industrial progress threatened to transform traditional warfare into something much more horrifying.

Lieutenant **William A. Glassford** (1853–1931), who had joined the Signal Corps in 1874, was assigned to develop a balloon section, where the operator of a captive balloon worked with the crew on a train, collecting military intelligence and transmitting it to the ground via a telephone cable. To learn about European operations and to buy equipment, General A. W. Greely sent Glassford abroad. The Corps' first balloon, named the "General Myer" in honor of the first Chief Signal Officer Albert J. Myer, was part of the Corps' exhibit at the World's Fair in Chicago; this is where Glassford and Chanute became acquainted.

To lobby for the Signal Corps, Glassford submitted an article to the widely read *Journal of the Military Service Institution of the United States* [73], "It is the tactics of this new engine of the art of war which we must study." He firmly believed that aeronautics was of utmost importance if the United States were to prevail in the coming century. Hearing of Lilienthal's flying activity, Glassford thought the Signal Corps should become more familiar with this kind of equipment. He approached his superiors and explained that, "The new soaring apparatus resembles a gigantic butterfly, with large curved, but fixed wings, provided with a flat tail, and upright keel projecting out behind, the only kind of flying machine, which has actually succeeded in transporting a man in free air" [74]. His superiors reluctantly agreed, but building such an apparatus was not as easy as Glassford had thought. Needing help, Chanute contacted Lilienthal, "I have been requested by Capt. Glassford, Chief Signal Officer, U. S. Army to ascertain on what terms you would furnish a 'Segelapparat' complete, with instructions, as advertised in the *Taschenbuch*." Lilienthal responded that he would gladly furnish one of his gliders with the latest improvements, but he wanted to combine it with the sale of his American patent [75]. The sale fell through, because the Signal Corps could not afford to buy the glider and the patent from Lilienthal.

As part of their communication, Chanute informed Glassford of his upcoming flying in the Indiana dunes and offered his Lilienthal-type machine to the Signal Corps, assuming it proved safe. He also invited a Signal Corps officer to join them in camp. Glassford passed this information along to

Lieutenant Joseph Maxfield, stationed at the Signal Corps Headquarters in Chicago, who visited the camp on June 24. He watched the multiplane glider fly and reported that experimenting with these heavier-than-air machines would be worth the military's interest.

Believing that the first nation to own a flying machine would undoubtedly possess an incalculable advantage in future wars, Chanute offered to build a man-carrying kite, if the Signal Corps wished to test its efficiency. "That style of kite has greater sustaining power than any other, and a chair for a man may be attached to it" [76]. Captain Baden-Powell, a Scots Guard in the British Military, had made many successful ascensions, and kite builders, like William Eddy and Charles Lamson, had demonstrated the safety of man-carrying kites. Maxfield offered to be the first military man in America to be lifted into the air and stated, "Mr. Chanute said that a kite could be made to take a man up to a great height with comparative safety and give him as great a freedom of movement as he would have in a balloon. This thing has been done in England and can be done in Chicago" [77].

When the Spanish-American war broke out in April 1898 [78], a Signal Corps balloon was sent to Santiago, Cuba, and Maxfield kept the military constantly and accurately advised of the movement of the enemy … until the balloon was shot down. Again, Chanute advocated using man-carrying kites to which Maxfield replied: "If a kite could be made that would carry a man's weight in safety, I would make the ascent. During Chanute's experiments in the dunes, it was shown that an aeroplane was capable of sustaining a man's weight in the air over a long distance from the starting point" [79]. Fortunately, the war lasted only a few months, and there was no time to test any aeronautical proposals in action.

3.13 New Year with New Ideas—Glider Flying in 1897

After receiving the latest news on the glider flying in 1896, Thurston suggested to deliver another lecture to his students. Chanute went to Ithaca on February 5, 1897, for his second talk at Sibley College, describing enthusiastically and without hesitation the team's flying activity [80]. He knew that more work was needed to develop an aeroplane, so he offered the use of either the biplane or the multiplane to anyone interested in continuing the experiments; he also offered to sell either glider for $300 each. The frontispiece of the *Sibley Journal* of June 1898 shows "Two Modern Flying Machines," Chanute's multiplane and biplane.

We know of two attendees who became involved in aeronautics after listening to the lecture: Charles M. Manly, majoring in mechanical engineering, who joined the Smithsonian Institute in 1898 to help build Langley's aerodrome and develop the light-weight internal combustion engine, used in the 1903 aerodrome, and Matthias C. Arnot, a banker from Elmira, New York, who had become interested in flight after meeting Chanute's son Arthur in 1891. Arnot attended the lecture as a guest and decided that he wanted to buy a glider; he contacted Chanute a few weeks later who gave the letter to Herring. Utilizing parts of the 1896 biplane, the improved "Herring-Arnot" Chanute-type glider was again built in Avery's shop. Arnot arrived in Dune Park on September 7, flew his glider several times, but became more interested in powered flight and left the glider behind.

Always looking for publicity, Herring had wired directions on how to reach the 1897 camp to several Chicago newspapers; a few reporters came and tried "tobogganing in the air" (Fig. 3.19).

When Chanute came for a brief visit, Herring asked him to finance a second week of flying. Even though no new knowledge would most likely result, Chanute agreed as he thought others might enjoy the experience. He invited fellow WSE members and wired an invitation to Means in Boston, who arrived on Wednesday, September 15, staying overnight at the Chanute house. The next morning, everyone met at the Englewood Station, which included WSE members Bion Arnold, William Karner, Warren Roberts, and Leland Summers, as well as Chanute, his son Charley, Avery, Means and

SOARING ALOFT ON AEROCURVE.

Fig. 3.19 With a short run and jump, any person can fly long distances and guide the flying machine. When a small motor is attached, the inventor declares that a trip through the air can be made from New York to Chicago at the rate of forty miles an hour. *New York Journal*, September 13, 1897

Herring, each bringing a tent and supplies for the next few days. It was probably a highlight in Means' life to join the team, but his diary entry only read: "At Camp Chanute" [81].

The weather was perfect for the outing; everyone enjoyed the primitive life style and the fun activity (Fig. 3.20). And Chanute recalled a few months later, "We had taken the machine to the top of the hill. A gull came strolling inland and flapped full-winged to inspect. He swept several circles above the machine, stretched his neck, gave a squawk and went off. Presently he returned with eleven other gulls, and they seemed to hold a conclave, about 100 feet above the big white bird, which they had discovered in the sand. They circled round after round, and once in a while there was a series of loud peeps, as if a terrifying suggestion had been made. The bolder birds occasionally swooped downward to inspect the monster more closely. After some seven or eight minutes of this performance, they evidently concluded that the stranger was too formidable to tackle, if alive, or that he was not good to eat, if dead, and they flew off to resume fishing, for the weak point of a bird is the stomach" [82].

Just like in the previous year, the glides were 250–350 feet in length at an estimated angle of 1 in 6. Some readers may wonder, how these relatively long flights of up to 14 seconds could have been achieved. We know that the dunes were almost 100 feet high, with Lake Michigan not too far away from the foot of the dunes. The biplane glider weighed about 23 pounds, with a 16-foot wingspan and a 135 square foot wing area [12]. This craft, just gliding down a dune in calm air descended rapidly, but gliding or soaring in strong 25–30 miles per hour winds from Lake Michigan against the relatively steep sand dunes made a big difference. "If we had a long straight ridge and a suitable

Fig. 3.20 Having fun flying the biplane with everyone watching. Chanute Photo Album, Manuscript Division, Library of Congress, Washington, DC (wb0039p-008-29 and wb0039p-011-41)

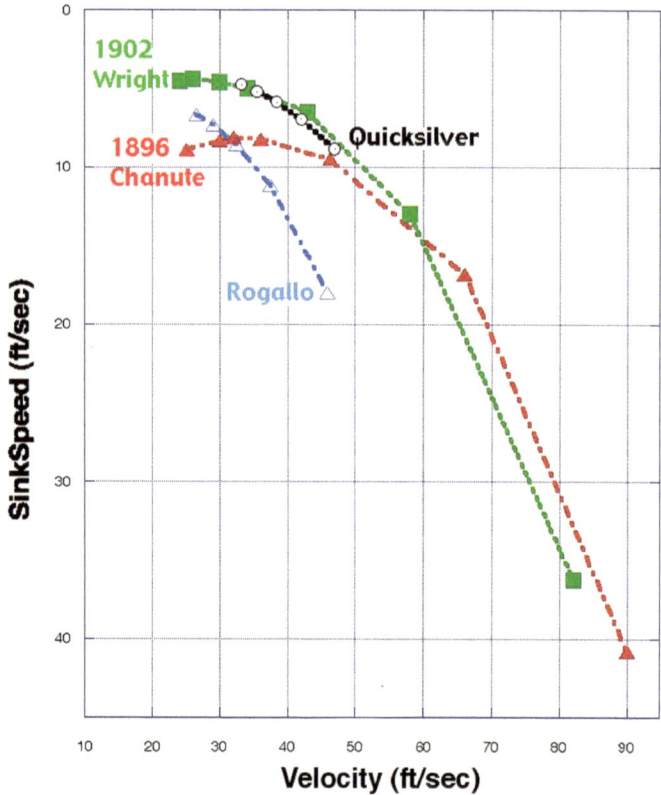

Fig. 3.21 The graph shows that the Chanute glider, with its maximum lift to drag ratio (L/D) of almost 5 was capable of soaring under the right weather and terrain conditions, while the Wright 1902 glider was a step ahead. To put the comparison chart into a modern perspective, the two modern hang gliders are included. Courtesy Albion Bowers

wind blowing at right angles thereto, we would have attempted to have sailed horizontally along the top of the ridge, transversely to the ascending current. This maneuver is frequently and easily performed by the soaring birds" [83] (Fig. 3.21).

The accompanying graph compares the sinking speed and velocity of four gliders, explaining the relative ability of the early gliders to ridge soar. The 1896 Chanute and the 1902 Wright glider exhibit a similar sinking speed and glide ratio. The graph shows that the Chanute glider, with its maximum lift to drag ratio (L/D) of almost 5, and a sinking speed of 5 ft/sec, was capable of soaring under the right weather and terrain conditions, while the Wright 1902 glider, with its maximum lift to drag ratio of almost 6, was a small step ahead. The Wright's improvement in L/D was primarily due to their much higher wing aspect ratio. Albion Bowers calculated the data using Ludwig Prandtl's theory as developed in his 1903 thesis [84]. To put the comparison chart into a modern perspective, Bowers included two well-known hang gliders of similar performance: the low aspect ratio Rogallo (maximum L/D of 5) and the rigid wing Quicksilver (maximum L/D of 7). Both hang gliders, flown in the 1970s, achieved soaring flight, giving credence to the claim that the Chanute and Wright gliders could soar as well [85].

3.14 How It Feels to Fly: Reporter Tries an Aerocurve

A nice first-person report about flying the Chanute glider was written by Harry Macbeth, published in the *Chicago Times-Herald* on September 8, 1897.

"Any man endowed with an average amount of nerve, a cool head, a quick eye and fair muscular development, can soar through the air nowadays, provided he is equipped with a machine like the one being used by A. M. Herring among the sand dunes near Dune Park, Ind. All that is necessary for him to do is to seize the machine with a firm grasp, say a prayer, take a running jump into space and trust to luck for finding a soft place when he alights. His chances of getting hurt are about one in a thousand in his favor, while having more sport to the second than he ever dreamed possible.

"Mr. Herring has been making flights with his machine for more than a week. Last year with a machine almost identical in construction he made daily experiments in flying for over a month. He made several hundred ventures into the air, and in none of them did he receive as much as a scratch. William Avery, who was of the same party to which Mr. Herring belonged, took the same risk fully as many times, and he, too, has yet to spend a cent for arnica or court plaster as a result of his seeming recklessness. With all these arguments before him in favor of the docility of the flying machine, a reporter for the *TIMES-HERALD* persuaded himself yesterday afternoon that he would like to hitch himself to the airy steed and try conclusions with a fish eagle that circled over his head a mile or so. It looked so very easy and it was a thousand per cent better fun to look at than shooting the chutes.

"One will never know what it is to sail through the air at a speed of thirty or forty miles an hour, sometimes at a height of ten feet and at the next moment three times as high, until he has tackled the aerocurve, or gliding machine. The first step is to get under the apparatus, and this is the most difficult part of the performance. The machine weighs only twenty-three pounds, but it is as big as the bay window of a cottage and has an alarming tendency to topple over on a man's head at a critical moment. With but two small upright sticks to grasp and a frail wooden bar under each arm on which to support the weight of the body one is not deeply impressed with the stability of the machine on coming into actual contact. Once underneath the machine one finds himself standing on a wide plank, which rests on the sloping side of a sand hill. The hill is about 100 feet high and steep enough to test the lungs and legs of the strongest man. You face the wind as squarely as possible and shift the machine to and fro until you feel that it is balanced fairly on your arms. You are suddenly aware that the broad expanse of varnished silk above your head is pulling on your arms, trying to get away from you with each gust of the freshening wind. At the same time, you remember to keep the front edge of the machine depressed until the instant of your departure from earth.

"It becomes necessary to start, of course, if one wants to fly. In the meantime, a sickening fear comes over one that he may lose his balance and plow a long and deep furrow in the sand with his nose. Somebody, one can't turn his head to see who it is, mutters that the wind is just right, and that it is a good time to start. Grasping the uprights with a grim determination to never let loose, and drawing a deep breath, one takes four or five running steps down the plank and jumps off, expecting to drop like a stone to the sand. To his surprise and pleasure, he experiences about the same sensations felt by a man when taking his first ascension in an elevator. There is the same queer feeling of being lifted from beneath and a corresponding exhilaration as the sense of motion is realized.

"As the machine mounts into the air, one sees the ground sinking beneath. He imagines he is a hundred feet in the air and begins to wonder if he will ever come down and be able to see his folks again. The thought no sooner comes, when the machine suddenly begins to descend with lightning speed. The wind rushes in the face of the operator like a hurricane and hums through the network of fine wire that forms part of the framework with a high, shrill note. There is a rustling sound, as of sand rushing over the white silk surfaces that sustain the machine in the air.

"Just as one stretches his legs out expecting to plant his feet on something solid, the wind suddenly lifts the machine again toward the sky. As it mounts upward, one's confidence returns. It is not so dangerous after all, just as Mr. Chanute and Mr. Herring and Mr. Avery said, and the possibility of flying

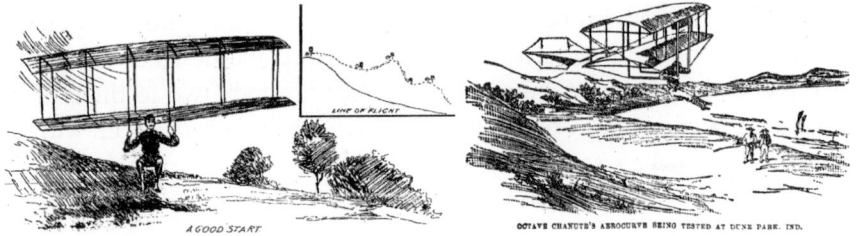

Fig. 3.22   Octave Chanute's progress with a flying machine and working out the manflight problem, *The Sunday Times-Herald,* September 12, 1897. William Avery achieved several very good flights in the "aerocurve" or the biplane. The *Chicago Record*, September 28, 1897

across the valley and returning to the starting point is mentally revolved. "The machine settles down slowly and steadily, and to the disappointment of the operator his feet strike the sand. His experience in the air is over. He turns around and looks up the side of the hill, feeling that he has traveled at least a thousand yards. When the tapeline is brought out, he is somewhat disgusted to find that he is only 110 feet away from his starting point. He wonders how this can be, when he was up in the air at least ten minutes. Then he receives another shock, when he is told that his flight lasted just five seconds. He still fails to understand, knowing positively that he was at least 100 feet up in the air, but some of the observers tell him that he was never more than thirty feet above the earth. This is the funny part of coasting on the air in the beginning. It's different when you know how" [86] (Fig. 3.22).

3.15 Elegant Engineering: The Chanute-Type Biplane Glider

The most creative innovations are often new combinations of old ideas. Innovative thinkers do not just create, they connect. Being familiar with the requirement for safe bridges and the ever-changing forces of the wind, Chanute applied his knowledge to a glider design that made a difference. He introduced the concept of vertical struts (under compression) and diagonal wires (under tension), the patented Pratt truss design, to airplane construction. In creating the biplane, Chanute had taken Wenham's idea of mutually reinforcing wings and added a masterly understanding of structural load paths. What he had learned over decades of designing bridges, he translated aloft, placing manned, powered, heavier-than-air flight within man's reach. This was a pivotal step in the history of aircraft design, ushering in the era of easily analyzed rectangular structures that quickly superseded the curves and framing of earlier attempts. It was a clean, logically thought-out, practical aircraft, introducing a pattern for subsequent airplane designs. Its stability was so good that the pilot only needed to shift his body two to three inches to maintain equilibrium. The elegant simplicity of the construction, with the direct transfer of the strains arising from loads through its members, to and from the points where those strains are concentrated are apparent, even to the non-technical eye.

The primitive Chanute-type gliders of 1896/7 were successfully flown by novices with no prior knowledge of flying. In later years, Chanute explained that his work simply consisted in the adaptation of the bridge truss between the wings to secure a maximum rigidity, never claiming it as an "invention". He just wanted to make it easier for others to build safe and strong, light-weight flying machines [87].

Looking for an entry on aeronautics in their next supplement, the editors of the *Encyclopedia Britannica* contacted Langley who recommended asking Chanute. To bring something new in this widely distributed publication, Chanute compiled a list of record flights with dirigible balloons and flying machines. Next, he provided instructions on how to design and build a glider using engineering principles and how to calculate the performance of the machine. As the Lilienthal table had been so helpful to him, he created an expanded version. The new table gave data of air pressures at various angles of incidence for flat planes (the Duchemin formula), originally published in 1894 in his *Progress* book, as well as the data for wings with a 1/12th curvature (the "Lilienthal data") in comparison [88]. Concluding his write-up, Chanute stressed that safety and stability must always be sought first.

Hearing about this literary project, Herring who considered himself the sole designer of what became known as the Chanute-type biplane, was concerned about not receiving credit [89]. He contacted Arnot, his current benefactor, who requested a review of Chanute's essay by the *Britannica* editors. They, in turn, forwarded Arnot's letter to Chanute who responded that he would take the matter up with Herring and that there was no reason to alter his write-up. A week later, Chanute asked Herring if he endorsed Arnot's statements. Herring took his time to respond. "I have never seen why you should claim the whole credit for the invention of the two-surface gliding machine, since its success depends wholly upon the efficiency of the regulating mechanism which was my work and furthermore, I made the design for the original two-surface machine alone at my home & put them on paper as a scale drawing in the course of one afternoon in your study. This first machine, it is true, was built with your money, and you paid me for my time. And you did suggest superimposing & trussing the surfaces together. But I do not think this is sufficient for you to take entire credit for the invention of the two-surface gliding machine" [90]. Chanute replied with Old World courtesy: "It is natural for every man to overvalue his own achievements. I do it myself." He then recalled the facts as he saw them and closed his letter stating, "Go on and demonstrate some important results and there will be no lack of appreciation in my writings" [91]. In a separate letter, Chanute suggested that Herring find someone to pay for the renewal of the communication with the British Patent Office or let it lapse. There was no reply to either letter. Several years later, Chanute clarified: "In point of fact, I made the design and Herring made the working drawing under my instructions, but what is the use of entering into a controversy about it now" [92].

With the public becoming interested in the people who had flown, Carl Dienstbach discussed "*Herring's Work*" in the *American Aeronaut* in May 1908 [93]. Chanute thought the facts were a bit distorted but said nothing. The next issue brought an article by Albert Zahm, commenting on who had invented the biplane glider: "It is questionable whether the type often described as the 'Chanute glider' can be properly regarded as the invention of one man. It seems rather the development of several minds ... However Mr. Chanute was the first to build such a glider, to operate it and to patent it" [94]. Reading Zahm's clarification, Chanute responded, "My work simply consisted in the adaptation of a bridge truss between the surfaces so as to secure maximum rigidity. A thing obvious enough, although new."

Sensing a controversy, the editor of *Aeronautics* asked Chanute to give his side of the story. This sounded like a good idea, and he was ready to recall the overall development. Chanute credited Wenham as the originator of superposed surfaces in an aeroplane, patented in 1866. Two years later, Stringfellow produced a model based on Wenham's principles. In 1878, Linfield tested an apparatus with twenty-five superposed surfaces, towed by a locomotive, but it proved unstable. Chanute witnessed Renard's gliding model at the 1889 Aeronautical Congress in Paris traveling against the wind and changing direction by using a rudder. In 1893, Horatio Phillips patented variously shaped wing sections, deducing what he thought was the best shape to provide lift. Hargrave had invented the cellular kite, described at the 1893 aeronautical conference, and, incidentally, Chanute had experimented with a similar gliding model, which however did not perform to his satisfaction. Then in the summer of 1896, Chanute's team first flew the Lilienthal-type glider but discarded it. Next his men experimented with the *Katydid*, learning with each alteration of the wing position. Any engineer had to foresee the variety of ways in which technology could fail or succeed, so Chanute sketched a triplane design on a sheet of cross-barred paper and calculated the air resistance in front of Herring, and asked him to build the glider in Avery's shop. "Being a builder of bridges, I trussed these surfaces together to obtain strength and stiffness. When tested in gliding flight, the lower surface was found too near the ground. It was taken off and the remaining apparatus consisted of two surfaces connected together by a girder composed of vertical posts and diagonal ties, known as a Pratt truss. Then Mr. Herring and Mr. Avery together devised and put on an elastic attachment to the tail. This machine proved a success. Over 700 glides were made at angles of descent of one in six or one in seven" [50].

In the 1930s soaring became a nationalistic and competitive sport. Discussing the historical background as they looked into the future, engineers and glider pilots like Robert Kronfeld [95], but also today's historians like Tom Crouch and John Anderson, agreed that this strong, light, and straight rectangular structure, was the first modern aircraft structure, but its origin was clouded in controversy.

There is no question that Chanute's engineering and mathematical knowledge in wood and iron bridgework formed the basis for the biplane glider. The truss bracing and the tail, consisting of a fixed vertical tail and horizontal stabilizer, originated with Chanute. Herring introduced the "regulator", the connection between the tail and the trailing edge of the wings, and Avery shaped the uprights aerodynamically, always stating that he built the machine at a carpenter's salary of $0.60 an hour [96]. With both men being employed by Chanute, one can theorize that the over-all design was a team effort under Chanute's leadership. Each member was knowledgeable in his respective field, and the words and thoughts of one person most likely triggered the thought process in the mind of his coworker.

On the occasion of the 40th anniversary of Chanute's glider flying experiments in 1936, the Western Society of Engineers dedicated a five-ton glacial boulder with a brass plaque in Millers, Indiana, to honor "Octave Chanute, the eminent engineer, Father of Aviation, who made the first successful flights in an heavier-than-air craft from these dunes in 1896" [97]. Another sixty years later, on the 100th anniversary of the gliding experiments, attention was focused on Chanute's role in shaping the community of engineers, scientists, experimenters who gave birth to the airplane [98]. The dune area at Millers, now an eastern suburb of Gary, Indiana, became a National Landmark of Soaring, sponsored by the National Soaring Museum and the Chicagoland Glider Council (Fig. 3.23).

Fig. 3.23 The Western Society of Engineers dedicated a bronze plaque to honor the eminent engineer in 1936. The National Soaring Museum and the Chicagoland Glider Counsel recognized Chanute's role in sharing knowledge with other experimenters and dedicated the dune area as a National Landmark of Soaring in 1996. Author's collection

3.16   Flying Flivver Is Assured

To share the excitement of flying, Chanute wrote *"Recent Experiments in Gliding Flight"* [12] and Herring discussed *"Recent Advances toward the Solution of the Problem of the Century"* [99]. Both reports were published in Means' 1897 *Aeronautical Annual*.

A few months later, Chanute addressed the WSE on the glider flying of the past two summers [100], as he was now convinced that man would make his way through the air by dynamic means. Highlighting his talk with lantern slides, he explained: "It was found that by moving the operator's body backward or forward, an undulatory course could be imparted to the apparatus. It could be made to rise several feet to clear an obstacle, or the flight might be prolonged, when approaching the ground, by causing the machine to rise somewhat steeply and then continuing the glide at a flatter angle. It was interesting to see the aviator on the hillside adjust his machine and himself to the veering wind, then, when poised, take a few running steps forward, sometimes only one step, and raising slightly the front of his apparatus, sail off at once horizontally against the wind; to see him pass with steady motion and ample support 40 or 50 feet above the observer, and then, having struck the zone of comparative calm produced by eddies from the hill, gradually descend to land on the beach several hundred feet away" [100]. And Chanute asked

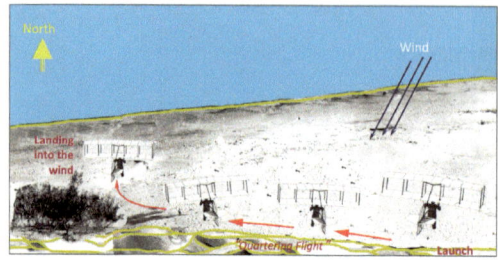

Fig. 3.24 "Quartering" or flying at an angle with the wind in order to make use of the ascending current over the slope which furnished both support and propulsion. The machine faces north but advancing west of northwest in a wind coming from the north-east. William Avery Papers, NASM 7A-09329

Herring to describe the final slides. Maybe he felt uncomfortable lending credence to Herring's long flights or he may have wanted to give him full credit. Both men described "quartering" as flying "at an angle with the wind to make use of the ascending current over the slope, which furnished support and propulsion. The machine faces north, but is advancing west of northwest in a wind coming from the northeast" (Fig. 3.24).

The WSE published Chanute's presentation in their *Journal*, and Moedebeck published it in the first issue of his new *Illustrirte Aeronautische Mittheilungen* in English and in German [101]. A slightly different version appeared in the British *Aeronautical Journal* [102] and the German weekly *Prometheus* [103].

By systematically approaching problems, Chanute moved solutions forward; he linked his correspondents into an informal network, provided motivation and leadership and served as the clearinghouse for the world's knowledge on aeronautics. He was quite sure that sooner or later, the right person would step forward, use his biplane glider design and take the manflight problem to the next level. In hindsight, it was the stimulation of Chanute's freely given counsel and experience that enabled the many pioneers world-wide, including the Wrights, to become proficient in gliding before adding an engine to their glider.

The effect of his efforts and leadership became obvious during the next decade when aviation technology improved quickly. His practice of working in goal-oriented information-sharing settings became a hallmark of cooperative development in the early part of the twentieth century (Fig. 3.25).

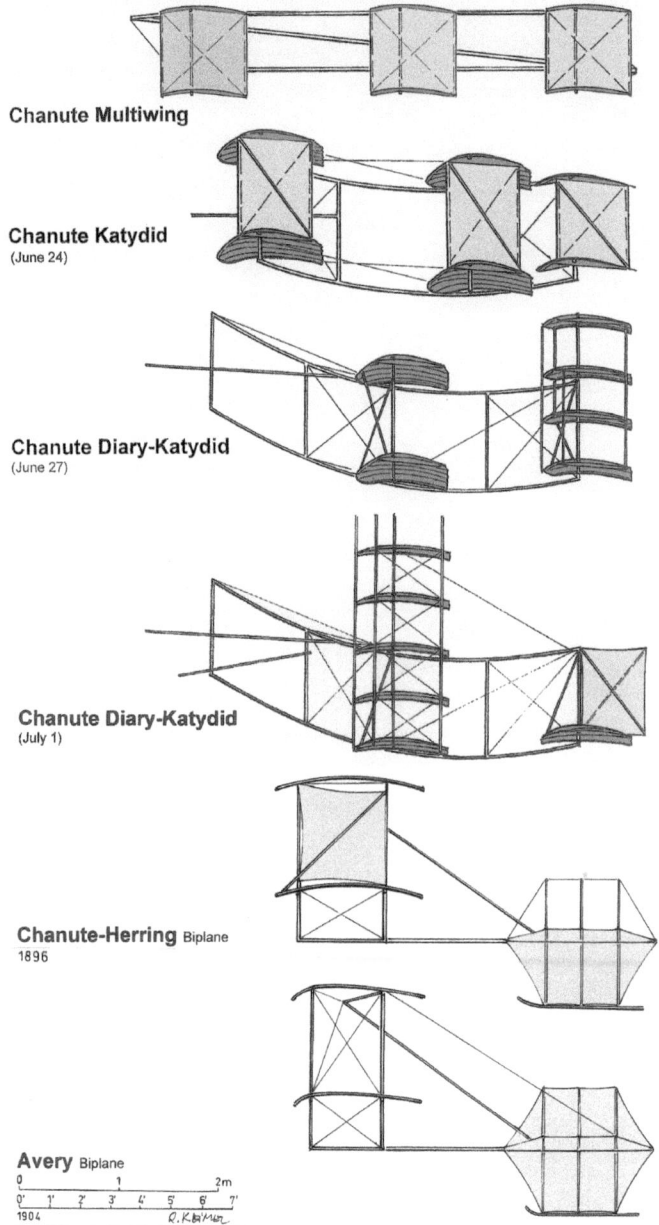

Chanute Multiwing

Chanute Katydid
(June 24)

Chanute Diary-Katydid
(June 27)

Chanute Diary-Katydid
(July 1)

Chanute-Herring Biplane
1896

Avery Biplane
1904

Fig. 3.25 Schematic of the development from the ladder kite design in 1895 to the biplane flown in 1904. Courtesy Reinhard Keimel

4

Collaboration for Progress in Flying Machines

It is safe to prophesy that the flying machine of the twentieth century will be analogous to a sailing vessel with an auxiliary screw, rather than to a mastless steamer. This is the prospect that makes flying worth-while to search after. It is the effortless soaring of the condor, not the fussy flapping of the sparrow, that must be taken as a model.

Garrett Fisher, The Art of Flying (1900)

The editors of the *New York Times* had anticipated manflight to be the crowning glory of the nineteenth century, but greeted the twentieth century in a less optimistic mood. "Everything thus far attempted in the way of aerial navigation has assumed a mathematical impossibility as its starting point and a mechanical paradox as its objective ... The magic carpet of the oriental fairy tale had a distinct advantage over anything in the line of modern invention; it charmed the fancy without inviting disappointment as the result of calculation" [1].

Reading about the many different visions about the future of aerial navigation, Chanute submitted his thoughts to the *Independent* magazine in early 1900. "Notwithstanding the pessimistic opinions of some scientists there is at least one sound reason to believe that man will eventually navigate the air. Within the last ten years artificial motors have been produced which approach closely, in relative lightness, the proportions which are obtained in the muscles of birds ... Success is likely to be achieved in two directions: First, with dirigible balloons, which will chiefly be used in war; and, second,

S. Short, *Flight Not Improbable*, Springer Biographies, https://doi.org/10.1007/978-3-031-24430-8_4

with flying machines, which promise greater speeds. Neither form of apparatus can possibly compete in reliability and cheapness with existing modes of transportation, but they will have uses of their own. Flying machines will not fill the heavens with commerce, abolish customhouses or revolutionize the world, for they will be expensive for the loads, which they can carry, and subject to weather contingencies. However, success is probable, as each new experimenter has added something to previous knowledge, which his successors can then avail of" [2].

4.1   The Boston Aeronautical Society

After graduating from the Massachusetts Institute of Technology (MIT) in 1884 and being interested in exploring the atmosphere, advertised to be a "wealth promoting science," **A. Lawrence Rotch** (1861–1912) erected the Blue Hill Observatory on the summit of the Great Blue Hill. Starting in early 1885, his group began to measure the height and velocity of clouds, using tethered balloons. To learn what others were doing, several members of this station attended the Aeronautical Congress in Chicago in 1893. Listening first to **William A. Eddy** (1850–1909), who had experimented with kites, and then to Lawrence Hargrave's work with box kites, left a big impression. To learn more about kites, Rotch invited Eddy, a journalist at the *New York Herald* newspaper, to join his team at the Observatory. On August 4, 1894, they flew a modified thermograph and a barograph with five combined "Eddy" kites to an altitude of 1,430 feet above Blue Hill. This was the beginning of regularly scheduled kite ascensions with recording instruments at the observatory [3]. But Eddy's kite-flying work was short-lived, as the rectangular Hargrave box kites, as suggested by Chanute and introduced by the staff of the Weather Bureau under Willis Moore as chief, worked more efficiently to study meteorology.

Remembering Chanute's urging to establish a national "Society for the Promotion of Aerial Navigation," several Bostonians decided that their town should become the center of the science of flight. The Boston Aeronautical Society (BAS), was formed on May 2, 1895 by James Means, Albert A. Merrill and William H. Pickering [4]. The goal was to advance the science of aerodynamics, to encourage experiments with flying machines and to collect and disseminate knowledge on solving the problems of aerial navigation. Activities were to include lectures with discussion, but also competitive trials of designing and flying kites as well as machines like Lilienthal's glider. Needing help with building kites and gliders, they first hired Albert Horn as a mechanic and then the German immigrant Gustav Weisskopf, who had claimed to have worked for Lilienthal.

Several members were in close communication with Chanute, who was elected an honorary member of the BAS.

In the early 1880s, **James Means** (1853–1920) became interested in aeronautics "after he had liberated himself from what he regarded as the shackles of the shoe business in Boston," his son wrote in the 1960s [5]. Means self-published in 1891 a 30-page pamphlet entitled "*Manflight*," which any interested person could order by sending four cents in postage stamps. The last sentence in the pamphlet read: "If you want to bore through the air, the best way is to setup your borer and bore." Chanute ordered five copies and wrote: "I will mail you some of my writing and would be pleased to correspond with you." This started a lasting friendship and collaboration.

Receiving a copy of Chanute's *Progress* book, Means wondered if he could take aerial navigation to the next level and encourage the general public to become interested. Means' first *Aeronautical Annual* [6], a 172-page softcover book, was published in December 1894. He clarified in the introduction that it contained not much new information, just things that deserved to be reprinted. In his editorial he stated, "I have become more than ever convinced that the soaring machine is the instrument by which we must, for the present, acquire knowledge."

Several Bostonians had used the drawing from Lilienthal's 1893 German patent to build a glider, but none was very successful. To learn and improve their flying skills Means suggested to stage a "Convention of Aeronauts" on the treeless Cape Cod peninsula, where every aerial craft would be submitted to a test under the direction of a board of skillful aeronauts. To make this event a success, Means contacted the only person able to fly repeatedly, Otto Lilienthal, in early August 1895. In a write-up for the press, he explained: "The Lilienthal aeroplane is a remarkable clever device, and while it is not yet of practical value for making protracted journeys in mid-air, it is a hopeful indication of the progress made. Its purpose is to imitate the soaring of birds, as well as their ordinary flight, which is executed by the flapping of their wings. The improved machine comprises two wings, which, after the manner of birds' wings, are slightly vaulted upwards. The wings are fixed by two rods laid crosswise, one upon the other, and firmly connected together, thus forming a carrying frame, or a part of a carrying frame, to which the person intending to fly may hold, as to be suspended between the two wings" [7].

Lilienthal was pleased when asked to participate in this gathering, but his business did not allow an extended absence; he suggested that several BAS members come to Berlin for a six- to eight-week training course. The cost would be $7,000 for five people, and Lilienthal would teach not only the building of his patented glider, but also how to safely fly [8]. Neither the flying camp at Cape Cod nor the expedition to Berlin materialized and the want-to-be aeronauts from Boston continued to teach each other.

In January 1896, Means published the second *Annual*, highlighting contemporary work of Lilienthal, Maxim, Chanute and Herring, along with shorter pieces by BAS members Lamson, Millett, Rotch, Pickering and Fergusson, discussing their work with kites. The third volume, published in 1897, was dedicated to Lilienthal and was devoted to the encouragement of experimentation and the advancement of the science of aerodynamics. Langley, Chanute, and Herring discussed the results of their experiments in 1896; and Means published articles by Lilienthal, Pilcher and Huffaker, but nothing from BAS members as he had encountered personal difficulties with their management and had resigned.

The *Aeronautical Annals* did much to encourage enthusiasts at a critical time of growth. Complimenting Means a decade later, Chanute wrote: "It was portentous service to aeronautics which you rendered in 1895, 1896 and 1897 by republishing some of the classics, describing promising experiments and formally giving data which have led to final success" [9]. Devotion to encourage, document and share the work of others, as well as his own, secured Means a special place among the promoters of aerial navigation.

An 18-year-old high school student, **Albert A. Merrill** (1874–1952), delivered an essay on aerial navigation, predicting that man would fly in small ships with light, powerful engines within 10 or 15 years. Hearing that Augustus Herring had repeated Lilienthal's soaring experiments in late 1894, the now 21-year-old Merrill also wanted to give it a try; William Eddy submitted the news story to the *Railroad Journal*. "The apparatus was a two-winged aeroplane, having a total spread from wing tip to wing tip of 22 ft., and a width of 7 ft. The wings consisted of two rectangular bamboo frames of canvas connected by a continuous pole. The wings were not hinged at the center point where there was an opening for the body of the experimenter. The guiding movement was made by slanting upward the whole apparatus at an angle, so that when one wing extended upward the other extended downward, thus resulting in a circular flight. Mr. Merrill told me that his apparatus weighed 21 lbs., but said that Herring's aeroplane weighed only 19 lbs. By running with the outspread wings against the wind, Mr. Merrill was supported clear of the ground for a distance of 15 ft. down the incline of a steep hill. Next, Mr. H. P. Clayton, of Blue Hill Observatory, made two jumps of about the same length, and in one instance, the aeroplane, owing to a puff of wind, raised Mr. Clayton 4 ft. or 5 ft. above the ground, clearly demonstrating the supporting power of the air" [10]. We do not know who built this glider and what happened to it in the years to come.

Fig. 4.1 Merrill sent this photo to Lawrence Hargrave on October 15, 1898. Lawrence Hargrave Papers, Museum of Applied Arts and Science, Sydney, New South Wales

One year later Merrill helped form the BAS, was elected secretary, and established contact with MIT, where Gaetano Lanza operated a "blowing tube" in the Mechanical Engineering department. This wind tunnel, about 3-feet × 3-feet with an air current of 10–12 mph, was used by several seniors and by Merrill.

To gain experience in flying, Merrill thought of buying a Chanute glider in spring 1898, but neither Herring nor Avery was interested in building one. He then used the drawings from Means' *Aeronautical Annual* to build his "soarer", using the same 1/12th wing curvature as Lilienthal and Chanute, but introduced a different tail configuration. He made flights from fifty to one hundred feet in length, and realized that the problem of stability was a valid one; a more efficient wing curvature was needed (Fig. 4.1).

Merrill's next project was a small whirling table to experiment with wing shapes of different cross-section and curvature; he discussed his findings in a stream of letters with Chanute. The results of his airfoil research were published in the British *Aeronautical Journal* in the summer of 1899. As part of his experimentation Merrill learned that the existence of a wind vortex under a bird's wing was an important factor to consider in soaring flight. Actually, Francis Wenham had noticed that a birds' wing is quite rough on the underneath, which might possibly have a direct bearing on flight [11]. Merrill closed his report with, "I am strongly inclined to think that Mr. Wenham is right, and that this roughness is important, though not in exactly the way he imagined it to be" [12].

At about the same time, Merrill wrote an utopian novel, *The Great Awakening. The Story of the Twenty-Second Century* [13], published in 1899. The book cover shows a pilot, flying a Chanute-type glider, trying to shoot a bird in flight (Fig. 4.2). This technical utopian described the author who was reborn in the year 2153 to an entirely new world in Boston; a Professor Harding helped him learn about his new life. Chapter nine tells of his ride in a flying machine, "And now we began to circle. The motors were stopped, the propellers disconnected, and we commenced soaring. We glided in great circles of, I should say, a mile in diameter and as we went with the wind we would drop, and then, turning into a puff, we would with our increased

Fig. 4.2 There is nothing in Merrill's letters to Chanute, why he selected this image for the cover of his utopian novel, *The Great Awakening*

velocity regain our former altitude; but each circle was a little farther from the wind than the former; in other words, while always at the same average height, we were gradually drifting to leeward. This, indeed was the triumph of man's skill, of his control over nature … Thus ended my first ride through the air, and I can truthfully say that never, either before or since, have I enjoyed anything as I did that sojourn in the skies. My ride back with the professor in his horseless buggy seemed tame, and I wondered how I ever could have been captivated with things that move on the rough, uneven earth, when I could soar forever with the eagles in the deep blue sky. My friend must have read my thoughts, he said, 'To fly is a pleasure, to ride in streets a necessity'".

Back to reality and wanting to fly for longer periods of time, Merrill thought of launching his glider with the help of a derrick and a falling weight. This idea sounded reasonable, and Chanute tried to calculate how heavy the weight should be and how high it should be lifted to give the glider the momentum to become airborne. But he was concerned about the other end of the wire after releasing: "As you neither say how you propose to control the speed at the drum, nor how you propose to attach the wire to the gliding machine, I cannot judge how safe your arrangements will be. Horses or boys running towed Pilcher, they could be stopped instantly, so he tied the cord to the machine. I am inclined to think this would be safer and simpler than your proposed falling weight unless you can devise a perfectly reliable trip hook, or hold the cord between fingers. I shall be glad to hear from you and to find all the fault I can" [14]. As part of their ongoing communication, Chanute shared the Wright's work with Merrill, just as they learned of Merrill's work. Is it possible that the idea of using a derrick appealed to the Wrights to use with their Flyer one year later?

Instead of building a derrick, Merrill considered a windlass next. "Your plans look sound and I hope you will test them. Be very sure however to have a man at the brake constantly, so as not to be dragged over the ground in case your gliding machine does not rise when you raise the front edge. As for myself, I prefer starting from a sand hill" [15]. As a friendly reminder, Chanute finished his letter with, "Not knowing the construction, I cannot judge the safety of your experiments. Remember, while your backer will only risk a little money, you will be risking your limbs in the attempt to advance knowledge."

For the next several years Merrill did not spend much time on aeronautics, he focused on his banking career and was satisfied watching the aeronautical developments around him [16].

Samuel Cabot (1850–1906) earned his living as a manufacturing chemist, owning Cabot's Creosote Shingle Stain. In his spare time, he was an avid amateur bird watcher, built and flew small flying models, and discussed his various aeronautical problems with Langley. Attending the Conference on Aerial Navigation in Chicago, he introduced himself to Chanute and started a friendship.

For several years the Boston Camera Club had sponsored photo competitions focusing on scenes around the house and garden. Cabot now proposed staging a competition on aerial navigation, or "*Caught on the Wing*," offering prizes for photographs showing birds who soar. He explained that the word soaring meant the attitude of the bird in the air when no wing motion is apparent [17]. Chanute's middle daughter Lizzie submitted her snapshot of a seagull in flight in this photo contest, winning second place, or $50.

Cabot knew from Langley that power was an absolute necessity when designing a flying machine; so he developed a power plant with an aerial screw, or propeller, mounted between two bicycles and operated it by foot power near his summer home at Cape Cod. "When the big fan began to turn, away went the bicycle, and the farther it went the higher became its rate of speed, until it was bowling along at the rate of ten miles an hour" [18]. Believing to have solved one problem, Cabot now looked for a glider.

After spending the summer in Europe, also meeting Lilienthal and watching him fly, Cabot was full of ideas about flying machines and hired Weisskopf, the BAS mechanic, to build a glider for him [19]. The letter exchange with Chanute provides an interesting story. On May 7, 1897, Cabot wrote: "… Weisskopf so far has made a conspicuous failure with his apparatus, and I fear that he is a pure romancer with a supreme mastery of the gentle art of lying. I feel, however, that it is perhaps premature to make a final condemnation of him …" [20] In the original typed letter, Cabot left a space and inserted Weisskopf's name by hand prior to mailing. Chanute responded a few days later: "… I had a very hearty laugh over the way in which you put Mr. Weisskopf's accomplishments, but I hope he may yet show you some good glides" [21].

A week later Cabot mailed his next report: "… I have had occasion, since last writing you, to modify my views about Weisskopf's failure so far. He foolishly attempted his experiments in a place, which would have been very dangerous had he got a flight, and it was so lacking in the proper conditions for a quick fall at first that he was, in my opinion, condemned to disappointment from the beginning. He is now in New York doing some kite-flying for a firm who are making Millet's kites" [22].

Continuing the work on his glider, Cabot explained to Chanute: "I am modeling mine upon the one described by Herring in the last *Aeronautical Annual*, although I am planning to put my cloth below the ribs, to put the cross-sticks above the ribs, and to lash it entirely with elastic cord. I also plan to use flat wire instead of round, and to stiffen it in such a way that it will always stay edge to the wind. For that reason, I am particularly anxious to hear from you in regard to the necessary strength of the wire, that I may get my flat wire of the same strength that the round should be." Chanute, the civil engineer, explained how he determined the wire strength: "These are proportioned just as upon a bridge truss, with this difference, however, that the weight is considered as applied at the center while the support (air pressure) is uniformly distributed. Having decided upon the weight to be supported, we calculated the amount coming on each post at the panel point, and its diagonal resultant along the wire. This gives us diameters varying from two to five hundredths of an inch, with a factor of safety of five, and we take the nearest commercial size to that. We use steel wire with a breaking strength of 100,000 to 150,000 lbs. per square inch" [23] (Fig. 4.3).

With the help of BAS members and much advice from Chanute, Cabot's glider was finally ready to be flown. Reporting on July 26, "… I have practically finished my scaling [soaring] apparatus and sent it to Chatham, where I expect to greet it about August 3rd. It is like yours except that the tail is a little different in shape. It is hung elastically and is so attached that the instinctive reaching out of the leg in the direction of the dip of the machine, will raise the opposite corner of the tail and thereby raise the machine onto an even keel again …" [24] In mid-August 1897, Cabot offered his final comment on Weisskopf: "… I am just back from Chatham and have a good deal more to write to you … I had Weisskopf under my eye at Chatham (I paying expenses) and I think he is completely unreliable and much doubt if he ever scaled [soared] over fifty feet in his life." This apparently was the end of the Cabot—Weisskopf or Whitehead relationship.

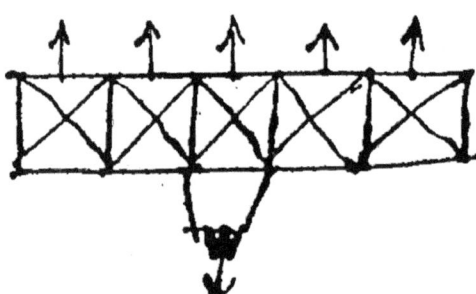

Fig. 4.3 Chanute mailed this sketch with his explanation on the selection of wire size when building the glider to Cabot on January 5, 1898, Manuscript Division, Library of Congress, Washington, DC

Cabot kept up his friendship with Chanute, always willing to support aeronautics but was not interested in owning or operating a flying machine anymore.

Charles H. Lamson (1847–1930) from Portland, Maine, was a jeweler by training but manufactured bicycle accessories for income. Becoming interested in aeronautics, he joined the BAS and built Malay and Hargrave box-kites. Wanting to experiment with a "single surface air sailer," he built a monoplane glider, using the drawing from Lilienthal's patent, with a wing span of 23 feet, 7 feet wide and weighing about 40 lbs. The ribs were made of spruce, covered with balloon fabric and braced with piano wire.

About six weeks after Merrill flew the Lilienthal-type near Blue Hill, Lamson made his first flight from the ridge on the west side of Great Diamond Island, near Portland, Maine [25]. The *New York Tribune* reported, "Lamson is at last sure that he has achieved success. He fastens two wings on his shoulders, runs a few steps, and the first thing he knows he is floating gracefully in the air. Several persons in Portland have seen him do this; but if he would come to this city [New York] and fly from the top of one of our high buildings, people generally would be more inclined to take stock in his invention" [26]. Lamson, however, was not so sure about being successful and reported to Means: "After a limited experience in trying one of his soaring machines, Lilienthal's apparatus seems to the writer to have a number of serious defects. First, it is too difficult of control for the average operator, when in the air. Second, it has no elastic rear edges to assist in forward propulsion. Third, the ribs and frame-work, as well as the body of the operator, present a too large and too rough surface to the air, impeding the necessary forward motion against the wind" [27]. And writing to Langley, "My first and only experience with the machine was not encouraging and I laid it aside." He then offered his "air sailer" to the National Museum in late 1898, but Langley was not interested.

With the knowledge gained, Lamson built a large box-type kite and traveled to Chicago to show it to Chanute. The *Chicago Chronicle* published details. "Lake captains who came into Chicago harbor on a twenty-five-knot breeze at 6 o'clock last evening turned their glasses on two airships which were floating in the sky not far from the north pier lighthouse at Ohio Street beach. The airy craft, whose keels were 100 feet above the lake, were the ingenious inventions of some American engineers. Charles Lamson, an inventor of Portland, Me., arrived in town a few days ago with an airship in his baggage. There is a fraternal feeling among designers and builders of airships, and Mr. Lamson at once sought Octave Chanute, who is Chicago's chief representative in aerial navigation. Yesterday afternoon the two inventors went out on the lakeshore to have a friendly trial of their airships. Lamson's ship had a yellow keel and four black wings. It was made of muslin, light woodwork and wire. It was ten feet long, and its wings spread of about six feet. Its owner kept it from sailing up to the new moon by holding fast to it by a string. Mr. Chanute was represented by an airship, designed by his chief assistant, A. M. Herring, an engineer from New York City. His airship was fashioned like a bird. It was made of white silk, wire and wood. It had a tail, which balanced it in the air. In extreme length it was about ten feet, by six feet in extreme breadth, and it weighed only four pounds. Mr. Herring also retained possession of his property by a string. Both models maintained an even keel and showed great stability in the gale, which was blowing twenty-five miles an hour. 'If my model was only a little larger,' said Mr. Herring, 'a sailor could ride on it with safety'" [28].

Traveling back to Portland, Lamson stopped in several cities to demonstrate his kite and reported back to Means, "I have recently discovered that my multiplane kite flies best the pointed end forward. Comical that I did not think to try it so before."

The next year, Lamson built a much larger kite, 26 feet in length with a 26-foot wingspan. The framework was shaped like a boat, and a wheeled undercarriage facilitated the takeoff and landing, but also the transportation from place to place. Seeing reports in the various papers, Chanute found it fascinating that Lamson could fold and move the large kite like we move gliders on a trailer decades later [29] (Fig. 4.4).

The highlight of that summer's flying season was probably the visit of a New York *Journal* reporter, taking a joy-ride in Lamson's kite. Ms. Grace Gould, better known as "The New Woman," had an exciting time, her report was featured on the first page of the Sunday's magazine section with two large engravings. She started her write-up with a simple one-liner, "I have flown," proud to be the first of her sex to venture on such a daring journey to test this new mode of progressing through the air [30].

Lamson submitted his kite design with five surfaces, three in the front and two as a tail, for patent, which was granted in January 1901 [31]. He mailed photos and drawings to Chanute, who responded: "The triplane surfaced one is especially interesting to me, as it is closely alike in outlines to a gliding machine with five surfaces which I have designed. I wish I had your skillful fingers to build a full-sized machine. Can you suggest some way to bring this about?" [32] Maybe the rocking or oscillating wing principles that Chanute envisioned could be introduced into this kite design to achieve automatic equilibrium?

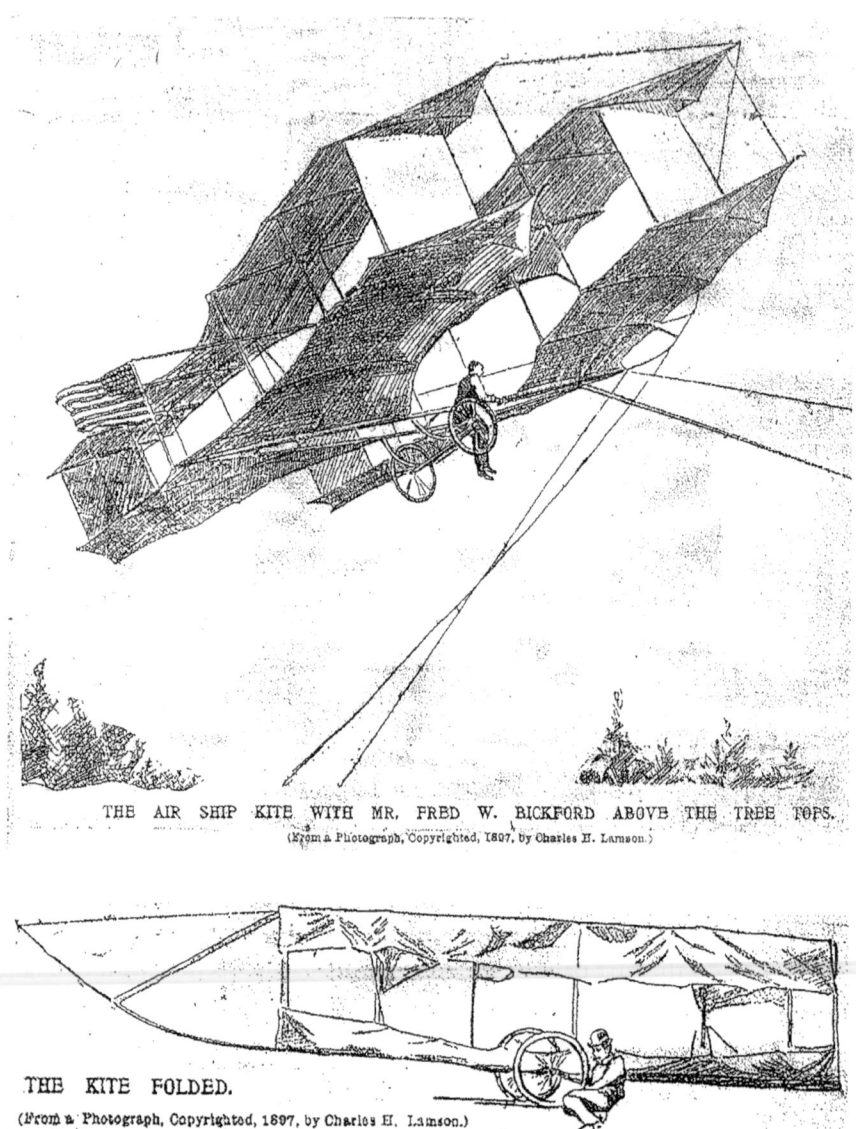

THE AIR SHIP KITE WITH MR. FRED W. BICKFORD ABOVE THE TREE TOPS.
(From a Photograph, Copyrighted, 1897, by Charles H. Lamson.)

THE KITE FOLDED.
(From a Photograph, Copyrighted, 1897, by Charles H. Lamson.)

Fig. 4.4 Charles H. Lamson's airship kite carrying a man towards the clouds, and disassembled or folded for transportation. *Portland Times*, June 27, 1897. Lamson Scrapbook, Chanute Papers, University of Chicago Library, CrMs176

4.2   Herring—Back to Chanute—Back Alone Again

After leaving Chanute's employment in mid-September 1896, Herring was ready to design his powered airplane. To test his theories, he built a triplane and mounted a sandbag between the wings in lieu of an engine and flew the craft in mid-October just north of the Dune Park Station. He reported glides of up to 927 feet in length, "all while 'quartering' on the wind. In a few of the flights it was found quite safe to turn the apparatus and it would have been possible to land on a higher point than the starting one" [33]. No one witnessed these flights, and most historians have doubts that they happened. But Herring could have been "quartering" or "crabbing" over a distance of 927 feet for 48 seconds and making partial turns as long as a strong wind

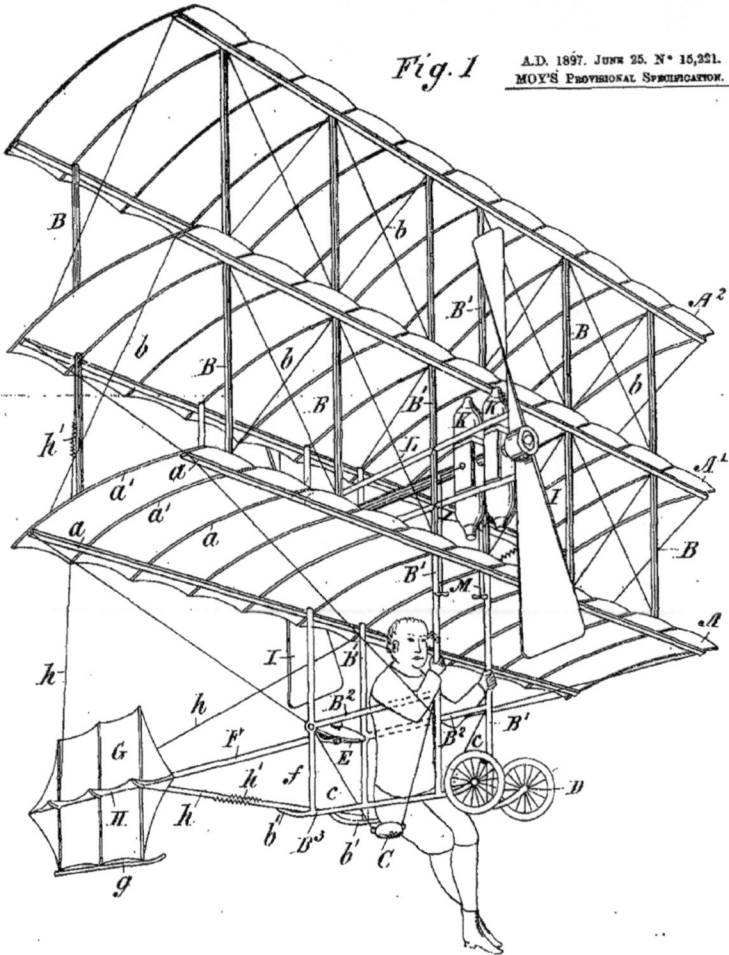

Fig. 4.5 Letter Patent combining Augustus Herring's triplane and Chanute's biplane design, mailed to Thomas Moy to be submitted to the British Patent Office as Communication, "*Improvements in or relating to Means and Appliances for Effecting Aerial Navigation.*" Patented on June 25, 1897

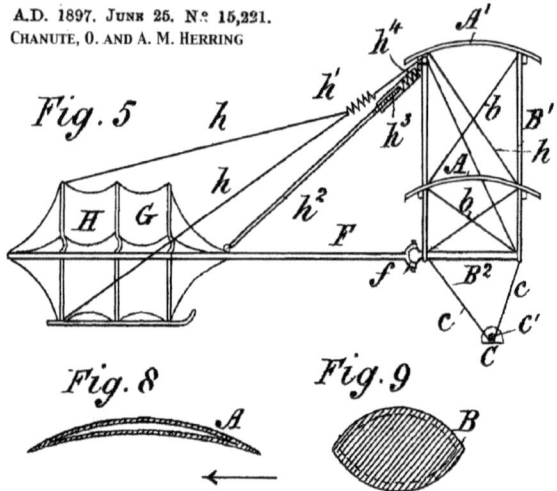

Fig. 4.5 (continued)

was blowing in a right angle to the sloping dunes [34, 35]. Six weeks later, Herring submitted his claims for a powered triplane to the United States Patent Office, they were rejected the following month; he then revised his claims, and they were again rejected. Looking for funding to build his aeroplane, he tried to "fly a kite" and contacted several businessmen, including the Barnum and Bailey Circus, newspaper magnate William Randolph Hearst and James Gordon Bennett. Slowly, one negative reply followed the other. In frustration, Herring approached Chanute again, who reluctantly rehired him in early June 1897.

Still thinking of automatic stability, Chanute envisioned a glider with wings shaped like those of a soaring bird, or "Pline's curve." Building or experimenting however was not what Herring had in mind; he only wanted his motorized triplane patented, and Chanute was the only one who had the influence and resources to get this done. Knowing that Herring's claims had been previously rejected, Chanute suggested to combine the claims for his biplane and Herring's powered triplane in one patent and forward them to Thomas Moy as a "Provisional Communication" to the British Patent Office. They were accepted a year later [36] (Fig. 4.5).

Seeing an ad for an engineering position at the Truscott Boat Yard in St. Joseph, Michigan, and having the patent applied for and almost accepted, Herring moved his family to this small community. This seemed like a great opportunity, as the boat yard had a good machine shop that might be helpful for building a powered aeroplane in his off-time and the beach along Lake Michigan could serve as a "runway."

Still receiving a monthly allowance of $200 from Matthias Arnot to develop the powered aeroplane, Herring copied an 8-h.p. Otto gasoline engine, but it did not perform as hoped. Arnot then shipped his compressed air motor design which did work. Now Herring built a biplane, instead of a triplane, and installed one propeller in the front and another one in the back, driven by Arnot's compressed air motor. In mid-July 1898, Chanute reported to Means that Herring had assembled the aeroplane to be photographed, but found it somewhat unmanageable. "It struck me that he was weakening in his faith of success, and a little afraid to try it" [37]. Finally, on September 30, 1898, Herring reported to Chanute with pride: "For the first time today I succeeded in getting full pressure on the tank after having eliminated the majority of the leaks." A few days later, Herring wired: "Blew inlet valve. It will take 2–3 days to make the new valve for the 2nd cylinder and perhaps as much or more time to do the other necessary puttering." On October 10, Herring wired again: "About ready today. If you come, don't bring any strangers." Anxious to see such tremendous progress, Chanute took the Graham and Morton night boat, leaving Chicago at 11:30 p.m. and arriving in St. Joseph at 4 a.m. the next morning, on October 11. Herring met Chanute a few hours later and told him that he had made a 50-foot flight the previous day over the level beach. However today, there was no wind, the compressor would not charge, and Herring could not repeat his flight.

Disillusioned, Chanute returned to Chicago and reported to Langley: "Herring invited me to St. Joseph on Oct 10 to see him fly. He was unable to perform the feat in my presence but said that on the preceding day he had flown ~50 ft. by running against a strong wind and simultaneously setting his propellers in motion with a compressed air motor" [38].

Doing more tinkering, Herring thought to be ready and invited Zahm to come and witness a powered flight, but he could not get airborne. Looking at the motorized glider critically, Zahm commented that "both a glider and a dynamic aëroplane should be controlled entirely by steering and balancing surfaces, and that the lateral balance should be controlled by changing the inclination of the wings on either side, while the double tail should be used to steer and steady the aëroplane sidewise and vertically" [39]. Herring had no objection, but right now he only "wanted to make a short flight with the machine as it stood, for the purpose of enlisting capital, and add the controlling devices at leisure."

Doing more work, Herring wanted Arnot to come. This time everything worked, and he made a straight flight over 73 feet on October 22, 1898. There is one photo, credited to Arnot, showing the powered machine with spinning propellers just touching ground, with Herring on his knees, holding

Fig. 4.6 The *Chicago Evening News* of November 17, 1898 reported that Herring's Airship actually flew and that the inventor took a flight of seventy-three feet against a twenty-five miles wind

the machine. According to their contract Arnot was to make the second flight, but he returned to Elmira without having flown (Fig. 4.6).

The successful accomplishment of such a flight covering a distance of seventy-three feet in eight or ten seconds, against a wind of thirty miles an hour, was reported in the Benton Harbor *Evening News* two days later on "today's experiments;" another report was published two weeks later in the Chicago *Record* [40], which was reprinted in the Elmira *Daily Advertiser* [41] and the *Chicago Evening News*. Reading the news, Zahm tried to verify the report but he was not able to ascertain the reporter's name, or any other witness to the event [39].

Maybe it is not really important if Herring flew once or twice; these "hops" in a powered Chanute-type biplane glider, with weight shifting for control, can hardly be considered a significant flight [42], but Herring's aeroplane was another step toward finally achieving powered flight.

4.3   Chanute's Trip to Europe in 1899

After spending about $10,000 on flying experiments in the past few years, Chanute became aware of the fact that his two tie preserving works were losing money. To return his business to profitability, he decided to visit Europe and learn about their timber treating procedures. To meet the new travel requirements, Chanute mailed his old passport to the State Department to be renewed, after updating his age and the color of his hair, plus $1. As his youngest daughter, twenty-eight-year-old Nina, wanted to come along, he added her name to the passport.

Father and daughter left Chicago on October 31, 1899, crossing the Atlantic on the "*Kaiser Wilhelm der Grosse*", the newest and fastest ship of the North German Lloyd, and the first German steamer to win the "Blue Riband," traveling at 22.4 knots. They stayed for two weeks in Hamburg and then traveled to Berlin visiting several wood treating plants. But then he wanted to spend time with aeronautics. The next stop was Wiesbaden to meet Moedebeck, who had published Chanute's articles with a German translation, "*American Gliding Experiments*" in early 1898 [43] and a year later "*Conditions of Success in the Design of Flying Machines*" [44]. Moedebeck now thought of updating his *Taschenbuch* (Pocketbook of Aeronautics) and Chanute offered his help: "It will give me great pleasure to assist in an addendum to the chapter on artificial flight. This should, in my judgment, consist in an appreciation of Lilienthal and an account of the accident in which he lost his life, to be written by yourself, and then a brief report of gliding experiments made since, for which I will furnish the notes. If you want to include also the experiments made with motors (Maxim, Langley, Ader, Tatin, etc.), please advise me" [45].

After an obligatory stop in Paris to finalize the estate of his mother who had died in 1893, Chanute traveled to London, where he met Hiram Maxim who informed him that the 32-year-old Percy Pilcher had crashed and died a few weeks earlier.

Next he met Francis Wenham, living northwest of London, who had designed in 1871 the first "wind tunnel," built by John Browning. Their tunnel was an 18-inch-square and 10-foot-long wooden trunk, with a blast of air from a fan, driven by a steam engine, blowing through its chamber. Even though this was a crude set-up, the results obtained using flat planes established the relationship between the velocity and the pressure of the wind, opening up new venues for the study of aerodynamic phenomena. As part of the conversation during the visit, Wenham stated that "the whole secret and success of manflight depends upon the proper curvature of the wing surface"

[46]. He explained "Due to the baffling eddies and ever varying currents I could not come up with meaningful results. I have therefore since rigged up a fan-blower running at 1,700 revs per minute and giving a current of 25 miles per hour. I could not get beyond this as it absorbed all my strength, but still the current was definite and steady and with proper arrangements to measure the lift and drift." Apparently, the 65-year-old Wenham used manual foot power to drive his fan-blower at 1,700 revolutions! In a follow-up letter Chanute thanked Wenham for a delightful visit and "I am eight years your junior and hope that I may be able to advance the question equally with you before I step out" [47].

Thomas Moy, the next person to meet, had just submitted the claims of his "Aerial Vessel" to the British Patent Office. An interesting letter exchange took place two years later, when Moy asked for financial help to build his machine. Chanute replied: "I am well disposed to assist you, but my ready money is limited. First, I want to know how much there would be to spend before returns could be expected. Give me also as much as you deem prudent as to the design of the machine, so that I may use my own judgment." Moy responded without giving details, so Chanute wrote: "I might be willing to advance money provided it was not too expensive, and I could get my money back. Singular as it may seem, I do not care to increase my present fortune, I have divided it up among my children and wife, leaving each a moderate competence, which I deem the happiest and safest. I am prepared to spend a couple of thousand pounds upon devices of my own without hope of return, but if I spend it on another man's devices I should want to get it back" [48]. A few months later Chanute mailed Moy £ 20 as a good-will token.

Father and daughter arrived back in Chicago in time for the Christmas holidays.

4.4   The Busy World of Aeronauts in Central Europe

To solve the problem of aerial navigation, Jules Verne and Félix Tournachon, better known by his nom de plume, Nadar, had formed in 1863 a heavier-than-air society in Paris where inventors could experiment with machines. "Thus, the impetus was given, inventors invented, calculators calculated all that could render aerial locomotion practicable." Thirty some years later, on October 20, 1898, several well-to-do Parisians formed the Aéro-Club to promote aerial navigation on a broader base. Knowing that the future of petroleum sales depended on the development of small internal combustion engines, Henri Deutsch, began to promote aerial navigation with vigor. In April 1900, he established an extraordinary prize of 100,000 francs to be awarded to the inventor of a machine, capable of flying from the grounds of the Aéro-Club to the Eiffel Tower and back in thirty minutes.

This high monetary prize was very tempting, and several Aéro-Club members gave it serious thought, with **Alberto Santos-Dumont** (1873–1932) being the most interested. Born and raised in Brazil, Dumont had come to Paris in 1892 to finish his engineering education; however, flying, and all the showbiz surrounding it, was much more exciting. The dapper Brazilian pioneer took off with his dirigible No.6 on October 19, 1901, rounded the tower and landed back at Parc Saint Cloud in twenty-nine minutes and thirty seconds. A jubilant crowd greeted him, but Aéro-Club judges squabbled over the particulars of the timing. While the judges deliberated, Dumont announced that he would give the entire prize money to his staff and helpers and to the poorest Parisians. This publicity stunt worked and the prize was made official (Fig. 4.7).

Born in the Ukraine, **Stéphane Drzewiecki** (1844–1938) had moved in 1859 to Paris to study mechanical engineering. Chanute had met him during the 1889 Aeronautical Congress and corresponded with him for several years; they both believed that flight depended on the principle of action and

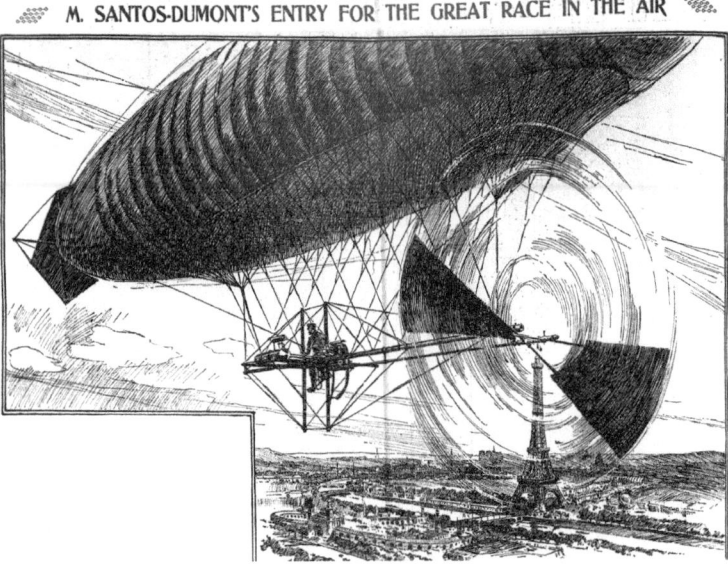

Fig. 4.7 Santos Dumont described his clever invention of a navigable balloon, with which he hoped to round the Eiffel Tower in the competition over the Grand Prix of the Aéro-Club in Paris, France. *New York Herald*, June 26, 1900

reaction, where the reaction of the air set in motion by the wings would continually overcome gravity. After studying birds in flight, Drzewiecki then focused his effort on propulsion and the screw propeller, first for submarines and then for aeroplanes. He presented his propeller theory at the Academy of Science in Paris in December 1892 [49]. His propeller was a twisted airfoil where each segment represented an ordinary wing, traveling in a helical path. In 1900, he presented a follow-up paper to the Naval Architectural Congress in Paris [50], and read his third paper in 1901. Drzewiecki mailed publication reprints to Chanute, who shared them with some of his correspondents. The topic of propellers was not well known or understood at that time, but it was of interest to Chanute who had researched propeller efficiency on canal boats and sea going vessels decades earlier. He agreed with Drzewiecki that the information was necessary for the development of any power plant in the future aeroplane.

The simple "blade-element theory" is usually referred to as the Drzewiecki theory, as he was the first to sum up the forces on the blade elements to obtain the thrust and torque for a whole propeller, and he was the first to introduce the idea of using airfoil data to find the forces on the blade elements.

The next World's Fair was again held in Paris (April 15 through November 12, 1900). The theme was to celebrate the accomplishments of the nineteenth century and the exciting prospects of the twentieth century. To publicize the Fair, French reporters interviewed Tournachon and Verne, as they were ardent supporters of the heavier-than-air principle or "*Le Plus Lourds que L'Air*." And Aéro-Club members, "the idle rich and a few industrialists who savored the joys of ballooning," triggered the public's interest in flying by offering daily balloon ascensions. Chanute was invited, but did not attend.

While the public's attention was focused on the events in Paris, **Count Ferdinand von Zeppelin** (1838–1917) piloted on July 2, 1900, his first airship, the *Luftschiff Zeppelin* LZ-1, over Lake Constance in southern Germany, reaching an altitude of almost 1,300 feet with five men on board. The airship traveled 5.25 miles in 17.5 minutes, answering to ascending and descending and steering, until a rope became twisted in a gear [51].

Interviewed by a Chicago reporter, Chanute thought that the flight of LZ-1 was not really an advance in aerial navigation. "While I regard Count Zeppelin's experiment an important step forward, it does not give us any encouragement that airships will be made to transport any great weight. Zeppelin's ship will be good only for use in making aerial observations" [52]. And thinking back forty years, Chanute told the reporter that von Zeppelin came as a military observer to the United States in 1863, and served with

Federal cavalry units in Northern Virginia. At the end of his military assignment, he decided to see more of this big country, just like other European
aristocrats. Starting in New York City, von Zeppelin traveled by train, boat
and coach all the way to Minnesota. Seeing a balloonist making captive
balloon ascensions and offering rides to the public in St. Paul, he decided
to give this a try [53]. Traveling back east, he spent a week in Chicago and
may have met Chanute who was at that time Resident Engineer at the Pittsburgh, Fort Wayne & Chicago Railroad, always looking for opportunities to
meet people who shared his interest in flying.

4.5   Two Brothers from Ohio Enter the Scene

In the late 1880s, **Orville** (1871–1948) and **Wilbur** (1867–1912) **Wright**,
had started a printing business and published a local newspaper. Looking
for interesting things to do, Orville bought a Columbia safety bicycle with
pneumatic tires and Wilbur purchased an Eagle safety bicycle in an auction.
In 1892 the brothers opened a bicycle sales and repair shop which was to be
their life business and income. But then, the horseless carriage caught their
attention and Orville suggested that they go into the automobile business.
The older brother did not agree, he thought the automobile would never
be practical. "To try to build one," declared Wilbur, "you'd be tackling the
impossible. Why, it would be easier to build a flying machine!" was a statement Orville shared with Fred Kelly, the Wright Brother's biographer, many
years later [54].

What caught their interest next was an article in *McClure's Magazine* entitled "*The Flying Man*" [55]. This nine-page write-up showed spectacular
photos of a human being, flying like a great bird with wings strapped to
the body. The brothers may have chuckled reading Otto Lilienthal's explanation that "Years ago the most distinguished professor of mathematics in
the Berlin Industrial Academy sent me word that of course it could do no
harm to amuse myself with such pastimes as flying, but warned me against
putting any money into them. A special commission of experts, organized by
the state, had, in fact, laid it down as a fundamental principle, once and for
all, that it was impossible for a man to fly." It is not known if either brother
realized that this man described in *McClure* was the same person, "*A German
named Lilienthal*," whose mini write-up they had printed in their West Side
newspapers four years earlier [56].

To learn about flying, but not knowing anyone in the engineering or academic community, the 32-year-old Wilbur wrote to the Smithsonian Institution in May 1899: " ... I wish to avail myself of all that is already known and then if possible add my might to help on the future worker who will attain final success" [57]. He did not have to wait long for an answer; Assistant Secretary Richard Rathbun mailed free-of-charge four Smithsonian Institution publications, "*The Empire of the Air*" by Mouillard, "*The Problem of Flying*" by Lilienthal, "*Experiments on Mechanical Flight*" by Langley and "*On* Soaring Flight" by Huffaker. Rathbun also suggested to buy Chanute's book "*Progress in Flying Machines*" for $2.50, Langley's "*Experiments in Aerodynamics*" for $1 and the three volumes of Means' "*Aeronautical Annual*" for $1 each. Wilbur purchased the recommended literature and read them with much interest. His lack of mathematical background guided him into the direction of mechanical tinkering, and the Pratt truss method of structural rigging appeared to be a workable solution. In this sense, Chanute provided the link between Wenham's biplane and Stringfellow's triplane to the Wright's powered flight, just another example of the flow of ideas in aeronautical engineering.

Having developed some technical understanding in their bicycle shop, Wilbur went to work [58]. "One evening while studying the movements of a little square paper tube, I noticed that the upper plane could be moved forward or backward with reference to the lower plane, which would be useful in controlling the fore and aft equilibrium of the apparatus ... I then constructed a little model made out of bamboo and proceeded to construct a kite with two wings. The connections of the standards to the planes were made flexible so that the top plane could be thrown forward or backward with reference to the lower plane." Wilbur probably had read in Chanute's *Progress* book how Albert Bazin controlled the "bi-polar kite" in flight with two strings, allowing the operator to vary the angle of incidence [59] [p:185]. Thus, his idea of "warping" the planes on the kite should work using Chanute's biplane design.

Almost one year after contacting the Smithsonian Institution, the budding aeronaut thought it best to consult the recognized authority on flying machines before venturing into building his full-size glider. Wilbur's five-page letter to Chanute, written on bluish stationary, was the beginning of the long road in the Wright brothers' eventual success. "For some years I have been afflicted with the belief that flight is possible to man. My disease has increased in severity and I feel that it will soon cost me an increased amount of money if not my life." Wilbur realized that it would be possible to fly without motors, but not without knowledge and skill. He then described the glider he intended to build. "In appearance it is very similar to the 'double deck' machine with which the experiments of yourself and Mr. Herring were conducted in 1896–7. The point at which it differs in principle is that the cross stays which prevent the upper plane from moving forward and backward are removed, and each end of the upper plane is independently moved forward or backward with respect to the lower plane by a suitable lever of other arrangement. By this plan the whole upper plane may be moved forward or backward, to attain longitudinal ~~stability~~ equilibrium, by moving both hands forward or backward together. Lateral equilibrium is gained by moving one end more than the other or by moving them in opposite directions." He continued, "The problem is too great for one man alone and unaided to solve in secret." By writing to the person who had the most up-to-date information on global efforts to crack the flying machine problem, Wilbur not only looked for confirmation on his thinking but also for advice on his design ideas [60].

Reading the letter made Chanute feel sympathetic toward this enthusiast. He agreed that no financial profit was to be expected from such investigations for some time to come. Following his standard handling of inquirers, he gave some advice on experimenting, listed some literature for further study and suggested areas where Wilbur might begin his gliding experiments. "The two most suitable locations for winter experiments which I know of are near San Diego, California, and St. James City (Pine Island), Florida, on account of the steady sea breezes which I have found to blow there. These are deficient in sand hills. Perhaps even better locations can be found on the Atlantic coasts of South Carolina or Georgia." In closing, Chanute wrote "I shall be glad to have you call on me, and can perhaps better answer questions that have occurred to you … If you do not expect to come to Chicago soon, I shall be pleased to correspond with you further and to have a detailed account of your proposal" [61]. Wilbur agreed to many of Chanute's suggestions and responded "… It is important that more persons should be intelligently interested in this subject."

Starting the construction of his glider, Wilbur wrote his third letter to Chanute, asking where to buy spruce of the proper length and what type of varnish to use for the cloth. Chanute responded quickly; after all, here was another newcomer who needed advice.

It did not take long for a special relationship to develop; Chanute with his devotion to scientific methods assisted Wilbur with documentary material, personal counsel and guidance. His answers were at times evasive, like a good mentor, trying to make his student think harder to come up with a better understanding and solution. In other cases, Chanute may not have fully comprehended what Wilbur was trying to do. Reading the hand-written letters, it appears that Wilbur was delighted to have found a friend in the much older man who always wanted to hear the latest development and who would comment, positive and negative.

Having selected Kitty Hawk as their experimental flying ground, Wilbur explained to his father in early September 1900, "It is my belief that flight is possible. While I am taking up the investigation for pleasure rather than profit, I think there is a slight possibility of achieving fame and fortune. It is almost the only great problem which has not been pursued by a multitude of investigators" [62]. Writing again a few days later, "My machine will be trussed like a bridge and will be much stronger than that of Lilienthal, which, by the way, was upset through the failure of a movable tail and not by breakage of the machine. The tail of my machine is fixed, and even if my steering arrangement should fail, it would still leave me with the same control that Lilienthal had at best … I have not taken up the problem with the expectation of financial profit. Neither do I have any strong expectation of achieving the solution at the present time."

Wilbur arrived at the Outer Banks on September 13; the glider was assembled and ready to be flown ten days later. In the meantime, the family had decided that Orville should join his older brother; he arrived on September 28. In the next few weeks, the brothers learned about wind and weather and how to fly the glider, either as a kite or manned. They found it "a rather docile thing" and noted that it flew better backwards than forwards. The glider, which had been damaged by a strong wind, was left behind at Kitty Hawk when they returned home on October 23 (Fig. 4.8).

Back in Dayton, Wilbur wrote a lengthy report to Chanute. "In October my brother and myself spent several weeks at Kitty Hawk, North Carolina, experimenting with a soaring machine" [63]. They had disregarded the fashion which prevails among birds and had placed the tail in front of their apparatus, calling it a front rudder, and they placed the operator in a horizontal position instead of sitting or standing upright. They had made a dozen free flights, and the total time in the air was about two minutes. "It is a magnificent showing, provided that you do not plow the ground with your noses," Chanute replied. Reading the letter again, he thought that the experiences of the brothers would be of interest to other enthusiasts, so he asked: "I should like your permission to allude to your experiments in a brief and guarded way as you may indicate." The reply came by return mail, "It is not our intention to make a close secret of our machine," but they wanted to test their machine's full possibilities first.

Fig. 4.8 (LH) 1900 glider flown as kite (00556u) and (RH) crumbled by the wind (00544u). The Wright Brothers Papers, Manuscript Division, Library of Congress, Washington, DC

Continuing their correspondence, Wilbur mentioned their plans for further experiments at Kitty Hawk [64] and asked the Chicagoan to come to Dayton to comment on his envisioned principle to control their glider. Being curious about his newest correspondent, Chanute agreed to stop in Dayton on his next trip east. When asked, Wilbur stated that Sunday would be a good day, as they would not be working in their bicycle shop. However Sunday was Chanute's day with the family. He arrived on Wednesday, June 26.

Some time back Wilbur had wondered if there was an instrument to measure and record lift, drift and wind velocity at any given moment while flying. And yes, Chanute owned two anemometers, one made by Richard in Paris and the other by Robinson. To help the brothers succeed, he brought along his French anemometer and explained how the instrument worked. The brothers used it during their 1901 and 1902 flying season.

It must have been an interesting visit. The brothers, being sure that their next glider was a complete success [65], proudly showed it to Chanute and explained their control system. Wilbur wanted to "warp" the biplane structure so that the trailing edge of one wingtip would flex upward while the other one would flex downward. This differential "warping" caused the lift on one side of the plane to be greater than on the other side, making the plane bank, or turn, toward the side with the upward twist. Chanute, with his railroad background and work with Mouillard, believed that by lowering a flap of material on the trailing edge of one wingtip, the increased resistance on that tip would turn the plane toward that side. This "braking effect" on one side is what Chanute thought would turn the plane. The ideas of warping to create lift and braking to create drag were, in effect, opposite approaches. At that time, no one knew which method would eventually bring success, and Chanute certainly did not see it the Wright's way.

The visit of this celebrity was seminal for Wilbur, who wrote to his elder brother Reuchlin, "He [Chanute] was very much astonished at our views and methods and results, and after studying the matter overnight said that he had reached the conclusion that we would probably reach results before he did. He will visit us at Kitty Hawk and will send one of his machines, just about completed, for trial. He also offered to send a couple men at his own expense to assist us in our experiments. This we refused but invited him to camp with us so that we could be of mutual assistance" [66].

Chanute's flying experiments in the Indiana Dunes had provided much previously unknown information, and he now wondered if the two mechanics from Dayton could take manflight to the next level. Would they add to the general aeronautic knowledge and—hopefully—some day—help solve the puzzle of manflight? They had written him many letters and had been absorbing all the available information. They had built and flown a glider, based on his biplane design, and incorporated their own ideas. They still corresponded with him practically every week about some aspect of aeronautics. They did not want his financial help, but they genuinely wanted all the knowledge and time that he was willing to give. As far as he could tell, the Wrights were currently the most promising duo to solve the problem of sustained, controlled flight. Now they had asked if he could join them in camp, this would be special (Fig. 4.9).

The Wrights went to Kitty Hawk for a second time in early July 1901. Chanute happily accepted their invitation, but he also wanted his two protégés, Edward Huffaker and George Spratt, to come along, so that they could learn and share what they knew.

Huffaker arrived on July 18. He was to help the Wrights and compare the glider, that he had built for Chanute, in flight with the latest Wright glider. Only one week after Huffaker arrived in camp, Wilbur wrote home that their visitor from Tennessee was intelligent, but lazy, and the glider was a total failure [67]. The culture clash between the three men became more obvious with every passing day. The 45-year-old Huffaker had grown up in the mountains of Tennessee, while the 30-year-old Orville and the 34-year-old Wilbur came from a medium size mid-western town. One man had an academic education, while the others had none. One believed that it is unimportant to dress in the field as one would in the city, while the others believed the opposite. Naturally, each one was convinced to be correct, and Huffaker with his academic background had a hard time taking the Wrights and their flight experiments as serious as they would have liked.

Fig. 4.9 Top: Chanute-type glider at the dunes along Lake Michigan, Indiana, in September 1897. Chanute photo album, Library of Congress (wb039p-34-009). Bottom: Wilbur Wright in 1901 glider, Kitty Hawk, North Carolina. The Wright Brothers Papers, Library of Congress, Washington, D.C. (00570)

The 31-year-old Spratt arrived on July 25. He happily assisted the Wrights, first to assemble and then launch the 98-pound glider plus 140-pound pilot. Chanute described the launching process a few months later. "The machine was taken upon a sand hill about 100 feet high, sloping at an angle of about 10°, two assistants, each grasping one corner, ran forward against the wind until a sustaining pressure was obtained. When they were told by the operator to 'let go,' the latter then guided himself, gliding straight ahead and landing at the bottom of the hill, by striking the sand at a flat angle and sliding thereon on the skids attached to the machine" [68].

Chanute arrived on August 4, shortly after Huffaker's glider was destroyed in a rain storm. Being at camp with a group of younger enthusiasts was exciting. He had not seen a heavier-than-air aircraft fly since his own experiments ended four years earlier. What a thrill to be part of the activity now! Everyone helped launch and then watch Wilbur fly. As before, Chanute logged the distances flown and flight times (down to the split second), the wind conditions and the angle of the slope of the dune for each flight, just like he did in 1896 and 1897 with his team. He also brought along his Kodak camera with preloaded negative film to document the Wright's glider in flight. Unfortunately, several photos were failures [69], but the group photos clearly capture the human face of the airplane's development: the 69-year old Chanute, with his distinguished small, white beard, wearing a striped dress shirt with winged collar, black bow tie and a felt hat, sitting with the leisurely dressed Huffaker on a camp stretcher in the opening of the shed. Other photos show Wilbur and Orville, dressed in white shirts with stiff collars and dark ties, and Spratt, wearing the same clothes as he did working on his farm [70] (Fig. 4.10).

The 1901 Wright glider, with a 22-foot wingspan and a 7-foot chord, was tested in gliding flights and as a kite. Wilbur made several glides up to 335 feet, but then on one flight, he skimmed only a few feet above the sand and plunged into the ground. No one understood why it happened. Wilbur wrote in his diary that he had to use the full travel of the front elevator just to keep it from diving into the ground or climbing until it stopped flying. This loss of pitch control was baffling. While drinking many cups of coffee, the team discussed the cause of the glider's erratic behavior from every angle. Spratt was quite sure that this was caused by the sudden reversal of the center of pressure, and he explained that the center of pressure was reversing as the glider approached level flight. Huffaker agreed with Spratt, and they decided

Fig. 4.10 Edward Huffaker, Octave Chanute, Wilbur Wright and George Spratt at the Wright Brothers Camp in 1901. Chanute Papers, Manuscript Division, Library of Congress, Washington, D.C. (00583)

to test the theory. Working as a team, they removed the top wing and flew the monoplane glider as a kite. When the angle of attack was reduced, the wing started to pitch and buck, proving that the center of pressure was moving, thus causing the problem. They changed the wing shape to a less cambered airfoil, which improved the performance, but the glider still showed less lift than calculated [71].

After a week of camp life with mosquitoes, Chanute returned to Chicago but reminded Wilbur: "Please take plenty of snap shots. You will want them to illustrate whatever you write." Next, Spratt had to return home and Huffaker left with the remains of his glider.

With everyone gone, the brothers proved to themselves that warping the wing to effect lateral control worked, but the machine experienced significant adverse yaw, which resulted in an undesirable slipping action and a tendency to turn into the opposite direction [72]. They were convinced that the data, supplied by Lilienthal, translated and published by Chanute, was wrong. They considered this year's experiments a complete failure.

Chanute had sensed that the brothers had reached a certain crisis in their confidence to achieve flight. He needed to reenergize Wilbur and developed a subtle plan.

4.6   Western Society of Engineers Presidency in 1901

The Western Society of Engineers (WSE) of Chicago, founded in 1869, was one of the most prestigious engineering societies in the United States; members from around the country and Canada came to Chicago to attend the monthly meetings. Being elected president in 1901, Chanute wanted the society to become even more prestigious and expand the members' horizons. To do so, he invited out-of-town and even non-members, to discuss timely and thought-provoking topics at the monthly meetings.

Professor Hermann von Schrenk, a long-time friend of Chanute, delivered his talk in February on the "*Factors, which cause the decay of wood.*" Beginning his presentation, von Schrenk asked for "your hearty co-operation in this matter; our opinions are likely to clash in some respects. What the scientific man may see on the one side may seem practically unfeasible to the engineer. On the other hand, what may seem practicable to the engineer may seem very unfeasible to the scientific man, and unless the two come together and look at the problem from both points of view, we will often find that certain economic problems cost a great deal of effort and a great deal of money and time to attain results which might have been saved, because of ignorance of some of the fundamental facts that underlie the whole problem." This topic was close to Chanute's business as well as his "side issue," aeronautics; he appreciated the thoughtful comments by von Schrenk.

A controversial talk on the "*Ethics of the Engineering Profession*" by Victor Alderson, Dean of the Armour Institute of Technology (now Illinois Institute of Technology), was much to Chanute's liking. He agreed with the speaker, that "pure professional success, as distinguished from mere money getting, depends upon acting in harmony with the fundamental principles. The professional man must be a broader man, must have a wider grasp of relations, must have the ability to solve new complications, and must be the leader and thinker as well as the doer." Several Chicago papers discussed Alderson's talk, as it surely hit a nerve in the engineering community.

The architect Frank Lloyd Wright from Chicago talked on "*The Art and Craft of the Machine*," and Bernard Fernow from the New York State College of Forestry, Cornell University, presented a paper on "*The Forester, an Engineer*" next. Now Chanute wondered if flight might be of interest to members. After discussing the next available time slot with the WSE Secretary, Chanute invited Wilbur to come to Chicago to address that august group and share what they had learned about flying.

Not being accustomed to public speaking, Wilbur was not in favor of accepting Chanute's invitation, but Katharine nagged him until he agreed. Hearing about his talk to be a "Special Meeting" or even "Ladies Night," Wilbur responded tersely, "I must caution you not to make my address a prominent feature of the program as you will understand that I make no pretense of being a public speaker. As to the presence of ladies, it is not my province to dictate, moreover I will already be as badly scared as it is possible for man to be, so that the presence of ladies will make little difference to me, provided I am not expected to appear in full dress, &c." To help make the presentation a success, Chanute offered his lantern slides, and asked Wilbur to ship some negatives to be made into slides.

Wednesday, September 18, 1901, was the big day. Wilbur spent the afternoon with Chanute, most likely discussing how to address the best engineering minds in America. Dinner was served at 6:30 sharp and then the two men took a carriage to the 16-story tall Monadnock building at Dearborn and Jackson, arriving at 8:00 p.m. They took the elevator to the rented rooms of the WSE headquarters where about fifty-five members and guests had gathered (Fig. 4.11).

Being to a large degree responsible for the brother's early work, Chanute stated in his introduction, "Engineers have, until recent years, fought shy

Fig. 4.11 The Monadnock Building, designed by the Burnham and Root Architects, at the southwest corner of Jackson and Dearborn streets, has 980 feet of frontage and is furnished with eighteen elevators. Its sixteen stories and 180 feet of height make it seem almost as slim as a chimney. Burnham and Root Architects, Chanute Papers, University of Chicago Library

of anything relating to aerial navigation. Those who ventured, in spite of the odium attached to that study, became very soon satisfied that the great obstacle in the way was the lack of a motor sufficiently light to sustain its weight and that of an aeroplane, upon the air." He explained that two gentlemen from Ohio had made some interesting experiments with remarkable results. "I thought it would be interesting to the members of this society to learn of the results accomplished, and therefore, I have the honor of presenting to you Mr. Wilbur Wright."

Stepping forward, Wilbur began, "The bird has learned the art of equilibrium, and learned it so thoroughly that its skill is not apparent to our sight.

We only learn to appreciate it when we try to imitate it." He continued in a more serious tone. "We went into camp in the middle of July 1901 and were joined by Mr. E. C. Huffaker, of Tennessee, an experienced aeronautical investigator in the employ of Mr. Chanute, by whom his services were kindly loaned, and by Dr. G. A. Spratt, of Pennsylvania, a young man who has made some valuable investigations of the properties of variously curved surfaces and the travel of the center of pressure thereon. Early in August, Mr. Chanute came down from Chicago to witness our experiments, and spent a week in camp with us." Wilbur then described their machine and mentioned that they achieved lateral control "by a peculiar torsion of the main surfaces, which was equivalent to presenting one end of the wings at a greater angle than the other." Comparing their results with those of other investigators, he indicated that previously published data relating to air pressures on curved surfaces appeared to be in error. Wilbur also tried to explain "soaring flight, by which the machine is permanently sustained in the air by the same means that are employed by soaring birds. What sustains them is not known, though it is almost certain that it is a rising current of air."

In the next letter to her father, Katharine described Wibur's trip [73]. "We had a picnic getting Will off to Chicago. Orville offered all his clothes so off went Ullam, arrayed in Orv's shirt, collars, cuffs, cuff links, and overcoat. We discovered that to some extent 'clothes make the man,' for you never saw Will look so swell." She also wrote that Wilbur declared Chanute's study ten times dirtier and more cluttered than their father's ever was. He had models of flying machines and stuffed birds suspended from the ceiling so thick that one could not see the ceiling. However, Katharine was not sure if she believed that story.

As Chanute had hoped, preparing for the talk and listening to comments from attending engineers reenergized Wilbur. His presentation, "*Some Aeronautical Experiments*," was published in the December WSE *Journal* [74] and Chanute ordered 200 reprints to share with his correspondents but also with Wilbur to give to his friends.

The 32nd annual WSE meeting took place on January 7, 1902, with about 160 members and guests attending. In his address as the retiring president [75], Chanute was proud of the papers presented during the past year, but especially the five talks by non-members, "men who were experts and who could elucidate subjects new to us and who had done novel things." Following the usual feasts, several members shared stories about Chanute and his love for flying, praising him for being an outstanding civil engineer with a vision and now Past-President of the Western Society of Engineers. William

J. Karner of the Illinois Central recited a poem without title, recalling his time flying in the dunes in 1897:

> … Well, I frankly confess there was a time
> In days of yore, when time was young,
> And birds conversed as well as sung,
> On a Chanute glide I soared aloft
> Until some people said that my brain was soft …

The feature of the after-dinner talks was the announcement by incoming President Finley that the retiring president, Chanute, had presented the society with a check for $1,000 to be used as an endowment for three engraved medals to be given annually for the best papers on civil, mechanical, and electrical engineering, presented by members during the past year. These medals were to encourage members to publish their achievements, and thereby grow the profession and the development of its members. This award was an important milestone for the engineering community. The criteria for the "Chanute Medal" changed over the years, but the WSE still presents them more than a century later. Chanute's leadership as president cannot be underestimated.

4.7 An Early Aerodynamic Researcher

After graduating in May 1894 from the Medico-Chirurgical College of Philadelphia, **George A. Spratt** (1870–1934) first joined his father in his medical practice, but having developed health problems, he moved to Coatesville, Pennsylvania, to manage his father's farm.

Believing that there is some fundamental principle that gives flying creatures an assurance of security, Spratt began to research flight. He compiled an essay in 1896 and mailed it with a two-page cover letter to Chanute. "Being very much interested in flying machines and fully believing in their economical practicability, I have had my views on the subject type written down and send you a copy. I wrote them down primarily that I might have clearer grounds for experimenting, but am getting discouraged accomplishing so little for various reasons, principally lack of sufficient funds. With the discouragement, boldness makes itself felt and I take this liberty of addressing you. The flying machine must come and it will soon come. Studying the subject principally from observation of birds, etc., in complete isolation from other interest, I am ignorant of the advance made. Will you do me the favor of reading and criticizing the premises and conclusion? Am I on the line of thought generally accepted as correct? How can I keep in touch with the advances made? I want to know more—I want to do more" [76]. As a courtesy, Spratt even included return postage, something that not many inquirers thought of doing.

The letter arrived on Christmas Eve, and Chanute responded right away. "As there is some of it which I fail to understand, it would not be fair to criticize it in detail … My suggestion is that you observe soaring birds when you have the opportunity and that you experiment with kites and gliding models to work out automatic stability in the wind. You may chance upon something that none of the other researchers have found" [77].

To better understand the laws of flight, Spratt began to experiment with air currents, induced in a conduit (or wind tunnel), to determine the most efficient wing shape. Initially Chanute discouraged him from spending much time on this project. "Such experiments have been found to be very delicate and perplexing, as they are affected by the sides of the vessel and the varying velocity and directions of the currents. Also, many of these experiments were tried by Horatio Phillips in England in 1884/5 where he tested several curved wing shapes in artificial currents of air, produced by induction from a steam jet in a wooden trunk. Others had experimented since, none were very successful and no practical results had come of them" [78].

To encourage Spratt to experiment further, Chanute mailed him two paper wings with a metal strip folded in along the leading edge. "I enclose herewith a couple of aeroplanes, such as described in my book, page 73. If they have not gotten out of adjustment in the mail, (and very little change will do it) they may exhibit a very simple way of getting propulsion out of gravity" [79] (Fig. 4.12).

Spratt followed the instructions given by Chanute in the *Progress* book and commented back on the paper wings' flight characteristics a few days later.

To measure the forces of lift and drag of wing shape models requires identical test conditions. Spratt explained to Chanute in March 1900 that measurement of lift and drag could be made more accurate if they were measured as a ratio, rather than separately. A drag-to-lift ratio could be determined by attaching a small wing shape vertically at the end of an arm. If the arm was free to move, a flow of air would cause it to rotate to the angle at which the forces of lift and drag would balance each other. This angle (called the tangent angle) would be the drag-to-lift ratio for that particular wing shape at that particular angle of attack. One could measure the tangent angles for a whole series of wing shapes and develop a complete table of ratios for any shape of wing. If the researcher then made a separate measurement of lift, he could use the ratio to calculate the drag. It would not be too difficult

Fig. 4.12 Among the many interesting letters in the Chanute Papers are two curious objects. Chanute wrote to Spratt in January 1900, enclosing two wings made of paper with a metal strip folded in along the leading edge for Spratt to experiment with. Collage assembled by Lewis Wyman. Chanute Papers, Manuscript Division, Library of Congress, Washington, DC [80]

to produce a set of graphs that would accurately show the characteristics of any wing shape at any angle of attack. It was, as engineers like to say, a simple but elegant solution to a difficult problem.

As the Wrights had encountered problems with their 1900 glider design, Chanute asked Spratt to share his findings with the brothers: "Your center of pressure experiments will do you great credit. Your method is very ingenious and I see no fallacy in it. The results show a reversal of the movement of the C.P. for the sections less than a semi-circle, but agree with my general understanding of results obtained by Mr. Huffaker for Prof. Langley. Of course, the 'Pline's curve,' or soaring birds wing curvature, is the most stable and was the one used by Messrs. Wright in their experiments. These gentlemen leave on the 8th for Kitty Hawk, off the North Carolina coast, and Mr. Huffaker is to join them with a gliding machine, which he has been building for me. If your farm work will permit of your leaving it, and you want to visit us, I will defray your expenses down and back, and perhaps you can assist in putting the machines together and in gliding. All the information you obtain to be treated as confidential, as Messrs. Wright may want to patent" [81]. This sounded interesting, so Spratt mailed a quick note to the Wrights and joined them in camp.

Returning home after watching the Wright's 1901 glider's erratic behavior, Spratt decided to study further. He then mailed a photo and resulting data from his wind tunnel experiments to Chanute, and a similar letter with a drawing of his measuring device to the Wrights.

In mid-January 1902, Spratt informed Chanute that he was to get married. "I congratulate you upon having found a nice girl to go into partnership for life. This will make a great change in your purposes. You will have to devote more thought and energy to material cares and to the accumulating of a competence, and any scientific experiments must be relegated to your leisure. If you need a little help, I shall prefer to furnish it, imposing no obligation whatever upon you, providing you eventually publish whatever you find, so as to help others" [82], were Chanute's sincere fatherly comments.

The three-way correspondence between Chanute, Spratt and the Wrights in the next few years showed a detailed interchange of scientific thoughts and information on wing aerodynamics and airplane stability; they clearly were all working on the general advancement of the aeronautical science as advocated by Chanute.

4.8   The Unwelcome Assistant

After leaving the Smithsonian Institution in late 1898, the 42-year-old **Edward C. Huffaker** was at first without a job. Feeling sorry Chanute offered him $50/month to build models and experiment with automatic stability, on the condition that all results would be communicated back to him so that they could be published [83]. The first project was to build a glider with five wings, stacked one above the other, and to build it as light as possible, consistent with strength. But then Chanute changed his mind, he now wanted Huffaker to build a monoplane glider, foldable like Lilienthal's machine or Lamson's big kite. "Let me know how soon you expect to have the machine ready for trial and whether you think it best to try at home or to come to Chicago. In the latter you will have the benefit of Avery's instructions about its use which you will find awkward at first" [84]. Sending his next monthly paycheck, Chanute stressed: "I desire to have your machine completed in time to test this summer" [85].

After visiting the Wrights in Dayton in June, Chanute continued his travel to Chuckey City, Tennessee, to see Huffaker and the glider he was building. Seeing the structure was disappointing, and Chanute wrote to Wilbur: "The mechanical details and connections of the gliding machine which Mr. Huffaker is building for me are so weak, that I fear they will not stand long enough to test the efficiency of the ideas in its design" [86].

For his visit to the Wright's camp a few weeks later, Chanute suggested to Wilbur: "I will send Mr. Huffaker and his machine to your testing grounds and pay his share of camp expenses. He will assist you in your experiments in exchange for your assistance in testing his machine. The latter I expect to be brief. He was for three years Prof. Langley's assistant, and is a trained experimenter, but lacks mechanical instinct. Kindly advise me of your decisions." Even though Wilbur was not keen in having many people join them, he did mail Huffaker instructions how to reach the camp. And Chanute mailed Huffaker $100 for expenses and wrote "I do not think that you will need shelter for your machine, but you better buy a cot and a blanket in Norfolk" [87].

Huffaker arrived with his glider at Kitty Hawk on July 18. Two weeks later, he recorded in the diary [88] he was to keep for Chanute, "My own single surface machine was about ready for trial with the Wrights machine. But then, a long and heavy rain late in the day almost ruined my machine, a number of the tubes yielded under the strain of wind and rain," basically destroying the glider. It is not clear why the structure collapsed so quickly, as the tubes, made from a paper composite, were varnished for additional

Fig. 4.13 Huffaker's glider, built with reinforced paper tubing collapsed during a heavy rain storm. Special Collections and Archives, Wright State University (15-5-20)

strength and two coatings of pyroxylene were applied to stiffen the cloth covering of the wings (Fig. 4.13).

Huffaker's final entry in the diary reads, "the work at Kitty Hawk seemed to still further emphasize the importance of equilibrium whose complete solution alone is lacking to secure the success of mechanical flight" [88]. He then mailed the diary to Chanute. This was the end of Huffaker's employment.

4.9   More Knowledge on Air Resistance is Needed

In his first article on the progress of flying machines, published in October 1891 in the *Railroad Journal*, Chanute stated that someone needed "to bring order out of chaos in aerodynamics and reduce its many anomalies to the rule of harmonious law." He was sure that all discrepancies about air resistance could be clarified by some explanation that had been overlooked so far.

Albert Zahm had presented a paper at the 1893 aeronautical congress discussing air resistance experiments, in which he attributed the unreliable data to the fickleness of the wind. He joined the Catholic University in Washington, DC, in 1895 as Associate Professor of Physics, while studying air resistance for his doctoral degree at John Hopkins University. To gain practical results, he shot four-inch hollow spheres from a gun along a dark room into a box of wool. Three thin ribbons of sunlight, crossing the bullet's path and focused in the camera, served to measure its velocity and deceleration, this was to help calculate the resistance. Receiving Zahm's manuscript of his thesis, Chanute found it interesting and wrote "I trust that your thesis will receive proper recognition from scientific men." The Rector of the Catholic University then agreed to provide some funding for an aerodynamics laboratory.

In August 1899, Chanute received a letter from Hugo Mattullath, who had designed a large seaplane to transport goods and passengers commercially. But to obtain a patent and secure funding for the construction of his aeroplane he needed to prove the feasibility of the design. Chanute suggested that he meet Zahm in Washington. Mattullath visited the laboratory, was impressed and offered financial resources for the operation. In April 1901 the first complete wind tunnel laboratory in the United States was fully operational, equipped for a wide range of aerodynamic experiments using instruments capable of accurate measurement [89]. Zahm's wind tunnel research, together with a description of his instruments, was presented at the Philosophical Society of Washington in May 1902 and published in *Science*. [90] Unfortunately

Mattullath died unexpectedly in December 1902 and funding for Zahm's lab stopped. The Smithsonian Institution provided a grant of $500 in late 1904 to experiment on atmospheric resistance to moving bodies, something Chanute had studied almost thirty years earlier when trying to speed up the railroad. Zahm proved that the best shape for the balloon part of the airship was that of the submarine torpedo, and that skin friction was a major contributor to drag, which was in direct contrast to Langley's statement that friction drag was negligible [91].

The Wrights had used the Lilienthal table to calculate the supporting power (lift), the resistance of the surfaces (drift), and the propelling power (tangential) to determine the aerodynamic performance for the 1900 and their much larger 1901 glider. But their practical results did not match their calculated data. Trying to understand, Chanute reread Lilienthal's article with the help of a German-speaking engineer and wrote to Moedebeck, "It follows therefore, that in applying coefficients to an aeroplane, not only should the arching be taken into account, but also the form, its proportions of length to breadth, its direction of advance, and the movements of the center of pressure, which will be found to vary greatly on different arched surfaces" [92].

To help Wilbur understand what was required, he mailed him his copy of Lilienthal's book, as well as copies of several translated chapters which he had received from Langley. As usual, Chanute played his role as the premier disseminator of aeronautical information.

Instead of experimenting with down feathers as Chanute had done in 1896, the Wrights built an "apparatus" and a device to measure the "rectangular pressure" of the external forces on a variety of wing shapes in early October, similar to what others had been experimenting with. Not obtaining repeatable data, they decided to follow Spratt's suggestion [93]. Success! they now had reproducible results on the relationship of wing shape and lift, and Wilbur shared sketches in a stream of letters with Chanute and Spratt, who was glad that his explanation helped. And Katharine wrote to their father, "The boys are working every night on their 'scientific (?) investigation.' Will says he'll have to eat crow for after all he has discovered by accurate tests, that Lilienthal's tables are not far off" [94]. And Wilbur had realized that no table is of universal application.

Sending a mini-report of his latest trip to the Outer Banks to Langley, Chanute wrote, "He [Wilbur] has done what neither Lilienthal, Pilcher nor myself dared to do: sailed on a machine of 300 square feet surface with the man prone on the frame work. His angle of descent has not been flatter than my own, but the results discredit the Lilienthal coefficients ... Have you a set of your curved surfaces?" [95] Langley obliged, mailed a drawing and data on the cambered airfoil that he was using, but asked Chanute to consider it confidential. As Langley's airfoil was similar to some of the shapes that the Wrights were testing, Chanute responded, "I will show this to no one save Mr. Wright, who is puzzled with the result of some of his tests in the wind" [96].

A few days later, Chanute mailed Wilbur data on windmill blades that he had researched in the early 1870s, and Langley's letter with the drawing of his airfoil. "I enclose it and beg that you will return it when you have absorbed his data for a surface very similar to yours. It is as he says confidential" [97]. The brothers made a model of Langley's airfoil and reported back to Chanute, that it was not as efficient as some other surfaces they had tested, especially the wing shape of the gull, or Pline's curve, which was superior in lift and much better in tangential [98].

4.10   Changes and Progress in 1902

Chanute celebrated his 70th birthday among family members in Pasadena, California, with a bottle of red wine from the Napa Valley and a fresh package of his favorite cigars. Only two months later, the 68-year-old Annie, Octave's wife of almost 45 years, died of pneumonia, leaving a large vacant spot in his heart and life. Trying to cope with his new life, the lonesome widower briefly considered stopping all aeronautical work, but the interaction with the many younger enthusiasts was simply too exciting; he felt needed, and he wanted to give (Fig. 4.14).

The daughters hired additional servants, which brought a different kind of problem. "My den was invaded by a house cleaning cyclone during my absence, and I may need a little time to find things" [99], Chanute wrote to von Schrenk.

And considering his next steps in life, Chanute wrote to Wilbur. "I think I will not experiment any more myself, but I desire you to test the comparative merits of what I have done in the hope that you will get some good out of it" [100]. Maybe the Wrights could rebuild his multiplane and the biplane gliders, when their regular work slacked up, and he would give them to the brothers as a present. Wilbur questioned the wisdom of such a present and responded. "It is not certain that we would be able to find opportunity for such extended use of the machines as would justify so great an expenditure on your part. Our use of the machines ought to be an incident rather than

Fig. 4.14 Octave and Annie Chanute (late 1880s). Courtesy Chanute Family

the primary purpose of their construction. A friendly spirit of emulation is a spur to progress in every line of human endeavor" [101].

Being without a job again, Herring stopped at the Chanute home and asked to build any glider "to beat Mr. Wright" [102]. With Wilbur not interested in building the multiplane or the biplane, Chanute accepted Herring's offer and agreed to pay him $150 to rebuild the *Katydid*, and he would furnish what was still in storage in Avery's shop. "I agree with your proposal to make the wings 8ft. × 3ft., and am glad that the machine will be a little lighter per square foot than before" [103].

Meanwhile, the Wrights had incorporated the results of their wind tunnel research into their 1902 glider, now with a 32-foot wingspan and a 305-square-foot wing area. This was a significant step up from Chanute's biplane of 1896 with its 16-foot span or their first glider of 1900 with a 17-foot wingspan (Fig. 4.15).

The brothers traveled for the third time to Kitty Hawk in late August 1902, assembled their glider and Wilbur flew it a week later. To control the higher aspect ratio glider in flight, they had developed a hip-cradle to activate the wing warping, but the craft yawed into an undesirable sideslip when the wing warping was applied. Orville then suggested changing the vertical tail into a moveable rudder to counteract warp-drag. This was the needed breakthrough for a working three-axis control system [72]; now both brothers shared the excitement of flying.

As everyone enjoyed Spratt's company, Chanute wrote: "Have a new multiple-winged machine to be tested. If your engagements permit of your

1896 Chanute   *1902 Wright*

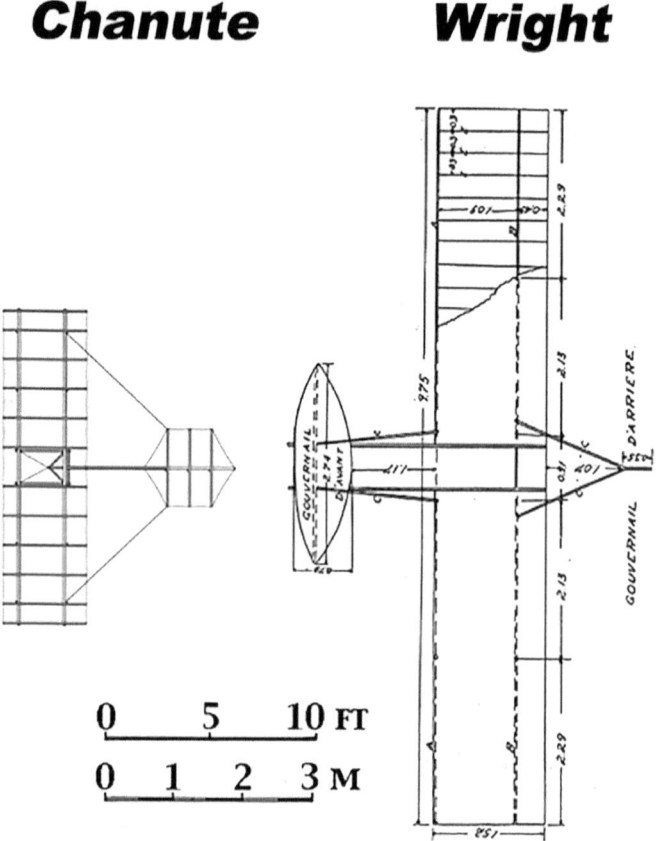

Fig. 4.15 Comparing the general size of the 1896 Chanute glider with the 1902 Wright glider. Courtesy Reinhard Keimel

being present, you may see how you can improve the design and the performance" [104]. And Wilbur wrote a quick note that three machines would be flown, so he ought to come and be part of the excitement. Spratt arrived in camp on October 1, staying for two weeks.

Chanute arrived with Herring on October 5 in the middle of a rainstorm. The next day Herring assembled "his" *Katydid*, or the "swinging wings machine," and made a short glide, but the new lightweight wings buckled and twisted in the wind. A second attempt ended after a twenty-foot glide, breaking the crossbar on landing. After a quick repair, Herring flew the glider as a kite, concluding that it did not produce sufficient lift. Spratt was anxious

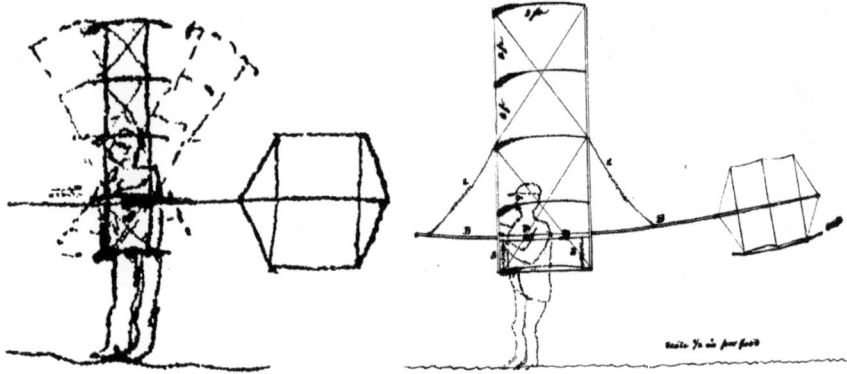

Fig. 4.16 Chanute's ideas on how to achieve automatic stability: (RH) Multiplane with "rocking" wings, letter to F. Wenham, September 24, 1897. (LH) Multiplane with "oscillating" wings, letter to E. C. Huffaker, December 22, 1898. Chanute Papers, LoC

to also fly the rebuilt *Katydid*, as Chanute had offered, but he did not have the courage to ask, as Herring made it very clear that it was his glider (Fig. 4.16).

Lamson had built the "oscillating wing" glider for $234. This glider with three sets of wings, one stacked above the other, could be folded for transportation. The crates had arrived at the Outer Banks late in September 1902, and Herring, Spratt and Orville now spent a morning assembling the machine with its 20-foot wingspan and a surface area of about 150 square feet plus the vertical tail. After flying it first as a kite, Herring made two gliding flights, but could not adjust the springs to activate the oscillating mechanism. When Lamson inquired, Chanute responded: "The machine was rec'd in good order. Workmanship is superb, but the joints frail. Owing to late arrival and bad weather, I could test it but one day during my 8-day visit. Left it for further trial with Mr. Wright. I am changing my mind about the expediency of adding a motor, as the gliding machine has been so much improved that it seems safe to advance a step forward" [105]. Even though Chanute's idea of "oscillating wings" did not work in Lamson's glider, the idea was successfully reintroduced in the early 1970s in modern hang glider technology (Fig. 4.17).

With his two glider projects out of the way, Chanute watched Orville and Wilbur fly over longer distances across the coastal sands. Despite the sluggish shutter speed and the negative film material in Chanute's box camera, he captured their gliding activities with remarkably vivid action at almost every stage: Wright's 1902 machine; the field of experiment; ready for a glide; beginning to run; gathering speed; good speed attained, the flight begun; skimming close to the ground; gliding downhill, were some of the captions

Fig. 4.17 Augustus Herring flies the Lamson-built "oscillating wings" triplane at Little Hill (October 1902). From *Revue Générale des Sciences,* November 30, 1903

Chanute assigned to his photos. The group picture shows the participants in front of the 1902 Wright glider, which was the break-through (Fig. 4.18).

Chanute and Herring left camp on October 14, both heading to Washington. Even though he had no appointment, Chanute saw Langley at the Smithsonian castle for an instant. Langley's assistant, Charles Manly, then received the main points of the Wrights' experiments. Some inconsistencies crept in relaying the news, but one point was clear: the Wright brothers were ready to install a motor. Next, Chanute met with Charles Marvin at the Weather Bureau who had helped figure out the scientific explanation of the horizontal and vertical forces in gliding flight. And the last person to meet

Fig. 4.18 The party at the Wrights' camp in October 1902: From L to R: Chanute with his Kodak box camera, Orville Wright, Wilbur Wright, Augustus Herring, George Spratt and Dan Tate in front of the break-through 1902 glider. Lorin Wright, photographer; Special Collections and Archives, Wright State University (15-7-3)

was Albert Zahm at the Catholic University. There was so much to learn and share!

Herring also stopped at the Smithsonian, but the secretary was not available. A few days later, Langley wrote Chanute, who replied: "I have lately gotten out of conceit with Mr. Herring, and I fear that he is a bungler. He came to me, said that he was out of employment and urged that I let him rebuild gliding machines 'to beat Mr. Wright.' I consented to building new wings for the multiple wing machine, but could give it no attention, as the work was done at St. Joseph, Mich. Herring adopted new forms of wings, but when the machine was tried in North Carolina, it proved a failure; he said he did not know what was the matter. Doubtless he got some ideas from Mr. Wright's machine" [106].

A very successful flying season ended for the Wrights. "We increased our records for distance to 622.5 feet, for time to 26 seconds and for angle to 5° for a glide of 156 feet" [107]. Of course, Chanute wrote to his many correspondents about this latest trip to the Outer Banks. Writing to Moedebeck, "Wright has made a great advance over last year. As soon as practice has been obtained to learn all the tricks of the wind, I believe that soaring flight will be performed under favorable conditions, and that the time has nearly come to introduce a motor. I am glad that this type ('type Chanute' as Captain Ferber is good enough to call it) is likely to prove to be one of the types of machine with which safety will be achieved" [108]. Writing to B. F. S. Baden-Powell of the Aeronautical Society of Great Britain, Chanute described the glider and explained: "It was equipped with a horizontal rudder at the front, and a vertical rudder at the rear, and it proved quite manageable, so that skill was quickly acquired to steer it up or down or sideways and to meet the turmoil of the wind" [109]. The following month, Baden-Powell mentioned the Wrights' progress in his presidential address, but did not include Chanute's comments about the rudders.

4.11   We Will Surely Navigate the Air!

As there were no publications in the early twentieth century focusing on the needs of those interested in aeronautics, William Edwin Irish from Glenville, Ohio, decided to fill the void. The first issue of his "*The Aeronautical World*" appeared on newsstands in August 1902; Irish hoped that his magazine "would sweep away all the difficulties and that the practical aerial machine would be available for rapid transit with the graceful ease and speed of a bird" [110]. But he stopped the publication of his magazine one year later.

A free-lancing reporter, William J. Jackman, who had covered the gliding experiments in 1896 for the *Chicago Inter Ocean*, wanted the public to know what had happened in the past five years. His interview was published in the *Chicago Times-Herald*. "Mr. Chanute had with his other possessions, frigate birds and kindred birds of the sea capable of long and sustained flight with little wing effort. Their flight over the waters in the region of their habitat had been watched time and again by the man bent on solving finally the flying problem." And Chanute stated conservatively: "I have confined myself to the problem of securing safety through automatic equilibrium. It is only after this has been secured that it will be safe to apply a motor and a propeller. The commercial machine will be developed by a process of evolution; one experimenter finding his way to certain definite results, the next overcoming some of the remaining difficulties, and so on, until some ingenious man shall add the finishing touches. He will probably obtain the credit of being the true inventor, but his predecessors who shall have pointed the way, may perhaps not all be forgotten. It is believed, however, after many centuries of failure, that we are at last within measurable distance of success in aerial navigation" [111].

Sitting at his big, flat desk and smoking his favorite cigars, Chanute did his best work late in the evening; he had created a remarkable information center on man's attempts to fly in his library on the top floor of his home on the north side of Chicago. Receiving clippings with new information, he either mailed the originals around, or he made copies in an extra letter-press book and tore the page(s) out, or he asked his daughters to transcribe the information on a typewriter, brought into the Chanute family in the early 1890s. One way or another, all new information was shared with his many acquaintances, as he continued his unofficial leadership by encouraging others through personal correspondence and visitation. Receiving letters from inventors who declared to have solved the problem of aerial navigation, Chanute studied their ideas and reminded them to work without expectation of financial reward other than being remembered: "For, in the usual course of such things, it would be the manufacturers who would reap the pecuniary benefits when commercial flying machines were finally evolved" [112].

5

Two Eventful Years of Intertwined Development

The airship, which shall be effective, responsive and safe, continues to be the aspiration of many scientists, the hope of many cranks and the means of notoriety for some adventurers.

Editorial, *The Evening Star*, Washington DC (1902)

In a well-researched paper on airplane wing trussing, Felix W. Pawlowski from the University of Michigan discussed in 1916 what was needed to create a successful aircraft structure. Based on his personal experiences, he admitted that designers frequently had difficulties breaking away from the forms, types, or examples that exist in nature. The first efforts to build a flying machine were attempts to imitate the flapping of birds' wings, just as there were attempts to build locomotives that moved on legs. The aeroplane, as finally created, was not a replica of a bird, and it did not imitate the bird-motions in flight (except to a certain extent in gliding flight), but it still retained the essentials of a bird's wing. The problem faced by the early pioneers was how to build wings large enough for a man-carrying aeroplane. They had to determine, how a big wing surface could be made lightweight and rigid at the same time, and how the members of the wing framing could be arranged. Francis Wenham in England was probably the first to realize the difficulty when building his multiplane kite in 1866. Prominent mechanical engineers like Sir Hiram Maxim (1894) with his huge multiplane craft, and Otto Lilienthal (1895) with his monoplane, could not produce a simple and statically clear structure to combine the wings of their machines. A bridge engineer was the first to do so; it was Octave Chanute who introduced the

© The Author(s), under exclusive license to Springer Nature Switzerland AG 2023
S. Short, *Flight Not Improbable*, Springer Biographies,
https://doi.org/10.1007/978-3-031-24430-8_5

bridge truss into the biplane, and the idea was adopted immediately by most airplane builders" [1].

5.1 Pas À Pas, Saut À Saut, Vol À Vol—from Step to Jump to Flight

One of the most important and yet little-known figures of the precursors of aviation was **Louis Ferdinand Ferber** (1862–1909), a powerful influence on the aeronautical development in Europe. He promoted the heavier-than-air movement in France when people still believed that manflight could only be achieved with dirigibles.

When asked in spring 1905, he explained to a reporter of *L'Opinion* how he became interested in aviation. "I believe that Jules Verne had a primordial influence on our country. The sons of the well-to-do classes, who devoted their efforts to literary and artistic matters, now occupy themselves with mechanics. Verne used several mathematical calculations in his novel "*From the Earth to the Moon*" that intrigued me, even though I was only fifteen years old. So, I begged my father to buy me a book on algebra; I studied it and soon was first in math. The teacher became my ally and told my parents: This boy must go to the Polytechnique and here I am."

An article on Lilienthal in a German paper in 1898 inspired Ferber to build a glider. The results were disappointing but he did not give up. A few years later, he read an extract of a lecture by George H. Bryan on aerial loco-motion in the *Revue Scientifique,* [2] stating that the most difficult question connected with the flying machine was its balance and stability, and stressing the importance of mathematics, as no accurate interpretation could be done without it. Bryan also gave the names of several pioneers. Ferber decided to learn from them and contacted Gustav Lilienthal first; he just warned him to be careful with unpredictable winds [3]. Next, he wrote to Chanute, who was pleasantly surprised to receive a letter from a French enthusiast. Responding, he apologized for possibly making mistakes in writing as he had no prac-tice with his French language, and he mailed Ferber a copy of his *Progress* book, the *Aeronautical Annals* of 1896 and 1897, and reprints of his and Wilbur Wright's WSE talks. In passing, he mentioned that "Quite recently, a Mr. Wright did even better than what I did, but he has not published anything yet" [4]. Receiving this package proved to be decisive for Ferber and the French aeronautical community.

Reenergized, Ferber designed his next glider by combining parts of Chanute's 1896 and the Wrights' 1901 design. The airframe was built of bamboo, and the wing covering fabric was loosely attached in separate pieces, assuming that the pressure of the wind would provide the necessary camber. The glider performed a bit better than his first ones, helping Ferber understand what he needed to do next.

Visiting Europe in early 1903, Chanute also visited Ferber; he described the area (Fig. 5.1) where Ferber did his flying in a letter to Wilbur. "Ferber has an abominable practice ground, being a series of rocky slopes with precipices above and below. He has accordingly built what he calls an 'aerodrome' with a mast, 59 ft. high and a rotating arm 98 ft. across, running on ball bearings, to test various forms of apparatus with motors. He now has a 6 hp gasoline engine, which weighs complete, with the shaft, inside of 200 lbs., and he estimates the screws to weigh about 22 lbs. He is now looking for the best

Fig. 5.1 Captain Ferdinand Ferber experiments with his glider. *L'Auto*, June 23, 1909 and flying his pilotless glider on his "aerodrome." *L'Auto*, January 21, 1903

and lightest construction of screws and is to let us know what he finds out. He says that he is not trying to invent a new system, but simply to experiment with the best that others have designed and is much of the opinion that you are ahead of all others. He says he is very tempted to go to America as he would like to buy your 1902 machine" [5]. Wilbur responded that they prefer not to sell yet.

Looking for input and advice, Ferber discussed many of his problems with Chanute who usually translated the letter and forwarded a copy to Wilbur. This stimulated a certain competitiveness and the brothers became more interested in Ferber's work. Here is one episode told by Ferber: "There is in my present apparatus a disgusting instability which I scarcely understand." The glider pitched up and down, and he wondered if he should add a tail or use only one surface with the Wright's horizontal rudder [6]. Forwarding an exact translation with a redrawn sketch, Wilbur responded that the problems "were due to too great a depth of curvature and the fact that the surfaces are not sufficiently held in shape by ribs. We never had such extreme insta-bility as he describes, but my first address gave some account of our troubles with a surface of too great curvature and the means we used to correct it" [7] (Fig. 5.2).

Being part of the military, Ferber accepted an assignment with Charles Renard, commander of the French Army Balloon School at Chalais-Meudon in early May 1904. Here he built his Type VI with a rear-mounted horizontal stabilizer, and featuring dihedral to obtain better lateral stability. But most of

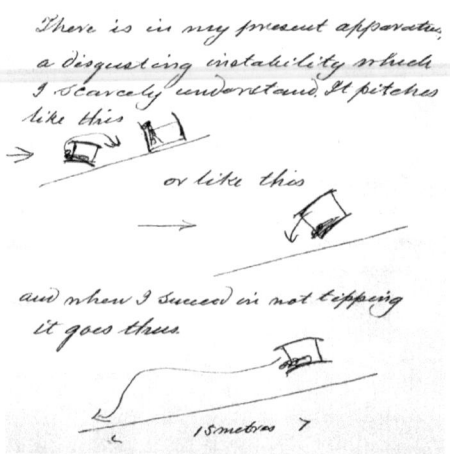

Fig. 5.2 Ferdinand Ferber consulted regularly with Chanute when he encountered problems with his flying machine, they were usually shared with Wilbur Wright. September 7, 1903. Chanute Papers, Manuscript Division, Library of Congress

the personnel at Chalais-Meudon believed that dirigibles were the only way to achieve flight, and Ferber was considered an outsider. He then asked for an extended leave from the military and joined the Antoinette company [8].

5.2   Spreading the Word as the Goodwill Ambassador

The fall season with its social gatherings in Chicago usually brought discussions on where to spend the winter months. Chanute's daughters had seen an advertisement for a tour to Egypt, which sounded exciting, and on the way back to America, they could visit southern Europe and spend springtime in Paris. The almost 71-year-old Chanute felt reasonably well, and his doctor saw no problems for an extended vacation, but gave a new heart tonic (Fig. 5.3).

Chanute then received an urgent wire from the Louisiana Purchase Exposition organizers in St. Louis; they were concerned that interest in their upcoming aeronautical competition was not forthcoming. As Chanute knew almost everyone in the European aeronautical community, they wondered if he could help. This idea sounded interesting and yes, he could combine the family's travel plans with visiting as many European aero clubs as possible to promote the aeronautical contests at the St. Louis Fair.

The Chanute family left Chicago on December 29 and embarked on the *SS Commonwealth* of the Dominion Line on January 3, 1903. Much

Fig. 5.3 Chanute and his wife Annie, and her brother James with his wife. Courtesy Chanute Family

to Chanute's disappointment this steamer had little luxury and sailed with a service speed of only 16 knots. After stopping briefly at the Azores and Genoa, they arrived in Alexandria, Egypt, on January 19. The family boarded at the luxurious Shepheard's Hotel and took daily sightseeing trips along the Nile River, just like other tourists.

One item on Chanute's personal agenda was the estate of Louis-Pierre Mouillard, who had died in 1897; he had heard that his papers were transferred to the French Consulate in Cairo to be claimed by whoever had the right of inheritance. Meeting the Consul, Chanute was told that no one at the embassy knew anything about this pioneer's estate, but he suggested to contact Mouillard's sister, who lived in Nice.

On February 1, the Chanutes ferried to Italy, staying for one week at the Grand Hotel in Naples. Chanute pointed out in his diary [9], that this hotel "was close to the sea, had an open and healthy situation, with a splendid view." Another week was spent in Rome, then Florence, Milan, Venice, and finally Nice, France, boarding at the Cosmopolitan Hotel, where each apartment had electric lighting and a hydraulic lift for guests to reach their rooms on the upper floors.

While in Nice, Chanute arranged to meet Madame Teillart, Mouillard's sister. She sent her visitor to her son-in-law, M. Camoin, who had worked for her brother in Egypt; he told Chanute that the family had no interest in the estate. Seven years later, members of the "League Aérienne" obtained Mouillard's papers, as they wanted to publicize his discoveries of wing warping and his patent regarding lateral control [10].

The next person to meet in Nice was the French artillery officer, Captain Louis Ferdinand Ferber. During their discussions, the two men realized that they both believed in the philosophy of the "School of Lilienthal." "To invent an airplane is nothing; to build one is something; but to fly is everything," acknowledging Lilienthal's influence in both their personal development [11].

On Friday, March 13, the daughters took the train to Paris, and Chanute traveled to Vienna to begin his missionary work for the St. Louis World's Fair. It was a bit coincidental, but before he knew that he was going to Europe, he had mailed reports of the Wrights' flying activity to his many European correspondents; these reports now appeared in print while he appeared in person [12, 13, 14].

Chanute had written a quick note from Nice to the four-year younger **Wilhelm Kress** (1836–1913) in Vienna, Austria, with whom he had corresponded since 1894. Arriving in the midst of a furious snowstorm, the two men then spent the afternoon talking about the latest aeronautical affairs. In the evening Kress hosted a dinner at the Hotel Bristol with eleven influential members of the *Flugtechnischer Verein* (Vienna Aero Club) in attendance.

Kress had designed a large flying boat, his *Drachenflieger I,* in the late 1890s. He used lightweight steel tubing for the structure and a heavy Daimler engine for propulsion, with two propellers providing the motive power. Three arched wings were mounted one above the other. The entire apparatus, reminiscent of Langley's aerodrome, weighed 1,870 pounds, including the aviator. When Kress launched his aircraft from the shores of the Wienerwaldsee, he had to swerve the craft quickly to avoid hitting a rock that he had not seen earlier; the pontoons filled with water and the machine sank close to the shore. To improve the design Kress added a fourth wing, which brought the machine's total weight to almost 2,000 pounds [15]. Seeing the *Drachenflieger* the next day, Chanute seriously doubted that it would ever become airborne, even if a more powerful motor could be obtained.

Victor Silberer, President of the Flugtechnischer Verein, had called a Board meeting for Sunday morning, at which Chanute presented facts about the St. Louis Fair, followed by a good discussion of the current state of aeronautics in Austria and in the United States. Later in the afternoon, Chanute socialized with the 86-year-old civil engineer Friedrich Ritter von Lößl, a senior board member of the club, who had been responsible for the development of the railroad system in the Austro-Hungarian Empire, much like Chanute in America.

The next stop was Breslau in southeast Germany, where he met Hermann Moedebeck, who showed him the manuscript of the updated *Taschenbuch für Flugtechniker und Luftschiffer* which was almost ready to go to the printer. Chanute was pleased to see his contribution on recent aeronautical events [16] and mentioned that it would be nice to have a translation for English speaking readers.

Continuing his travel to Berlin, Chanute met the Commissioner for the World's Fair, Mr. Lewald, and members of the Board of the *Berliner Luftschiffer-Verein*. They invited Chanute to the Officer Corps of the Airship Battalion at the military airfield at Reinickendorf-West where a special breakfast was served. Chanute happily shared information on aeronautical progress, followed by an interesting discussion on the pros and cons of airships versus aeroplanes.

Traveling to Strasbourg next, he met the Chairman of the International Aeronautical Commission and members of the *Oberrheinischer Verein für Luftschiffahrt*. Several members hoped to travel to St. Louis, providing they could receive government funding for the trip.

Chanute arrived in Paris on March 22, catching up with his three daughters, getting reacquainted with family members and friends and enjoying springtime in this excitingly social city. One of his correspondents, **Victor Tatin**, (1843–1913) invited Chanute to lunch at the Automobile Club and then gave him a tour of the St. Cloud airfield. Michael Lagrave, the French Commissioner for the World's Fair, also came to the hotel to discuss the rules of the contest at the Fair and how to grow aviation in France and the United States.

Thursday evening, April 2, was the big day; almost every Aéro-Club member attended the monthly meeting at the hotel of the Automobile Club. As Chanute had done now more than a dozen times, he began his presentation by first describing how he was introduced to aeronautics. Having learned of Lilienthal's flying activity, he wanted to do his own experimenting and take flying to the next level. He described with some humor the events of 1896 and 1897, where members of his team, but also fellow civil engineers and newspaper reporters, enjoyed tobogganing on the air. Everyone who wanted to try, succeeded and even the cook became almost an expert in a short time. "Our experiments, methodically conducted, will permit us, little by little, to learn completely 'the art of the bird', an art which, without seeming so, is extremely difficult. We will know it better in a year or two. Until then, it is useless, and even dangerous, to burden oneself with a motor, and I much prefer to use an untroublesome and simple motor like gravity as supplied by nature. Besides, these glides provide the most original and most enticing of sports; several of my friends have devoted themselves with enthusiasm to aerial glides!"

Believing that the only way to solve the flying machine problem was by helping one person, who would then help the next person, and so on. His invitation to repeat and improve on his work did not receive an answer until 1900 when the Wright brothers contacted him. He was impressed with their work as aeronautical workers, and the rest of his talk focused on the brothers' gliding experiments of 1901 and 1902. "In short," Chanute told his audience, "I can claim no other merit than taking Lilienthal's experiences and to perfect them as best I could, have others resume my work and bring them to the perfect result. Progress in the sciences, especially in aeronautics, takes place in successive stages. I would be too happy if I could have contributed, however little, to advancing the question, hastening the solution of this great and difficult problem, which is the passion of our time" [17, 18].

Combining his enthusiasm with showing fascinating lantern slides, Chanute stressed that flying was exciting and fun. He explained that sailing flight [or soaring] is aerial locomotion, using the energy supplied by nature.

Progress was tedious, but "the feat of flying had been performed in America, which should spur aviators to try again and again." In closing, Chanute reiterated: "It is necessary to use a great deal of prudence while conducting gliding flights. Do not try to beat existing records, it only prompts foolhardy adventures and causes accidents. Competitive runs where amateurs practice together are fruitful because the fliers learn from one another ... One must never forget that balance, control and safe landing are the main things, no matter the length of the distance flown. It can be hoped that when many fliers start to work seriously, progress will happen so fast that the time will be near when flying becomes practical" [19].

News of the talk by the *célébré aviateur américain* spread swiftly through the press. Ernest Archdeacon's report [18] was published nine days later in *La Locomotion,* showing four photos, taken by Chanute, of the Wright 1900 and 1901 gliders, but mis-labeled as Chanute gliders. He described the American engineer as being different from other inventors, as he believed in the necessity to share information to reach the goal of a practical flying machine. Archdeacon explained that the pilot is upright in the Lilienthal machine and shifts the body for control, while in the Wright machine the pilot lies horizontally. Perhaps taking license with the fact, Archdeacon also wrote: "To regulate the equilibrium in the transverse direction, he moves two cords, which operate by warping on the right side or the left side of the wing; and simultaneously by the displacement of the vertical rudder in the back." In closing Archdeacon mentioned that Chanute would send drawings of all the gliding machines, so that similar ones could be built in France. And he reminded his readers that aviation had to be perfected in Montgolfier's country and not by foreigners. The Secretary of the Aéro-Club published the minutes of Chanute's talk in the April issue of *L'Aérophile,* [20] again misidentifying the Wright gliders as Chanute's; several European publications were quick in pointing out the error.

On April 10 the Chanute family arrived in London. Having read **John Bacon**'s recently published book *Dominion of the Air,* and seeing his name mentioned a few times, Chanute arranged to meet the author and spent an afternoon at his home in Berkshire. Years later Bacon's daughter **Gertrude** wrote about "... the visit of the courteous, keen-eyed, white-haired American gentleman with a pleasant smile and eyes that looked into the future, sitting at our table and telling us, in modest, matter-of-fact tones, enthusiastically of certain wondrous experiments with kites and gliding machines that he was making in his own country, and of the rather striking work that was being done by two young American brothers of Dayton, Wilbur and Orville Wright. They had chosen as their experimental ground a remote stretch of the North Carolina coast where a steady wind blew constantly from the Atlantic, and the sand dunes gave admirable jumping-off places. We liked the old gentleman and we were mildly interested in what he told us" [21].

To entertain the guest, Gertrude showed her collection of "blindfolded pigs" and cajoled Chanute into adding his contribution. Seeing Maxim's masterwork, he agreed, was handed a black pencil and a sheet of paper and was blindfolded to draw his rendition of a pig. Remembering this from days long gone, drawing a pig with one's eyes closed gives a psychological insight into a person's character. Removing his blindfold, Chanute looked at his masterwork and added, probably with a smile, his firm autograph (Fig. 5.4).

Years later Bacon recalled, "Many a time since then have I looked with satisfaction upon his boldly sprawling animal and firm autograph in my book, because, as events subsequently proved, this Octave Chanute was the pioneer of a great discovery. His work was directly responsible for the first flights of the first aeroplanes and the first men who ever successfully flew; and long before he died he was enthusiastically acclaimed, both in Europe and America, as the father of modern aviation" [22].

A few days after this pleasant interlude, members of the Aero Club of the United Kingdom entertained Chanute and his daughters with a dinner at the Carlton Hotel. The next evening, Chanute delivered his final talk at the General Meeting of the Aeronautical Institute, held at St. Bride's Institute in London. "Mr. Chanute first gave a general outline of the St. Louis Aeronautical Competition of 1904, and then proceeded to deal with the more recent experiments of the brothers Wright in South Carolina. Chanute pointed out

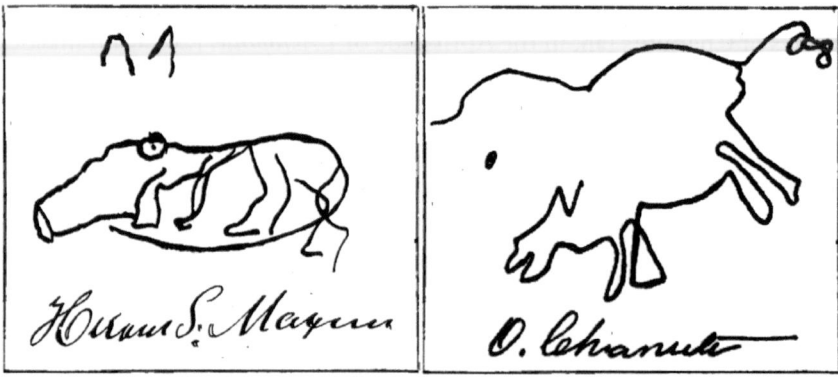

Fig. 5.4 Gertrude Bacon had asked several aviation enthusiasts to draw a pig while being blindfolded. Seeing Maxim's masterwork in her collection of "blindfolded pigs," Chanute agreed to contribute. *Strand Magazine*, March 1899 (Maxim) and December 1912 (Chanute)

to an interested audience that in all cases of soaring flight it is an indispensable condition that the center of pressure and the center of gravity should be in the same vertical straight line. In the machines of Lilienthal and others it was attempted to maintain this condition by shifting the body of the operator. Chanute introduced different ways to restore the condition of equilibrium, not by shifting the operator's position or center of gravity, but by altering the center of pressure by manipulation of a tail. In all experiments, however, the operator maintained a vertical position. The Wrights were the first to adopt a horizontal position, reducing the resistance offered by the operator's body to the air to about one-fifth of what it is in the vertical position." Those who listened to the talk will not forget Chanute's graphic expression in reference to the wind: "The wind is not a steady flowing stream, as many people imagine, but a series of whirling billows of air" [23].

The Chanutes left Southampton on April 29 on the recently launched *Kronprinz Wilhelm* of the North German Lloyd, arriving back in Chicago on May 9. Reflecting on the past six weeks, traveling through central Europe and listening to European aeronautical experts, Chanute was convinced that no other nation was ahead of the United States in regard to aviation at that time. He had established or reestablished contact with everyone who was someone in aviation and had inspired a new generation of experimenters. He had achieved his personal goal, and he had created interest in the upcoming fair in Saint Louis with its aeronautical contest.

5.3   The Busy World—Aeronauts Soar on Silken Wings

As promised, Chanute compiled his notes of the April 2 Aéro-Club talk and forwarded the text to Georges Besançon, secretary of the Aéro-Club and editor of *L'Aérophile*. And he included three-view drawings of his three gliders (*Katydid* multiplane of 1896, the biplane of 1897 and the 1902 Lamson-built "oscillating wing" triplane), as well as the Wright 1902 glider.

The drawings showed dimensions but no technical details, perfect for those enthusiasts who were inclined to use their imagination to take their ideas to the next level. Besançon translated the text and published it in their August issue [19] (Fig. 5.5).

In his introduction Chanute described the two philosophies to achieve manflight: use either motor-driven machines, à la Langley or Maxim, or motorless machines as advocated by Lilienthal. He then described his own gliding experiments in the Indiana Dunes and his less successful attempt at

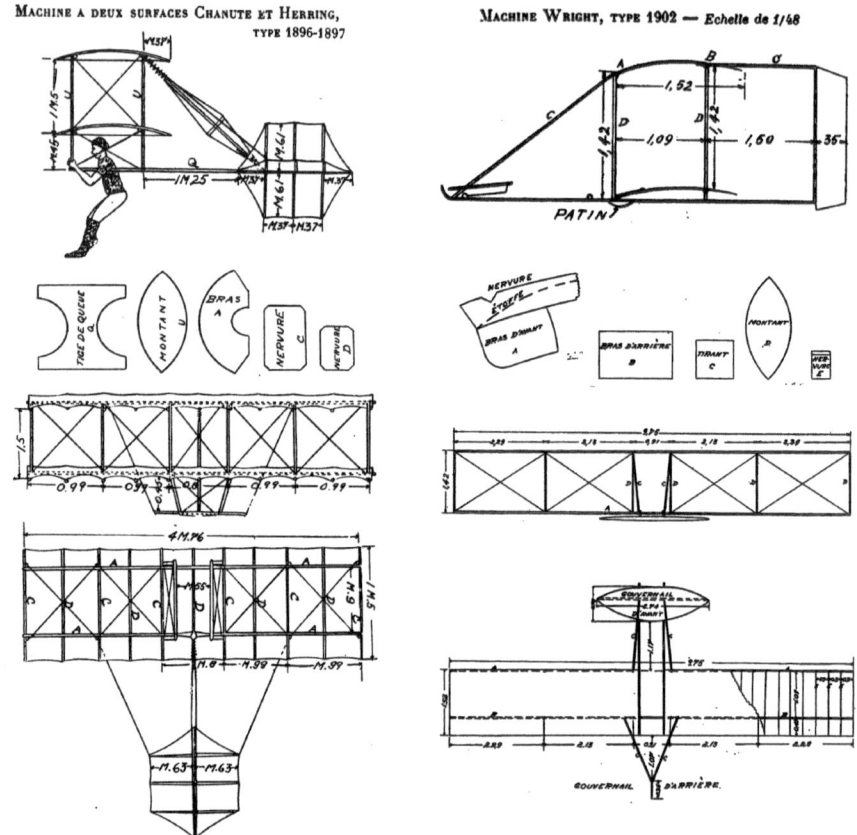

Fig. 5.5 The August 1903 issue of *L'Aérophile* published Chanute's Aéro-Club lecture, including three-view drawings of Chanute's 1896, 1897 and 1902 gliders and the Wright's 1902 glider. *L'Aérophile*, August 30, 1903

Kitty Hawk in 1902, reminding his readers that all experiments were made without accident. The remainder of the article discussed the gliding activity of the Wrights in 1901 and 1902.

The question of how the Wrights controlled their machine stirred much controversy in the next decades. There is no record that Chanute described the Wright's break-through concept. He did however point out their method of warping the wings, and indicated its significance, however most people interested in flying at that time did not catch the hints. The English-language text, which was translated into French for publication, only described the three improvements introduced by the Wrights, with one of them being the "warping of the wings to steer to the right or left," with no mention of any corresponding movement of a vertical rudder. Most flying machine designers at that time still believed that the aircraft would be controlled in two dimensions, just like a ship. To climb or descend, the operator would need to only increase or decrease propulsion, and a primitive vertical rudder was to control the direction of the aircraft in the horizontal plane. To most observers, the flying machine basically floated in space just like a ship floats on the water surface.

As Tatin wanted to experiment with a "type du Chanute" glider, he asked for help, and Chanute pulled the photos he had submitted to *McClure* [24] three years earlier, as they clearly showed how members of his team took-off, glided downhill and landed safely [25]. And he responded: "It is difficult for me to give all construction details in writing to enable any amateur to construct a biplane glider. There are however some details from my experiences which may be useful to you.

1. I think it is better to use wood for the arm rest than metal. I prefer fir, well dried.
2. We made the wing ribs of rattan or ash wood. Willow is also good if you can get pieces with the desired curvature and length from the basket maker. If ash is used it has to be soaked, or steamed, and dried slowly over a mold of the desired curvature. Several ribs may be formed in one section to be separated later on.
3. In the drawings, I showed the ribs a little heavy, so that there is no deformation of the wing. In case wings are desired to be flexible in the back, which I consider to be of benefit, flat ribs may be used and the connection at the leading edge may be made by means of wire at the point of deflection.
4. More stability is obtained by letting the root of the wing come 75 to 100 millimeters below the center. You may have noticed that the attitude of the sea gull is more stable than that of the vulture.
5. Connections may be made by means of string windings, or loops of very soft wire, or by means of light steel tubes as used on bicycles. I prefer string windings which give a certain elasticity to absorb landing shocks.
6. Wright uses unvarnished cotton cloth for the surfaces while I use varnished Japanese silk.
7. The cloth is attached in the front and pulled back tightly, so it becomes very smooth; then it is doubled up across the wire which connects the ribs in the rear; next the cloth is fastened with pins. Two layers of varnish were applied next, which glued the folds together and shrink the cloth so that it becomes tight as a drum. If desired the pins may then be removed.

8. We use pyroxylene as varnish which is prepared as follows: 60 grams of cotton powder, dampen with alcohol to make it safer. Combine with 1 quart of alcohol and 3 quarts of ether. To this are added 20 grams (2/3 oz) of Castor Oil and 10 grams (1/3 oz) of Canada Balsam. When ready to use, pour a very small quantity into a saucer and apply with a flat brush. It dries quickly" [26].

Tatin then submitted Chanute's explanation and photos to *L'Aérophile*, they were published in March 1904 as "*The practice of gliding*" [27].

Ernest Archdeacon (1863–1950) wanted to repeat, confirm and then improve the performances of the Wrights. Using the drawings from Chanute's article in *L'Aérophile*, Archdeacon asked if M. Dargent could build a 1902 Wright glider at the Chalais-Meudon military works. In late February 1904, the technical commission of the Aéro-Club of Paris inspected the "exact" copy of the Wright-type glider. Archdeacon then located an area near Berck-sur-Mer, with a fairly high dune, semi-circular in shape and oriented toward the west, without a single tree, shrub or clump of grass. At its feet lay the flat sand, and breezes came steadily from the sea (Fig. 5.6).

Starting in March, several Aéro-Club members were flying biplane gliders of various configurations, and newspapers reported on the latest craze of aerial tobogganing. "Gilded youth [rich kids?] is going in for aeroplanes,

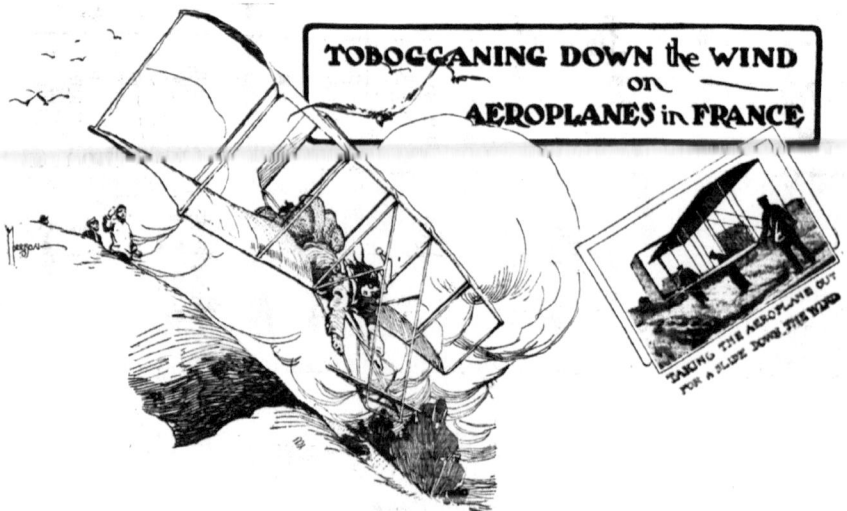

Fig. 5.6 Several American papers discussed this new and exciting sport of aerial tobogganing in France. *The New York Press*, March 20, 1904

and they call it tobogganing in the air or sliding down the wind at Berck-sur-Mer, in the dune country not far from Calais." Another headline read: "The aeromaniacs got the necessary hint from the Wright brothers successful flight and petroleum motors were to be added after the knack of the air is acquired." *Scientific American* reported in December 1904 on Archdeacon's experiments showing a photo of him lying flat in his glider. The text explained that it "resembles the Wright aeroplane in its general principles, but contains different modifications in details, which will no doubt make it an improvement over the former."

The write-up of Chanute's lecture in *La Locomotion* caught the attention of **Gabriel Voisin** (1880–1973). The simple fact that Chanute was born in France raised his impression of him even higher. And, "I set to work on a set of wings similar to those I saw in the illustrations. The construction was similar to our Hargrave glider that we had built in late 1902. On the first day of the south wind, we were behind the brickyard, and a few minutes later our machine took to the air … The Chanute is laterally more stable than our old Hargrave box kite; it lifted us with incredible ease and the day passed without incident. But in spite of trying different tethering points, we could not obtain a small angle of attack. Then I removed Chanute's tail and replaced it with a Hargrave cell from previous trials via a rough method. The effect was instantaneous. Our glider stopped flying tail-down and took up in the wind, even when loaded, in a nearly horizontal attitude" [28].

Needing someone to fly Archdeacon's Wright-type glider, Ferber recommended hiring Gabriel Voisin, who recalled in his biography [28, 29]. "On Easter Sunday 1904, I was on a sandy dune at Berck. Towards ten o'clock in the morning I took off. Archdeacon had chosen the site, which was probably perfect for duration trials with a fully prepared machine, but for a first attempt it was unsuitable. I had some good flights, but I could not circle over the area in graceful curves as the seagulls did above us … I brought my aircraft down to the ground, at the foot of the hill, without incident, which gave me, immediately, a certain amount of prestige".

Francois Peyrey discussed the activity from a different angle. "The tests on Sunday, April 8, 1904 were particularly interesting. The aeroplane lifted admirably on that day, it even raised up 7 to 8 meters above the starting point. But it did not make much progress, as the longest glide was only twenty meters." However, Archdeacon was thrilled: a Frenchman was seen by the public to fly in a French-built heavier-than-air machine between the earth and the sky over France [30]. Several photographers took photos of the flying activity, which were then made to postcards and mailed to every Aéro-Club member and the press (Fig. 5.7).

Fig. 5.7 Gliding in the dunes near Berck. Author's Collection

To improve the performance of the 70-pound Wright-type glider, Archdeacon suggested, "The ribs, perfect as fabricated, were, on the other hand, quite defective in their shape. The rear part, instead of being flat and almost horizontal, is, on the contrary, convex like the front, and the air, once entering under the curvature, does not escape quickly enough, thus lifting the device and preventing it from moving forward. I am going to make new ribs and build a new airplane which will differ from the previous one only by small details of which experience has shown us the primordial importance, and which were not indicated in the Chanute memo" [31]. The rebuilt glider was ready in early 1905. And Voisin recalled, "I made some detail modifications to my glider, which might be tested by being towed by a motorcar. I wanted to find out exactly what power would be needed for the future engine. My rich friend, E. Archdeacon, asked for a few days to think it over. I feared that our research was going to be stopped. But I was wrong. Archdeacon, reluctant to bear the new expenses alone, assembled a group of friends and formed a company with the object of providing financial support for our research. The *Syndicat d'Aviation* was formed with myself being nominated 'Engineer' at a salary of 190 francs a month. It was not much, but the thought of being paid to work upon the realization of my dreams made me forget all about my indigence."

Believing that water was the ideal shock absorber, as stressed by Chanute, Archdeacon decided to make trials with the next glider on the Seine River. In

June 1905, the rebuilt glider was towed by a speed boat at 40 kilometers per hour. Voisin piloted the "lovely glider" and rose instantly some 50–60 feet into the air and alighted gently on the water. "For the first time in the world, without the aid of ascending currents, a hydroaeroplane had left the water and returned to it after a flight of 2,000 feet, carrying a pilot in full control of its maneuvers, powered by an external mechanical power source. And everybody in Paris who was interested in heavier-than-air craft witnessed and photographed our trial" [28]. The second trial was somewhat less successful, and Voisin took a compulsory bath.

The experiments with *"Le Plus Lourds que L'Air"* or heavier-than-air craft continued, as several Aéro-Club members transitioned from balloons to aeroplanes. Ferber continued to push that the duty of French aviators was not to copy, but to study and improve the personal flying skills and the glider's performance. He was convinced that the "type du Chanute," or Chanute's biplane design with the tail in the back, was a better and safer performer than the Wright machine "type du biplan américain Wright" with the tail in the front. Eventually, most airplane designers abandoned the forward canard in favor of the aft tail.

Robert Esnault-Pelterie (1881–1957) was another young French civil engineer who had attended Chanute's talk. He too built a Wright-type glider and experimented near Wissant in 1904. Chanute wrote that the results were nearly equal to those obtained by Wright brothers in the first year they experimented. Pelterie demonstrated that it was possible to glide at an angle of 1 in 10, and "experience was gained which will doubtless be utilized later; for nothing but practice, practice, practice will produce adequate advance" [32] (Fig. 5.8).

Pelterie summarized his work in *L'Aérophile* [33] and included drawings of the updated Wright-type glider he had built. "The magnificent result from the other side of the Atlantic had left us somewhat skeptical. In science, skepticism has no value; to remove the doubt one had to repeat the experience. Therefore, we have built an aircraft by scrupulously following the instructions of the Wright brothers, and using plans published in *L'Aérophile*. Our device was absolutely similar to that of the American experimenters, both in its general dimensions, in the curvature of the ribs and in the arrangement of the rudders. The rectangular parallel piped shape, first applied by Chanute in 1897, is now classic; it is the form adopted by the Wright brothers and consequently by us … This first series of tests allowed us to see many interesting things; the rudder in front is an indispensable organ, contrary to what we had thought first. The twisting of the wings, recommended by the Wright brothers and tested by us, gives good results in maintaining the transverse

Fig. 5.8 Robert Esnault-Pelterie was fascinated about what he heard at Chanute's Aéro-Club lecture about the flying activity in America. Using the drawings from *L'Aérophile* and his technical engineering background, he designed and built a Wright-type glider. *L'Aérophile*, June 1905

equilibrium, but we consider this dangerous. It may cause excessive tension on the guy wires, and we feared to have structure failure in the air, which cannot happen with the standard rigging system. So, we decided to give up the twist. To act on the transversal balance, we employed two horizontal rudders at the front, independent and placed one at each end of the airplane. These two rudders were connected to a small steering wheel, within reach of both hands of the operator. When these two rudders were activated at the same time, they provided longitudinal stability; when they were activated in the opposite direction, they control lateral stability." Pelterie called them little wings or *aileron*.

Next Pelterie experimented with varying wing shapes, using sundry instruments to measure lift, drift and the center of pressure. His updated glider appeared a few weeks later: the wings had less curvature, the span was reduced to 31 feet 6 inches, and there was no canard. To takeoff from the level beach the pilot stood on a trolley holding the glider while being towed by a car. Following in Chanute's footsteps, Pelterie wrote, "The transverse equilibrium may perhaps be obtained automatically by a form of appropriate surfaces, but we believe that the vertical balance will have to be left in the hand of the airman, as well as the direction in the horizontal plane. However, all these difficulties can only be solved by a long practice. It is therefore necessary to arm yourself with patience."

Even though his research pointed the way to eventual success, Pelterie changed direction and designed a very successful monoplane next.

While Aéro-Club members had fun with their gliders, the now 30-year-old **Alberto Santos-Dumont** continued flying his Airship No. 9 over Paris. An editorial in the *New York Times* looked at his activity critically. "His airship may possess little real value, and he may have brought to light no new scientific principle of importance, but it cannot be denied that he has done much to arouse public sentiment regarding the question itself. The importance of the subject is sufficient to warrant public encouragement and with such support the man with the idea may be brought into view" [34]. But reading about the biplanes flown by Chanute and the Wrights inspired the Brazilian pioneer to invent a real flying machine that would rise without a gas bag. In collaboration with the Voisin brothers he developed his 14-bis aeroplane. To become airborne, Santos-Dumont attached his aircraft to a 60-meter-long steel cable stretched between two posts, similar to the zip-line system Ferber had used a year earlier. Success, on September 13, 1906, he made a flying hop of about 36 feet in his 14-bis biplane, powered by a 50-horsepower engine. One month later, on October 23, 1906, Santos-Dumont flew his biplane for nearly 60 meters (or 197 feet) at Bagatelle near Paris, winning the 3,000-franc Archdeacon Prize. And a few weeks later, on November 12, he made the first officially recognized, sustained, powered flight in Europe, flying in a straight line for 220 meters (about 720 feet), about 6 meters (about 18 feet) above the ground. Of course, Brazilians proudly proclaim that the world's first powered flight was made by one of their own, the fearless experimenter who exhibited the inventiveness, curiosity, and vision to succeed in doing what others thought could not be done.

An early pioneer from Denmark, **Jacob Christian Hansen Ellehammer** (1871–1946), had designed and built internal-combustion engines, which he installed in motorcycles. Seeing Chanute's article in *L'Aérophile* aroused his interest in aeronautics; he built and flew several Chanute-type gliders prior to tackling powered flight in 1905. His first aeroplane was driven by a three-cylinder engine with a four-bladed propeller, which he had designed. As the island of Lindholm, where he lived, was too small to attempt long-distance flights, Ellehammer built a circular device to which his craft was tethered, similar to Ferber's aerodrome, introduced a year earlier. On January 14, 1906, the tethered triplane spun pilotless around the pole until the fuel was exhausted. On September 12, 1906, Ellehammer made a manned, powered flight covering a distance of 138 feet, flying a straight line about two feet above the ground at a speed of 35.5 mph. Even though the flight was a success, his aeroplane is not credited as the first to fly under power.

The 30-year-old German inventor **Karl Jatho** (1873–1911) from Hanover, Germany, was another serious experimenter who had always sought new and different projects to experiment with. In early 1903 he became intrigued with flying machines and designed an aeroplane that showed some similarity to Chanute's 1902 triplane. His *"Motordrachen"* had six wings or sails, a horizontal steering sail, a horizontal mainsail, two vertical fixed sails and two vertical steering sails, a wind propeller and a basket carrying a 12-h.p. petrol engine. The equilibrium of the aeroplane was to be maintained by the horizontal steering sail, which controled the raising or lowering of the forepart of the machine. The two vertical steering sails served to direct the course by spreading or furling one or the other, while the two vertical fixed sails, or wings, keep the aeroplane, so to speak, resting on the air after it has ascended" [35]. Jatho managed to remain airborne for 18 meters (about 65 feet) on the very first flight with this triplane, with four people watching. He then changed his design to a biplane, *"Motordrachen* Nr. 2,"* in November. This time, he traveled about 200 feet in a straight line at a height of about 10 feet. Jatho then gave up because he could not visualize how to further improve his design.

When Chanute compiled his monthly articles on *"Progress in Flying Machines"* a decade earlier, he had mentioned the Irish born **John P. Holland** (1841–1914), who had claimed that it was possible to navigate the air using the screw principle and combine things already tried and proved by other experimenters [36]. Almost fifteen years later, the keen-eyed and now gray-haired Holland claimed, "Aerial navigation appears to be a much less difficult problem to solve than was submarine navigation. The simplest mode of locomotion is soaring in the air, and within twelve months people will be flying, and it will be the most popular method of locomotion" [37]. Holland believed that the true flying machine would simulate the bird, flapping its wings, and he believed that one would reach complete success with this method. But flying like the bird, flapping its wings, was pronounced fallacious by several engineers. And the editor of the *New York Tribune* closed his report with, "To suggest that Mr. Holland may be no more fortunate than those who have tried his plan before may seem ungracious; but, if he demonstrates its practicability, the *Tribune* will accord him due praise" [38].

The British mathematician **George Hartley Bryan** (1864–1928) became interested in aerial navigation "by tossing model gliders into the air." In February 1901 he read his paper *"History and Progress of Aerial Locomotion"* to direct attention to the importance of further investigation rather than furnish any specific kind of solution. He stated that the main difficulty connected with any attempt to fly in a heavier-than-air machine was longitudinal stability. The very fluctuations of wind velocity, which may furnish a source of energy for birds in sailing flight, vastly increases the danger of experiments with artificial flight. He told his audience that of the three people, Lilienthal, Pilcher and Chanute, who have done most to solve this question of balance and stability, the first two met with fatal accidents just when their experiments were becoming most successful. "Both Lilienthal and Pilcher used machines with broad curved wings, adopting a single-surfaced machine, with the operator relying on the movements of his body to counteract the effects of any sudden gust of wind tending to overturn the machine. Chanute, on the other hand, experimented with narrow superposed wings, some of his machines having as many as eleven or twelve aerocurves, arranged in pairs. Instead of balancing himself by his own agility, the wings moved about pivots and were held in position by springs so that their displacements, caused by a sudden gust of wind, gave the machine a tendency to right itself. In the early forms the movements of the body in balancing amounted to five inches, in the later ones they were reduced to two inches, and finally to one inch. Many glides were made of from 150 to 350 feet, in winds varying up to 31 miles an hour (nearly 1½ times the greatest wind-velocity in which Lilienthal experimented), and no accidents occurred." In closing, Bryan stated, that "the problem of artificial flight is hardly likely to be solved until the conditions of longitudinal stability of an aeroplane system have been reduced to a matter of pure mathematical calculation" [39]. Extracted parts of his lectures were widely re-published and translated [2].

Two years later, Bryan and his student W. E. Williams submitted an updated version of their research, titled "*The Longitudinal Stability of Aerial Gliders*," [40] published in January 1904. Ten years after his first foray into the stability of flight, Bryan published his classic book *Stability in Aviation*, where he traced the roots of his early research on stability of an aircraft and his interests in thermodynamics and statistical physics, again acknowledging Chanute and his biplane glider.

5.4   The Smithsonian Pushes Ahead

Having heard from Chanute about the Wrights' success flying their glider, Samuel Langley wanted to know more; he asked his assistant Charles Manly to combine a family visit to Chicago with meeting Chanute in late November 1902. Manly reported back that "Mr. Chanute feels that the Wright brothers have made a real advance even over his own ideas and experiments." The details of the Wright machine became a bit scrambled as they passed from Chanute via Manly to Langley, but it was clear, that the Wrights knew more about controlling their craft than he did.

Langley then wrote to Chanute, "I should be glad to hear more of what the Wright brothers have done, and especially of their means of control, which you think better than the Pénaud. I should be very glad to have either of them visit Washington at my expense to get some of their ideas on this subject, if they are willing to communicate them." Talking about the Wrights and their flying with Manly was one thing, but receiving Langley's letter "seemed cheeky." So, Chanute forwarded the letter to Wilbur and asked how to respond. Wilbur simply wrote, "It is not at all probable that either Orville or myself will find opportunity to visit Prof. Langley in response to his suggestion. But we have a number of matters demanding our attention just now" [41].

Newspapers reported in July 1903, that Langley had resumed his aeronautical work; in early August, the headlines gave details: "*Langley's Flying Machine makes its first flight. Performance was a disappointment. Went 500 yards, then sank in the river*" [42]. Chanute was not surprised reading about this mishap, as neither Langley nor Manly had any flying experience in a full-size glider. Even though Langley wanted no audience, the ever-present press reported on the next launch in early September: "*Prof Langley foiled again. Valve on his airship broke and he could not fly*," or "*Airship still misbehaving. This time it is the propeller which spoils the launching*." That thing just did not want to fly, but Langley was convinced that he was on the right track (Fig. 5.9).

Later in September Langley requested additional money, explaining his reasons fully and convincingly. The Secretary had gone through the original $50,000 from the War Department and money from a special fund of the Smithsonian. He may have also obtained financial assistance from other sources like the Fortification Act of June 6, 1902, where Congress had appropriated $100,000 for tests of the most effective guns and other implements and engines of war. The War Department may have stretched the latter category to include work on Langley's aerodrome.

On October 7, 1903, all was ready again. Manly started the engine, raised its speed to maximum, took a last survey and gave the order to release. The machine rushed down its track, impressively and majestically, according to witnesses, but on reaching the end of the launch track, it plunged sharply down into the water. Manly managed to pull himself through the narrow space between the wires and was picked up by a boatman. An editorial comment stated, "The ridiculous fiasco which attended the attempt at aerial

Fig. 5.9 The press reported on an unpremeditated voyage by Samuel Langley and his crew. The machine was not damaged and experiments will continue at an early date. *The Evening Star*, Washington, DC, July 18, 1903

navigation in the Langley flying machine was not unexpected, but it ought to fly as well as the average hen hawk" [43].

On December 8, Langley made a final attempt with Manly at the controls; the aerodrome plunged again into the waters of the Potomac. The press took no pity and ridiculed the secretary for wasting government money. *"Aerodrome did a flip-flop. Langley's flying machine again in the mud,"* was just one headline [44]. With this latest mishap, official support for the project was withdrawn and Langley was widely derided. An editorial in the *New York Times* reflected the public opinion: "… We hope that Professor Langley will not put his substantial greatness as a scientist in further peril by continuing to waste his time and money for further airship experiments. Life is short, and he is capable of services to humanity incomparably greater than can be expected to result from trying to fly … For students and investigators of the Langley-type there are more useful employments" [45]. Apparently Langley's efforts to achieve powered flight hardened the convictions of most Americans that flying was a totally impractical idea. After all, if Langley with his academic background, scientific experience and prestigious backing could not do it, then who could? (Fig. 5.10)

Fig. 5.10 The front page of Washington's *Evening Star* showed this cartoon to catch the reader's attention: "The Birds' Chorus: Wonder which of us was the Model?" And a more detailed report was published on page 11, "Airship fails to fly. Prof. Langley's machine goes to river bottom. Prof. Manly aboard was rescued from perilous position. *The Evening Star,* Washington, DC, December 9, 1903

In late January 1904, members of the House requested a statement showing the actual amount of money disbursed for the promotion of flying machine experiments and construction, either under Langley's direction or by anyone else. Mr. Hitchcock from Nebraska, the author of the resolution, contended against extravagance and simply asked, "… why a modern-day Darius Green should be established at national expense," and he insisted that the information sought in his resolution should be furnished to the House of Representatives. As a follow-up, Mr. Robinson from Indiana stated: "We should strike directly at that queer scheme of aerial navigation whereby a scientific promoter, encouraged by executive officers, if not by Congress, raised the high expectation of the public, and his demands for money, only to have the venture fall flat" [46].

The criticism following the aerodrome performances was disheartening, and Langley wrote to those who offered assistance, "If the government no longer wants it, then I no longer feel obliged, nor do I desire to pursue it." The Secretary ordered the aerodrome to be repaired and put in storage on the second floor of the Smithsonian's South Shed, locked away from the public. In one way Chanute felt sorry that pesky reporters and politicians labeled his friend a bungler, but he believed that Langley, like Maxim, Ader and others, undertook too much at once by trying to produce a full-fledged flying machine before making sure that control, stability and the possibility of landing safely were achieved [47].

5.5   The Airplane is Borne ... The Wrights Did It!

Just before Chanute left for Europe in late December 1902, Wilbur informed him that they were ready to submit the claims on their flying machine, were building a larger machine and proceed to mount a motor. With the help of their recently hired mechanic Charles Taylor, all efforts went into building the power plant for their "whopper flying machine."

Chanute arrived for his second visit in Dayton on June 6, 1903, bringing a brand-new Richard anemometer that he had purchased while in Paris, as he really wanted the brothers to succeed with their goal. The discussion quickly turned to the Wrights' propeller and engine design. Most likely Chanute mentioned Stéphane Drzewiecki and his findings on screw propellers, as he had just received his latest publications.

A couple weeks later, Wilbur addressed WSE members a second time, describing their experiments with a controllable glider. He stated that a thousand glides are equivalent to about four hours of steady practice, "far too little to give anyone complete mastery of the art of flying. Progress is very slow in the preliminary stages, but once it becomes possible to undertake continuous soaring, advancement should be rapid." Further he explained the two different methods of soaring: "When the weather was cold and damp and the wind strong, the buzzards would be seen soaring back and forth along the hills or the edge of a clump of trees. They were evidently taking advantage of the current of air flowing upward over the obstructions [modern day ridge soaring]. But on warm clear days when the wind was light, they would be seen high in the air soaring in great circles [modern day thermal soaring]. This seemed to indicate that rising columns of air always exist, but that the birds must find them."

Always looking for news to share, Chanute asked Wilbur, "I am writing an article for the *Revue Générale des Sciences*. Should the warping of the wings be mentioned? Somebody may be hurt if it is not. If you have one of your computations of a glide to spare, please send it to me" [48]. Wilbur politely responded, "It is not our wish that any description of this feature of our machine be given at present." He argued that the control was too tricky for beginners and that experimenters should use machines of less than 20 feet spread to learn the longitudinal control before attempting complex methods of operation. A few days later, Chanute again asked, as he believed that the Wright's operation of the vertical tail should be explained in this article more clearly. No explanation came. Being under time pressure, he sent another note by special delivery. Wilbur hastened to reply: "The vertical tail is operated by wires leading to the wires which connect with the wing tips. Thus, the movement of the wing tips operates the rudder. This statement is not for publication, but merely to correct the misapprehension in your own mind. As the laws of France & Germany provide that patents will be held invalid if the matter claimed had been publicly printed we prefer to exercise reasonable caution about the details of our machine" [49]. Sensing a certain apprehension, Chanute responded, "I was puzzled by the way you put things in your former letters. You were sarcastic and I did not catch the idea that you feared that the description might forestall a patent. I believe however that it would have proved quite harmless as the construction is ancient and well known" [50]. Chanute's article was published without discussing wing warping and the simultaneous useage of the tail in November 1903 [51].

Having worked in business and technologically advanced settings throughout his life, Chanute knew how to balance the excitement of new technology with one's need to maintain confidentiality. He understood the need for protecting potentially valuable "trade secrets," but he sincerely doubted that a fortune could be made from the aeroplane.

The Wrights arrived on September 25 for the fourth time at Kitty Hawk and revamped the 1902 glider, which they had left behind. To launch the Flyer, they had designed a rail system with a dolly, which they now tested with their glider first, starting successfully five times out of six. The boxed-up Flyer with its small gasoline motor and two pusher propellers arrived at Kitty Hawk on October 8.

Spratt arrived in camp on October 23 and left on November 6, when work on the Flyer came to a halt due to the breakage of the propeller shaft. He offered to express-ship the shaft to Dayton from the train station, where he crossed paths with Chanute who was on his way to the Wrights.

The cold and rainy weather during the week while Chanute was in camp, did not permit any flying, but provided plenty of time for talking and drinking coffee. The older engineer assessed their Flyer and was quite certain that the brothers did not provide a sufficient margin of power, [52] but thought that they might succeed. The brothers were sure that Chanute was wrong, but they fine-tuned the motor and drive-train the best they could after Chanute left.

The repaired propeller shaft arrived on 20 November, only to crack again a week later. This time Orville travelled to Dayton to machine a new unit. Sitting on the train, he wondered if there would be enough time to launch their Flyer successfully. Might Langley achieve his goal and win? Similar thoughts crossed Wilbur's mind. Waiting for his brother to return with the new shaft, he wrote to some people that he felt close to. Writing to Chanute, "I see that Langley had his fling, and failed. It seems to be our next turn to throw now, and I wonder what our luck will be" [53]. So far, things did not look promising.

Orville arrived back in Kitty Hawk on December 11; they installed the new steel propeller shaft and everything was ready on December 14; the brothers tossed the coin; Wilbur won. The Flyer moved on the launching rail under its own power, picked up speed and lifted off the rail, but then went into a steep climb and stalled after 3½ seconds in the air. "Misjudgment at start reduced flight. Power and control ample. Rudder only injured. Success assured. Keep quiet," read the wire sent home. Writing a more detailed letter later the same day, Wilbur admitted that he had raised the front too high after the Flyer left the downhill track, but he "was sure that the machine would have flown beautifully."

Machine and weather were again ready on December 17. The launching rail was laid on the sand into the wind; a camera was set-up to hopefully take a photo of the Flyer as it reached the end of the rail; the machine began to move with Orville to take off first. This time, the first sustained, controlled, powered flight became reality. Four flights were made during the morning, the last one, piloted by Wilbur, covered 852 feet, flying for 59 seconds. They sent a telegram home, setting the prearranged news release in motion. Katharine took the telegram to her brother Lorin who was to notify the press. On the way back home, she stopped at the telegraph office to share the news with Chanute who was the first and only person outside the family to hear about their success: "Boys report four successful flights today from level against twenty-one-mile wind. Average speed through air thirty-one miles. Longest flight fifty-seven seconds." The telegram arrived in Chicago shortly after 8 p.m. and Chanute replied to Katharine right away:

"I am deeply grateful for your telegram of this date advising me of the first successful flights of your brothers. It fills me with pleasure." He also wired the Wrights at Kitty Hawk: "Immensely pleased at your success. When ready to make it public, please advise me." He felt like a proud godfather to the Wright brothers' achievements.

As so often, it is good to have a bit of luck of the circumstances on your side, as Harry Combs described in his book *Kill Devil Hill* [52]. On that cold December 17, the wind blew at 27 miles per hour. Although the brothers were apprehensive about being able to control the Flyer in this much wind, it was actually the wind that saved the day. Flying against that wind, the airplane needed an acceleration of only 3 miles an hour to gain a flying speed of 30 miles an hour, and consequently, only 40 feet of the 60-foot track was needed for takeoff. If there had been no wind blowing on that day at Kitty Hawk, the Wrights 12-horsepower engine would have required a 755-foot-long track to get the aircraft to accelerate to its flying speed of about 30 miles an hour. In other words, on their 60-foot track, with no wind, the craft would have needed an engine of 120 horsepower, or 10 times the power they had, to get sufficient acceleration to take off. Another advantage was the actual location of Kitty Hawk. Being at sea level at a temperature of 34° F. early on December 17, this produced a density altitude of 1,800 feet below sea level. Chanute was correct that the Flyer was underpowered; it is always good to have some luck on your side.

Mingled between further sarcastic discussions about Langley's latest flying mishap just a week earlier, several newspapers mentioned the Wrights' flight, usually with an imaginary drawing that did not match reality. Chanute forwarded the *Chicago Chronicle* clipping, published in late January 1904, with his comments to Wilbur. "My attention was called yesterday to a clever piece of 'journalism' by which you are made to appear to have given an interview to the *Chicago Chronicle*. You are lucky to have the plans of your machine made for you by the newspaper men. This gives you time to get your patents" (Fig. 5.11).

An editorial in the *New York Tribune* reflected the attitude of the public: "If the performance of the Wright brothers at Kitty Hawk are correctly reported, they have gone one step further ... Now let them enter the lists at St. Louis and show the world how their machine compares with the others, which are to be exhibited there in action" [54].

Reading about the Wrights achievement, Augustus Herring wrote a four-page "frank straight forward" letter to the brothers, congratulating them on their success and reminding them "... that my work has contributed a tangible amount to your success, because my work with the two-surface

Fig. 5.11 Successful test of flying machine. Aerostat invented by the Wright Brothers from Ohio in flight at Kitty Hawk. *Chicago Chronicle*, January 24, 1904

machine gave you your interest in the problem." He did not think that litigation would benefit either of them, but they could work together, as there would be enough money for all. He suggested that the profit should be shared in three parts, with him receiving one third. Herring also asserted that he had been offered a substantial sum for his "rights" to interference suits against the Wrights [55]. As could be expected, Wilbur forwarded the letter to Chanute, "This time he surprised us. Before he left camp in 1902, we foresaw and predicted the object of his visit to Washington, we also felt certain that he was making a frenzied attempt to mount a motor on a copy of our 1902 glider and thus anticipate us, even before you told us of it last fall. But that he would have the effrontery to write us such a letter, after his other schemes of rascality had failed, was really a little more than we expected. We shall make no answer at all" [56].

Herring's imprudence asking for part of their invention seemed incredible to Chanute. "While I could wish that you had applied for patents when first I urged you to do so, I think that your interests are quite safe. The fact that Mr. Herring visited your camp, in consequence of circumstances which I subsequently regret, will certainly upset any claims, which he may bring forth. I suppose that you can do nothing until interference is declared. If it is, please call on me, and in the meantime I will try to find out who his patent attorney is" [57]. Apparently Herring reconsidered his next step, and Orville recalled, "I remember Chanute telling us in 1904 that Dienstbach became very angry when Chanute called Herring a blackmailer."

5.6   The Wright's Flight: Truth or Myth?

The December *L'Aérophile* published the report of the Wrights flights as the first item in the news from around the world. "We will wait before commenting, but need to remind French aviators of the alarm cry uttered by Mr. Ernest Archdeacon: The homeland of Montgolfier should be ashamed to let this ultimate discovery of aerial science take place abroad." And Ferber stated: "We must not let the airplane be perfected in America! It is still time, but we cannot waste a minute."

Reading the many colorful stories about their flight(s) in the newspapers, Wilbur decided to write and submit a clarifying statement to the wire service on January 5, 1904. Reading the write-up in the *Chicago Tribune*, and then receiving a copy from Wilbur, Chanute puzzled about the last sentence: "All experiments have been conducted at our own expense, without assistance from any individual or institution." He thought that he had helped them, just as he had helped the many other people in their quest to solve the problem of manflight. So he asked Wilbur, "Please write me what you had in your mind concerning myself when you framed that sentence in that way" [57]. Wilbur responded a few days later, "The object of the statement, concerning which you have made inquiry, was to make it clear that we stood on quite different ground from Prof. Langley, and were entirely justified in refusing to make our discoveries public property at this time. We had paid the freight, and had a right to do as we pleased. The use of the word 'any,' which you underscored, grew out of the fact that we found from articles in foreign and American papers, and in correspondence, that there was a general impression that our Kitty Hawk experiments had not been carried on at our own expense, etc. We thought it might save embarrassment to correct this promptly" [56].

The brothers believed that their older friend had portrayed them as his "pupils" in his talks in Europe in the spring of 1903, and that he failed to appreciate the originality and importance of their ideas. However, Chanute saw things differently. He had shared his knowledge with the Wrights, just as he had done with the many other experimenters who had contacted him; the brothers picked up what they could and took the flying machine project to the next level. Chanute's patented biplane was the prototype for their machine, and he was proud of their success. It is a credit to Chanute's wisdom that he dropped all further effort to solicit from the Wrights some clarification of his contribution, as Wilbur had made it perfectly clear that none would be forthcoming.

Reading an unsigned editorial in the *Revue Sportive*, Chanute thought about "putting a flea in Wilbur's ear" and mailed him a translation. "The duty of French Aviators is not to try to copy the apparatus which the Wrights had designed, but to study others, with single surfaces or with several surfaces with a lower center of gravity. Let us keep our eyes peeled, for we are at a turning point in aeronautical history" [58].

The January *L'Aérophile* [59] published a translation of the Wright's statement, and closed with "… we must point out some obscure points. Among others, Mr. Orville Wright did not tell us about the altitude difference between the point of departure and landing. Nevertheless, the experience merits our warmest applause" (Fig. 5.12).

The discussion was heated at the February Aéro-Club meeting in Paris, as members were not willing to accept the fact that the Wrights had flown "using an entirely new system of control" [59]. Victor Tatin, who had just received a letter from Chanute, considered the various reports on the Wrights "incomplete and often contradictory" and he did not think that Aéro-Club members should copy the machine of the Americans. And Archdeacon reminded fellow

Carrying the machine uphill to the starting point

Taking a trial flight of a few yards

The aeroplane in position for "pushing off"

The machine in midair, traveling against the wind

THE FLYING MACHINE MADE BY THE WRIGHT BROTHERS,
WHICH SAILED THREE MILES AGAINST THE WIND

Fig. 5.12 "The flying machine made by the Wright brothers, which sailed three miles against the wind." As the Wrights kept their flying activity secret, reporters used photos of Chanute's glider to highlight stories of the Wrights. Only the last photo shows the Wright glider. *Collier's Weekly*, January 23, 1904

members: "We will soon fail to catch up, but we need to surpass the Americans. To succeed in this difficult work, it is essential that all of us help. So, I count on you, gentlemen, not only as potential participants in our contest, but to go around and attract subscribers. It is essential to preserve the glory of France, the home of Montgolfier, in the ultimate conquest of the air. We cannot let it slip into the hands of foreigners."

Hermann Moedebeck offered his comments on the achievements of the two Americans in his German language publication, "Lilienthal has shown us the way. Chanute, Herring, Wright, Ferber, etc. work as his students and they do so with the best successes. We need to organize gliding competitions, train, compete and offer prizes! This awakens the sport-lusty youth. And if one can pull in outstanding men of cold blood and greater skill, then there will be no shortage of knowledgeable people who can guide a real flying machine, with motor and propellers, through the air in a flash" [60].

Talking to members of the City Club of Chicago five years later, Chanute explained. "I was in Europe in 1903, met the leading aviators and told them what had been accomplished in this country in the way of gliding machines. The Germans, Austrians, Italians and English were not much surprised. They had kept track of what was being done, and while they saw future possibilities, they did not enthuse. The French, however, became very enthusiastic. Some of them said it would be a national shame to permit Americans to complete an invention for which the world had been waiting so long; they invited their people to begin experimenting at once so as to anticipate the Americans. That is the reason why the Wright brothers became the 'mysterious Brothers Wright.' The brothers were afraid that somebody would build a motor-driven machine before they did; they hurried, and on December 17, 1903, they succeeded in making the first heavier-than-air flight made by man with artificial power" [61].

5.7   American Association for the Advancement of Science in St. Louis

For his talk at the American Association for the Advancement of Science (AAAS) meeting in late December in St. Louis, Chanute had received confidential information from Manly on the final launch of Langley's aerodrome, but had received nothing from the brothers, only the telegram from Katharine. Reading now in the Chicago papers that they had flown more than three miles at an altitude of sixty feet, he wrote Wilbur on December 27: "I have had no letter from you since I left your camp, but your sister kindly wired me the results of your test of December 17. Did you write? It is fitting that you should be the first to give the Association the first scientific account of your performance. Please wire me at St. Louis." Wilbur wired back: "We are giving no pictures nor description of machine or methods at present." Wilbur must have had a change of heart a few hours later, as he composed a five-page detailed letter describing his attempt on December 14 and their successful powered flights of December 17 and mailed it to St. Louis.

Anxious to socialize with the many people he knew, Chanute travelled to St. Louis right after Christmas. Section D was to meet on Wednesday, December 30, 1903; one of the invited speakers was General **Edward W. Serrell** (1826–1906), whom Chanute had met in the early 1870s; the two civil engineers had collaborated on several ASCE projects, like the artificial production of rain or how to measure air velocities. At this meeting, Serrell discussed building the Mandingo Ship Canal/ Tunnel through the Cordilleran Mountain Range in Central America in the early 1860s, and the work of the ASCE committee to select the best location for the Panama Canal twenty years later. This was a hot topic, as Panama had just declared its independence and there was a move to finally get the canal dug.

The last talk of Section D was Chanute's "*Aerial Navigation*" where he told his audience, "The empire of the air is still to be conquered, but we have certainly got a further glance into the Promised Land than we have ever had before." He described what was accomplished with balloons and flying machines and speculated about the future: "Navigable balloons and flying machines will constitute a great mechanical triumph for man, but they will not materially upset existing conditions as has sometimes been predicted. There are now dawnings of two possible solutions of the problem of aerial navigation, a problem which has impassioned men for perhaps 4,000 or 5,000 years. Navigable balloons have recently been developed to what is believed to be nearly the limit of their efficiency, and after three intelligent but unfortunate attempts by others, a successful dynamic flying machine seems to have been produced by the Messrs. Wright." He discussed the failures of Langley's aerodrome, which only proved that a better flying machine had to be invented, and mentioned that the Wrights had acquired the necessary proficiency and had flown their aeroplane two weeks earlier. "How they accomplished this must be reserved for them to explain; they are not ready to make known the construction of their machine nor its mode of operation" [62]. On a more forward thinking note, he discussed the prospect of anyone winning the prize money offered by the World's Fair Commission and urged interested parties to step forward [63]. *Popular Science Monthly* published his talk in March, and Langley republished it in the 1903 *Annual Report of the Smithsonian Institution*.

After listening to Chanute's talk, Serrell suggested that the two civil engineers reminisce about aeronautics during the Civil War. Everyone knew then that the balloon could take off, but there was no way for it to come back to its home base. The flying machine, however, could take up observers and go where they wished to go, and could carry explosives to punish the enemy. With this in mind, Serrell had designed a tin toy that he wound up and spun like a humming top, raising itself to about 100 feet. He showed it to Major-General Benjamin F. Butler, Commander of the Army of the James who "immediately expressed the belief that a machine could be made that would navigate the air and drop explosives." Serrell envisioned his "volomotive" to be lifted into the air by twin rotary blades, one above and one below the fuselage, driven forward by blades at the front and rear. Two large flat copper wings, each nine feet in span, were positioned on either side of the fuselage. The two wings were connected to a crank running through the fuselage, so that the crew could incline or depress the wings to provide lift. The fuselage was to be a cigar-shaped copper shell measuring 52 feet in length, with runners to operate like a sled. A chamber at the bottom would serve as a reservoir for the boiler water, and a second chamber above it for the coal. A lightweight steam engine with a vertical boiler was to be housed in the rear of the shell. A moveable weight at the center of the body was to balance the "volomotive," also called the "reconoiterer." Toward the end of the Civil War, Serrell had large sections of the three-man craft under construction, but gave up, as there was no engine available to propel the craft.

Chanute now urged Serrell to write up this fascinating story, because most people had never heard of these early efforts by the military. *Science* published his write-up a few months later [64].

5.8   The St. Louis Aerial Sweepstakes and the 1904 Chanute Glider

To attract a large crowd "to view hundreds of never seen aerial feats," Louisiana Purchase Exposition executives had committed $200,000 for prizes and other expenditures. Some people thought that such enormous prizes were out of proportion to the practical value of aeronautics, [65] but their reason for offering high prizes was not to fix a premium on the development of any particular type of flying machine, but to assist in determining the very best method of construction to help solve the problem of navigating the air. A grand prize of $100,000 was to be awarded to whoever made at least three flights over a fixed ten-mile course at a speed of not less than 20 miles per hour. And there were two additional awards for gliders. "A $2,000 prize was offered for the gliding machine, mounted by an operator, which shall advance in a calm or against the wind at a horizontal angle most acute with the horizon. And a prize for $1,000 was offered for the gliding machine, mounted by an operator, which exhibited the best automatic stability in the wind during at least forty glides, not less than 400 feet in length each."

Organizers envisioned more than 100 contestants to step forward and that "there would be no lack of delighted observers, but the chief charm of the coming tournament would lie in the mystery in which the outcome will be shrouded, for it was by no means certain whether any form of flying machine already invented will win or whether one 'dark horse', as yet concealed in its stable, will trot out and bear away the prize. Never before was such a scheme projected, never before in the world's history was the time quite ripe for such a scheme, and it is no exaggeration to state that from 1903 on a new epoch in aerostatics will begin" [66]. Ninety-two aeronauts requested entry forms, but only eight paid the required entrance fee of $250 by late August. The manager, Carl Myers, was let go, and Chanute became concerned that American leadership in aerial navigation might go away, as "none were fit to compete" [67].

Visiting the Wrights in Dayton in late January 1904, Chanute encouraged the brothers to take a closer look at the high financial prizes. Maybe Chanute was correct that by flying in public and revealing the extent of their work, they would ultimately profit, but the brothers just continued tinkering on their machine.

	1896 glider	1904 glider
Wing area (total)	136 sq. ft	155 sq. ft. (14.40 sq. m.)
Cord	4 ft. 3 in	4 ft. 11 in
Camber	$\sim 1/10$ of the chord at center of wing (circular arc camber)	$\sim 1/12$ of the chord aft of leading edge (parabolic camber)
Wing span	16 ft	15 ft. 9 in. (4.80 m)
Gap between surfaces	4 ft. 3 in	4 ft
Weight	23 lbs	40 lbs. (18 kg)
Length	13 ft. 2 in	12 ft. 8 in (3.90 m)
Elevator & fixed tail	~19 sq. ft	~19 sq. ft
Glide ratio	~6:1	~7:1

Dimensions extracted from William Avery correspondence and Musée de l'Air records

The 35-year-old competition balloonist Jacques Balsan, a rich young *bon vivier* and aeronaut, had listened to Chanute's talk in Paris and now wanted to buy a glider. Hearing of the potential sale, William Avery suggested to sell the 1897 glider, which was still stored in the rafters above his shop. But Chanute wanted a new "soaring machine" built, where the wings were shaped like those of the soaring birds or like "Pline's curve." Avery saw no problem cutting the wood for the ribs into a deeply curved shape and got started.

But then Balsan canceled his order with Chanute on rather short notice, as he wanted to buy the Wright's 1902 glider right away [59]. Losing the sale put Chanute in a certain bind, as Avery had already built the glider and wanted to be paid. Feeling quite comfortable about the performance of this new "soaring machine," Avery was sure that he could win either or both prizes for gliders, as offered by the contest committee. "This was like picking up money from the street," he told Chanute.

With no promising flying machine entry for the St. Louis aeronautical concourse, the two men decided to enter the new glider. Chanute paid Avery for building the aircraft, entered it into the St. Louis competition and withdrew from the World's Fair jury. But to enter the glider to fly required some device to launch it with its pilot into the air. A decade earlier, Chanute had discussed the high-wheeled carriage with a drum, a windlass, used by the Italian military for controlling their captive balloons in Abyssinia [68]. Percy Pilcher from England used people or a horse to pull his glider into the air in 1898. Wilbur Wright had considered in 1900 a 150-foot-high tower and a rope over a pulley system to get airborne. Early in 1902, Albert Merrill had proposed using a derrick with a falling weight to launch his glider. Weighing the pros and cons of all the different ideas to launch the glider, Chanute then combined a windlass with an electric motor. Using a narrow railway track with a small flatcar for Avery to stand on, while holding the glider level, seemed the most efficient solution to pull the glider with its pilot into the air. He drew up his ideas for "*Means for Aerial Flight*" and submitted the claims to his new patent lawyer in Chicago. The "apparatus for facilitating the practical operation of a flying machine" [69] was patented two years later. Little imagination is needed to recognize the ancestry of today's winches and towing arrangements in this patent (Fig. 5.13).

Avery went to St. Louis in the middle of August to check out the stadium ground and the rented electric motor for the rope-reeling drum, spending almost $600 on equipment. All looked good, and he returned to Chicago. A few weeks later the glider framework and various winch parts were shipped to St. Louis and the two Avery brothers, William and Frank, arrived on September 3. Chanute mentioned to Wilbur: "Mr. Avery started for St. Louis

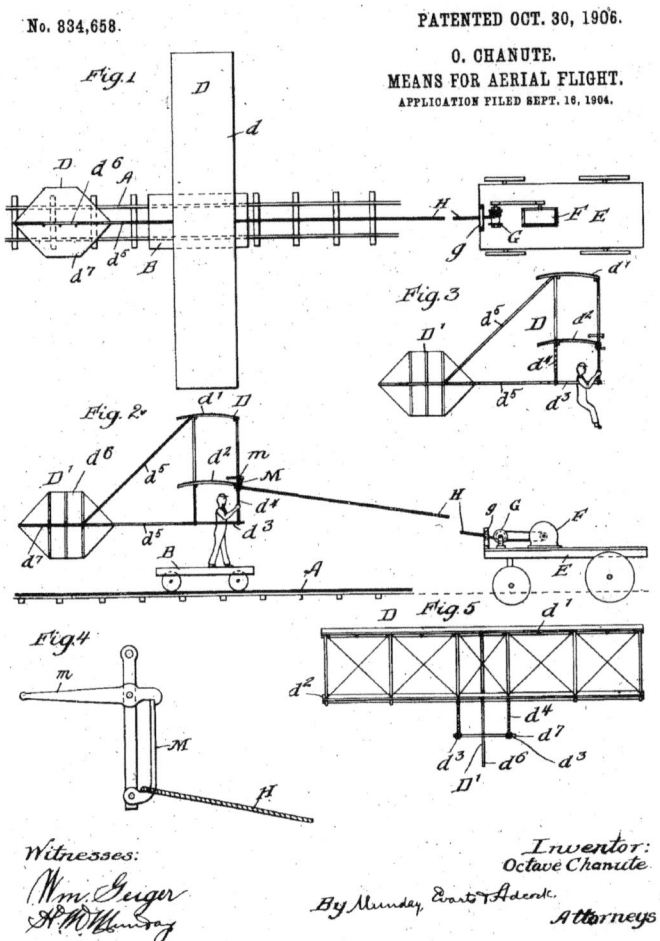

Fig. 5.13 Chanute's "*Means for aerial flight*" patent drawings, No. 834,658, patented on October 30, 1906

last night to make arrangements to compete for the gliding prizes. He is to use an electric motor and a portable railroad track and car. It will probably take him 2–3 weeks to get his plant together and make his preliminary experiments. As he is not well off, I am furnishing him the funds, and have therefore declined to act on the Int'l Jury, for that concourse at least" [70].

The frame of the glider was assembled a week later, and Avery reported back to Chanute, "It is a beautiful piece of work" [71]. After the photographer took pictures of the uncovered glider, the Avery brothers stretched the silk cloth, hemmed along the edges, tightly over the wings to take up bagging and laced it over the wire between the ribs. They then applied two coats of pyroxylene varnish to shrink the cloth and make it "as tight as a drumhead."

On Friday, September 23, Avery made his first launch at the aerodrome field, a seven-acre dirt field surrounded by a 30-foot-high fence to prevent non-paying onlookers from watching. As the winch mechanism was not yet ready, launching required manpower. One end of the rope was connected to the glider, held upright by the "steersman" Avery, and three men took the other end of the 1/8th inch diameter rope and started to run. While the men were pulling the glider, Avery, holding the glider steady, also ran down the field, until he felt himself lifted off the ground. At a height of about 30 feet (length of the rope), Avery released the tow hook and glided back, fighting the air turbulence created by the fence [72].

Avery was proud to report to Chanute: "I wish you could have seen the two flights. I know that they would have pleased you as much as they did me. I think that the curve of the ribs must be about right, for the last trial the machine was perfectly level and traveled at a very flat angle" [73]. Chanute was pleased but concerned: "Make your record by 30 Sep. Test efficiency of making line to the car instead of machine, as this leaves you freer. Also put rubber soles on your shoes to ease alighting," [74] and closing his letter with, "Be very careful in experimenting!"

On September 30, the original deadline to win the prizes, organizers extended the grace period to October 31 to help "dilatory aeronauts" in their race for the grand prize [75]. To win the prize for gliding, Avery needed more speed to pull the glider higher into the air to reach the required 400-foot distance in gliding flight. He gave serious thought to using an automobile, but the field was muddy and soft due to recent rains. Finally, fair organizers connected electric power to the 10 h.p. motor (Fig. 5.14).

On October 6, the electric winch was fully operational, providing a more powerful launching system. The team laid the 75-foot rail track into the wind; Avery stood on the rolling platform while two helpers held the glider level; the winch pulled him at a steady speed of about 10 mph into the air [76]. "After leaving the car, the aeroplane with its master rose gracefully into the air, reaching an altitude of about 35 feet, gliding 175 feet and allowing Mr. Avery to land with scarcely a jar," [77] was the report in the *St. Louis Republic newspaper* the next morning.

Now, weather permitting, Avery launched his glider almost daily in front of paying spectators, usually reaching an altitude of 70 feet, flying over a distance of 300 to 350 feet, at times even turning in flight before gliding back to earth and flaring out for a smooth landing on the concrete pavement on his thick rubber-soled shoes (Fig. 5.15).

OCTAVE CHANUTE   BARON VON TSCHUDI   WM. AVERY   PROF. CARL MEYERS

Fig. 5.14 William Avery is explaining how the electric motor will work to pull the glider with him into the air, just like today's winches. *Cherry Circle.* January 1908

Fig. 5.15 Ready to be launched by the electric winch, note the towing cable connection to the trolley and the glider. Reaching a height of 40 feet at Plaza of Saint Louis, and in the air at 60 miles an hour. National Air & Space Museum (NASM 79–10,566) and *Cherry Circle.* January 1908

A strong wind prevailed all day on October 10, and Avery decided to be safe and use manpower instead of the more powerful winch. The recently hired reporter for the *St. Louis Republic* newspaper, Arnold Kruckman, reported the next day: "As the aeroplane rose in the air and sailed through space a sudden gust of capricious wind caught the steering fin and whirled the machine at almost right angles with its course, causing the aeroplane to swoop sharply toward the ground. With remarkable coolness Avery managed to steer the machine so that beyond breaking one of the ribs in the steering fin the machine was uninjured and its occupant unhurt except for a slight jar and fall" [78].

Avery's flights were spectacular; every time he pulled the glider out to the field, some 6,000 people gathered to watch with incredulity. Organizers had to position about twenty guards to contain the crowd. One fair official told Avery, that "watching an aeroplane carrying a man into the air was the most wonderful thing that he ever saw." Milton Wright, who had come to the fair, observed Avery's performance, introduced himself and congratulated him on a fine job done. Another eyewitness was Captain Tom Baldwin; Avery wrote that "one could have hung a hat on his eyes." [79] Balsan, who did not buy the Wright's glider, had just arrived in St. Louis and watched Avery's performance. He now wanted to buy his glider right away.

Originally the international jury had ruled to award the glider prize for flying 400 feet; they now clarified that the distance was not from the point when air supported the glider but from the point of releasing the towing line. This could not be done on the stadium ground, so Chanute simply withdrew the glider from the contest and wrote Avery: "I regret that you should have no chance to win a prize, but it is better for me to lose my money than for you to lose your life" [80]. Talking things over, the two men then agreed that they would jointly decide when to stop the flying.

Exhibition management realized that the entrance-paying public wanted to see the only heavier-than-air machine in the air; they suggested moving the winch to the Plaza of St. Louis, an open space of almost 2,000 feet in length near the main entrance. Now using a 100–foot rope allowed higher and longer flights, but Chanute cautioned, "As you evidently desire to stay and try for the prize, you can do so as long as you are satisfied that they are perfectly safe. You need to examine your apparatus carefully after each glide and make sure nothing is broken or loose. Take no chances in tumultuous winds. When you have made the prescribed number of glides to win the prize, stop, and make your own arrangements with the administration about further exhibiting. It probably can afford to pay you a good price to secure large attendance" [81].

Date 1904	# Flights made	Comments
23 Sep	1	Three men pulled the glider, running down the aerodrome field, to get airborne
29 Sep	2	Same launching procedure
6 Oct	16	A 10-horsepower electric winch provided speed and altitude at the stadium field
7 Oct	3	Winch launch
9 Oct	5	Winch launch
10 Oct	1	Used manpower to get airborne
11 Oct	20	Winch launch
12 Oct	12	Winch launch. The last three flights on this day reached 600 feet distance
13 Oct	7	Winch launch. Distance to landing about 400 feet on each flight
14 Oct	10	Winch launch with longer towrope. Almost every flight reached about 800 feet distance and an altitude of about 70 feet
22 Oct	6	Winch launch. Flights went across the entire length of the Stadium Field
25 Oct	1	Winch launch at the Plaza St. Louis. Avery encountered rope break at an altitude of about 40 feet
	84	Total number of gliding flights made Estimated time in the air almost 30 min

Compiled from correspondence by William Avery to Chanute and to Percy Hudson

After almost one month of being launched in his glider daily and landing on his 1¾-inch-thick rubber-soled shoes, the inevitable happened. On October 25, Avery rose to about 40 feet when the hemp towing line broke. He came down rapidly and landed hard on the asphalt; holding the glider as level as he could, receiving only minor damage, but Avery twisted his ankle badly. The gliding flights stopped much to everyone's regret [79].

The unhappy and hurting Avery disassembled the glider and returned to Chicago. As soon as he could move without crutches, he replaced the one broken rib, and Chanute wrote to Balsan, "The glider is back here [in Chicago]. It is in good shape, but it will cost you 500 francs if you want it." He had originally thought of selling the glider for $75, which was about 300 francs, but he upped his purchase price to recoup some of the expenses in Saint Louis. This time Balsan sent money, and Avery crated the glider for shipment to France, the first American aircraft to be sold overseas. Balsan lost interest in motorless flight quickly and gave the glider to Charles Renard [82].

The much-heralded World's Fair with its aeronautic competition and its high monetary awards came and went. Organizers had originally announced that airships of various shapes and sizes were to make the heavens populous during the aeronautical concourse; that they did not do so is probably no fault of the exposition management, but the fact remains that the anticipated competitors were not there. American skill, ingenuity and experience did not triumph, and the airship invention was clearly still "in the air" [83]. However the highly publicized events did bring the public nearer to believing that a flying machine could become a reality in the foreseeable future. Chanute was sure that mechanical flight would be successfully worked out, "I fancy we will see progress in 1905, but the Aeronautical Concourse at St. Louis brought out absolutely nothing new and such progress as has been made outside is being kept secret" [84].

5.9   We Shall Yet Fly!

The Wrights did not enter the Saint Louis aeronautical contest, they established their experimental station near the Dayton-Springfield interurban tracks, known as Huffman Prairie, and continued working on their Flyer. They invited Chanute to come in early May to watch them fly, but his wood preserving business took priority. They also invited the press, but rain and no wind prevented a take-off. Most of the reporters had left when Orville finally got the Flyer into the air, breaking the propellers in landing, [85] Wilbur confided to their father on June 10, "we certainly are having more than our usual share of trouble."

To launch their Flyer without the wind of the Outer Banks, the brothers built a 20-foot derrick and used iron weights to provide the initial take-off speed, as they had heard from Chanute that Albert Merrill was experimenting with. Wilbur then reported to Chanute [86]: "The starting apparatus was tried for the first time on Sept. 7. Up to the present we have made eleven starts with it. It seems to operate perfectly." One of those starts ended in a half-mile flight, longer than any previous flight. And "On Wednesday, Sept. 15th, we made our first attempts to encircle the field but did not quite succeed, though on both trials a distance of half a mile was covered." Wilbur followed up two weeks later: "On 20 September we renewed the attempt and succeeded. Up to the present we have been very fortunate in our relations with newspaper reporters, but intelligence of what we are doing is spreading through the neighborhood. If your business will permit you to visit us this year it would be well to come within the next three weeks. In fact, it is a question whether we are not ready to begin considering what we will do with our baby now that we have it" [86].

The owner of the periodical *Gleanings in Bee Culture*, Amos I. Root, had witnessed the flights on September 20 and published his report with the Wrights permission in his journal on January 1, 1905, hoping to scoop other publications with his story. He recalled four years later, "Some years ago, I met Mr. Chanute, the man who had made experiments with gliding machines, even before Wright brothers had, and a man who is widely known all over the world wherever there is any interest in flying machines. When I was introduced to Mr. Chanute, he paid but little attention to me. As the party broke Orville Wright handed Mr. Chanute a copy of our journal, turned over to the pages that gave my story, and suggested that he might be interested in reading it. The next morning, Mr. Chanute's face had changed. He came up to me with a very friendly greeting. When I told him that I had been much disappointed to find that the article elicited so little interest, he replied something like: "Why, Mr. Root, your readers all supposed that it was a made-up story. The world did not believe you were telling the truth." Root responded, "Mr. Chanute, that article has the stamp of truth from beginning to end. I mentioned the locality, and the things that happened, in a way that would convince any reasonable person that what I related really occurred." Glancing at the pages again Chanute reportedly said: "Well, I guess that is so to a great extent; but what you are telling is too wonderful" [87]. This is a curious story, how factual is it?

Having received another invite to watch the brothers fly, Chanute stopped in Dayton on his return from St. Louis. "On the 15th of October, 1904, I witnessed a flight of 1,377 feet performed in 23⁴/5 seconds, starting from level ground and sweeping over about one-quarter of a circle at a speed of 39 miles per hour. The wind blew at some six miles per hour, but in a diagonal direction to the initial course. After the machine had gone some 500 feet and risen some 15 feet, a gust of wind struck under the right-hand side and raised the apparatus to an oblique inclination of 15 to 20 degrees. The operator, who was Orville Wright, endeavored to recover an even transverse keel, was unable to do so while turning to the left, and concluded to alight. This was done in flying before the wind instead of square against it as usual, and the landing was made at a speed of 45 to 50 miles an hour. One side of the machine struck the ground first; it slewed around and was broken, requiring one week for repairs. The operator was not hurt. This was flight No. 71 of that year."

Like all good stories, there is triumph, and there is tragedy; most importantly there are many fascinating and colorful people in aviation who did some amazing and challenging things. The story of this massive period of technological innovation and development at the beginning of the twentieth century is an exciting one, it deserves more recognition.

6

The Persistent Experiments to "Wing the Air"

Upon the whole, navigable balloons and flying machines will constitute a great mechanical triumph for man, hut they will not materially upset existing conditions as has sometimes been predicted. Their design and performance will doubtless be improved from time to time, and they will probably develop new uses of their own which have not yet been thought of.
Octave Chanute, *Popular Science Monthly* **(1904)**

Reading of Jules Verne's death in March 1905 made Chanute pull one of his favorite books from the shelf, *The Clipper of the Clouds,* [1] which can indeed be considered the ultimate manifesto of the nineteenth century's quest for heavier-than-air flight. Here Verne, or his fictional engineer Robur, shared with the reader his philosophy on manflight and how the heavier-than-air flying machine would conquer the world. When Chanute became interested in manflight, he studied Verne's explanations and then built up his own knowledge base. To him, learning from others and sharing information was part of growing in the profession; his philosophy about technical education was that a student should be well grounded in the general principles of science, taught where to look for information and learn how to use the tools of knowledge [2].

Receiving thoughtful questions after a lecture made Chanute feel like he had provided another tool. But at times he regretted that he did not have an academic title to affix to his name. "You will note that in 1850, when I began my career, there was not in this country the present opportunities to take a collegiate course, and knowledge had to be acquired from private

© The Author(s), under exclusive license to Springer Nature Switzerland AG 2023
S. Short, *Flight Not Improbable,* Springer Biographies,
https://doi.org/10.1007/978-3-031-24430-8_6

schools, tutors, and especially private study. This will always remain the most important way for college graduates who wish to succeed. I hope that the members of your fraternity will realize that while they are favored over the preceding generation by many opportunities to learn the state of knowledge and the use of mental tools, they will have to combine study with practice to become experts." [3].

Much to his surprise, he received a confidential letter in late summer that the University of Illinois Council had recommended for him to receive the honorary engineering degree, which was to be presented at the inauguration of the new president of the University of Illinois in Urbana. Chanute shared the news sheepishly with fellow Western Society of Engineers (WSE) members at the next meeting, as everyone intended to travel to Urbana to greet the new president on October 18, 1905.

Nearly 4,000 dignitaries from around the country crowded the Armory to the last available foot of space to attend the Inaugural Exercises. After the new president, Dr. Edmund J. James, had delivered his address, twenty-five honorary degrees were tendered [4]. Professor Ira Baker announced that the Doctor of Engineering degree was awarded to Chanute for his contributions in timber preservation. Stepping forward, he was pleased but felt a little embarrassed accepting his diploma, and later told his daughters, halfway joking, that the tassels on his mortarboard got in his way. In the years to come he occasionally added the academic title to his name, Dr. O. Chanute. This academic honor could very well have been one of the first such degrees awarded to a person who only attended a private school as an indigent child (Fig. 6.1).

6.1   Aviation Lures Youth

Being the informal center of aeronautical information, Chanute frequently received letters from enthusiasts describing that he or she had discovered the secret of manflight; reading some of these inquiries might have made him smile, but he usually took the time to critique and offer suggestions. A Mrs. Henrietta Belcher from Brooklyn wrote that she had a dream of what the perfect flying machine would look like, and she wanted Chanute's opinion. "It is most unlikely that a dream should reveal a practical airship, as the latter is a complicated apparatus, depending for success on many details. To gratify you I will examine a description and drawings of your vision and give you my opinion as to whether they are worth further thought. This I will do without charge." [5]. A few weeks later she mailed sketches and a description,

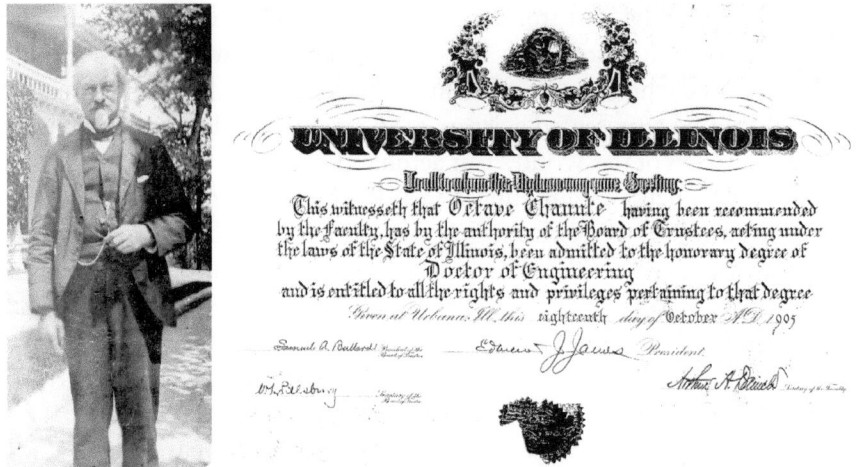

Fig. 6.1 Octave Chanute (1905) and his Honorary Doctor of Engineering diploma, received from the University of Illinois in October 1905. Courtesy Chanute Family

and Chanute responded, "So far as I can judge you ought to consider your dreams as worthless," and he explained, "To be of any service a vision should disclose the principals upon which a mechanical device works and should give enough details of construction to form some idea of its practicality." There was no follow-up, and there were no reports in the newspapers about Mrs. Belcher's airship.

While studying legal matters at the University of Michigan, **Israel Ludlow** (1873–1955) became interested in aeronautics. After graduating from law school in 1895, he moved to New York City to start his legal career. With flying still vividly in his mind, he contacted Chanute in July 1895, vaguely describing his idea of a flying machine; but he also wondered if someone in New York would like to work with him. Reading of Hearst's Lilienthal glider, Ludlow arranged to fly it and reported to Chanute that he did not have a good flight.

In the next few years, Ludlow built and flew models, followed by building a full-size tetrahedral kite. "There is no reason in the world why man shouldn't fly," Ludlow told a New York *Times* reporter [6]. "It is probable that my first flights will be short and crude, as the jumps and hops of a young bird are. In time, though, I will spurn the earth, and my movements in the air, like those of a full-grown bird, will be free and untrammeled." On July 25, 1905, Ludlow tried to launch his big kite-like glider, but could not get airborne. This did not stop him or his new found friend Charles K. Hamilton. They built another glider and took it to Ormond Beach, Florida, to demonstrate it at the automobile race. Ludlow, hanging in his glider, was towed by a car into the air, when a strong gust from the ocean collapsed the bamboo framework and the glider crashed to earth. The 33-year-old Ludlow broke his spine in the fall and spent the rest of his life in a wheelchair. His love for aeronautics remained, and he represented aviators as an attorney in the years to come.

When the news of the World's Fair aeronautical contest in St. Louis with its high monetary prizes was announced in 1901, the 27-year-old **Augustus Roy Knabenshue** (1875–1960) from Fostoria, Ohio, wanted to participate. His father, an editor at the *Toledo Blade*, had subscribed to various magazines, in which Knabenshue read about Chanute. He then decided to build a glider, become an aeronaut and wrote to Chicago. Receiving the letter from the teenager from Toledo, Chanute remembered reading in the *Chicago Tribune* about two teenagers who ran away with a big balloon, or, rather, the balloon ran away with them [7]. Was this the same young man who had a wild ride? If yes, this enthusiastic young man needed to know more about gliders. Chanute copied drawings and responded, "I recommend you begin with the two-surface machine which you will find much steadier than the Lilienthal. Mr. Wright has improved it by placing the rudder in front and has obtained still better results this year. If you adopt the upright position for the man, as I did, the machine can be built to weigh 25 pounds, and will glide at an angle of about 10 degrees. If you adopt the horizontal position, as did Mr. Wright, who has hitherto found this safe, the machine will weigh about 100 lbs. and glide at angles of 6 to 7 degree." Chanute closed his letter with "I shall be glad to furnish further information." [8].

Flying balloons appeared to be more exciting than building a glider, and Knabenshue travelled to the St. Louis World's Fair with a balloon outfit, found a place to live just off the exposition grounds and looked for income. A shoe company hired him to fly kites to advertise their products; when this project ended, he gave rides to the public in his tethered balloon. When Thomas Scott Baldwin arrived in August with his *California* Arrow airship, Knabenshue introduced himself and was hired to help assemble and then fly the dirigible. When his contract ended, he decided to start his own show business.

Looking for offers and places to perform, the 30-year-old Knabenshue joined friends in New York City who introduced him to the 25-year-old journalist Arnold Kruckman, who had recently joined Joseph Pulitzer and *The World*. Having reported on Avery's gliding at the St. Louis Fair the previous fall and the flying attempts of Israel Ludlow just one month earlier, Kruckman was convinced that the joy of speed and the sensation of flying, of overcoming space and distance, was bound to make air-sailing the sport of the future. He was ready to report on this activity to anyone reading the news-papers. Both men thought it beneficial to work with each other, Knabenshue needed publicity and income and Kruckman wanted publicity for his own career as aeronautical news writer.

A few weeks later, on Sunday, August 20, 1905, everything worked out as planned. Knabenshue's ascension in his airship was a great success, and the front-page write-up by Kruckman in the Monday paper said it all, "Crowds wild as airship skims over New York," and "For the first time New York saw a real airship in a real flight yesterday." The one-column text, highlighted with three images, gave good publicity to both of them [9]. Years later Knaben-shue stated that he gained a great respect for newspaper men, especially for Kruckman, from this experience in New York [10].

The White City Amusement Park on the south side of Chicago, between 63rd Street and South Park Avenue (today's Dr. Martin Luther King Jr. Drive), opened to the public in May 1905. Being a board member, Chanute, who had just received *The World* clipping about Knabenshue's flight over New York, suggested to invite him to come and perform in Chicago (Fig. 6.2).

Arriving in Chicago on Sunday, September 17, and seeing the "airfield" was a bit unexpected; Knabenshue became very concerned, "Flying here was difficult as our location was so completely surrounded by buildings with flag-poles, while in the center of the park was a very high tower. When I saw the location, I felt that nothing but disaster would result in trying to fly in and out of this small space with every kind of obstruction ready to snag the ship. However, we accepted the situation, hoping that at the time of the flight the weather would permit us to operate." [10]. Due to high winds, Knabenshue made no flights on the first two days and was called many uncomplimentary names. Starting on the third day, he fulfilled his daily contract, even though the wind was stiff (Fig. 6.2). White City management appreciated the large entrance paying crowds, and Knabenshue was pleased to see Chanute and his daughters sitting in the stands on several days.

Being grateful for all the help he had received from Chanute, Knabenshue thought of a special publicity stunt, or as he called it, "another thriller to spring on the aeronautic world." He wanted to take aviation's senior promoter as an honored guest on a flight in a balloon across Lake Erie in early January 1907. Chanute appreciated the thought and invitation, but being almost seventy-five years old, he did not think that he was up for such a ride in the middle of winter; he sent his regrets. In the years to come he enjoyed reading about Knabenshue and his many adventures.

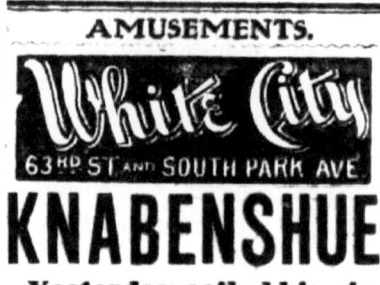

Young Man to Sail Airship Over Chicago.

AMUSEMENTS.

White City

63ᴿᴰ ST ᴬᴺᴰ SOUTH PARK AVE

KNABENSHUE

Yesterday sailed his airship from White City to 39th street and return, keeping it under perfect control and landing exactly at the starting point.

WATCH FOR THE AIRSHIP TODAY.

If weather conditions permit he will make an ASCENSION AT 1 P. M.

GATE ADMISSION 10 CENTS. Open from 1 p.m. Till Midnight.

Fig. 6.2 (LH) Roy Knabenshue from Toledo, Ohio, will fly over the city daily next week. *Chicago Daily Tribune*, September 13, 1905 and (RH) Advertisement in the *Chicago Daily Tribune*, September 28, 1905

The family of **William Bushnell Stout** (1880–1956) had travelled from Minnesota to Chicago to attend the World's Fair in 1893. Father and son spent much time in the Machinery Hall, where every kind of road vehicle and even airships were on display. Thirteen-year-old William dreamt of a flying machine to carry him, but this dream remained dormant.

Reading about experiments with flying machines a decade later aroused Stout's interest in aeronautics. Early in June 1907, the 27-year-old mechanical engineer contacted Chanute about flying and flying toys to use in teaching. Interested in helping but also knowing that he needed to clean up, Chanute responded that he would be glad to dispose of the toys that he still owned, "now that the problem of artificial flight is solved," [11] and he also offered to furnish directions on building a glider. This all sounded interesting, so Stout travelled to Chicago. Sitting and talking in his library, Chanute mentioned that the Civil Engineers' Society in St. Paul, Minnesota, wanted a talk on artificial flight, but his business kept him at home. Naturally, Stout volunteered and Chanute gave him his slides. "All in all, I had myself a time with a very much interested audience, including some of my old Phi Beta classmates. That little ego of mine swelled up just a bit and that occasion whetted my enthusiasm. With this start I went back to work with more intensity, scientific reading and a lot of engineering ideas." [12]. Chanute was most likely pleased that his mentoring had helped him and Stout.

Moving to Chicago in 1910, Stout became reacquainted with Chanute who gave him a letter of introduction for the *Chicago Tribune*; they hired him as Transportation Editor. Two years later, the Aero Club of Illinois asked Stout to edit their 26-page monthly *Aerial Age*, which started publication in June 1912, but lasted only for one year. Stout's main interest was teaching boys and girls of all ages to build and compete with model airplanes. He reportedly suggested that students should never resort to mathematics until they have exhausted the possibilities of two toothpicks and a piece of string (Fig. 6.3).

In 1919 Stout formed an engineering company in Detroit, Michigan, designed and built a lightweight automobile, the "cyclecar," and established in 1922 the Stout Metal Airplane Company to build the "Flivver of the Future." He sold his firm to the Ford Motor Co. in 1924, which began production of the first all-metal airplane in America with Stout being one of the engineers, remembering Chanute throughout his long aeronautical career.

Having caught the aviation bug, fourteen-year-old **Larry Lesh** (1892–1965) searched for information on building flying models in Chicago's public

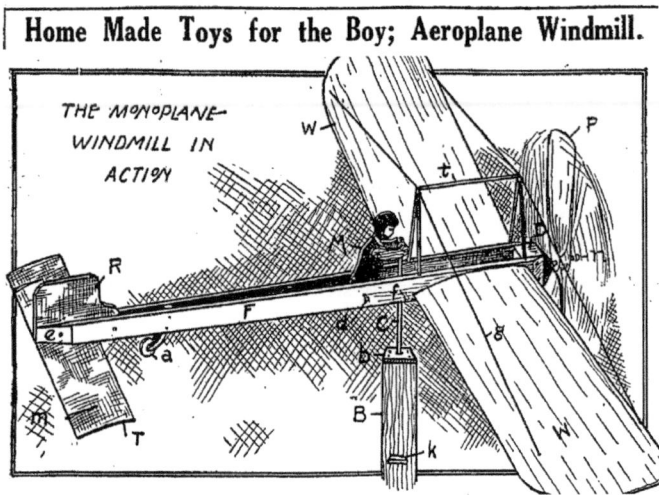

Fig. 6.3 William Stout admitted that not all flying models fly as perfect as hoped, this one is not intended to fly at all, but to merely be a curious and interesting windmill. *Chicago Tribune*, May 26, 1911

libraries. Reading of Chanute, and discovering that he did not live too far from his home in Oak Park, the teenager put on his best clothes and visited the elder statesman one day after school. Chanute was a little surprised seeing the teenager, but the two then went across the alley to meet Avery in his shop. Lesh later wrote: "I was relieved to find that Chanute was a very simple, cordial and delightful person. His head was high and broad, the repository of an amazing store of knowledge ... I left the Chanute mansion with an armful of books, papers and photographs. The next day, I received a letter with a check for $50 and instructions how to build my full-size glider." [13]. Lesh continued visiting Chanute on a regular basis until his parents moved to Kansas City.

Impressed by Lesh's enthusiasm and rapid learning progress, Chanute mentioned his young protégé to the Wrights, hoping that they might hire him. "I have been for the past year in touch with a 15-year-old boy, L. J. Lesh, who displays quite extraordinary aptitude for aeronautics and who has made a number of gliding experiments. I believe him to be thoroughly honest and discreet." [14]. Orville responded quickly, that they could not use the Lesh boy right now, but they might need help before the end of the year.

When the Aero Club of Chicago planed on staging an air meet at the White City Amusement Park in spring 1907, Chanute suggested to invite Lesh. Organizers offered a contract to pay the teenager $100 to bring and fly his latest glider. As the weather did not cooperate, organizers extended their air meet for a second week, and Lesh assembled his glider on Monday morning. "Mr. Chanute visited the show and saw my glider for the first time. He tactfully suggested that I keep it in the tent, pointing out that it appeared too weak in the joints." [15]. Feeling uncomfortable about young Lesh and his glider, Chanute wrote to the teenager's father in Kansas City that he saw his son assembling his glider. "This [glider] is so frail that I advised him strongly not to go up in it, and he promised me that he would refrain from the demonstration which some of the organizers had proposed." The proud teenager thought differently and was launched by an automobile the next day. All went well, but a rain storm later in the afternoon destroyed the glider. Lesh received his honorarium for making one flight in the only fixed-wing aircraft during the meet (Fig. 6.4).

The Lesh family then moved to Montreal, and in August 1907, the teenager performed the first flight of a heavier-than-air machine in Canada, when his glider was pulled into the air by a galloping horse. A few days later, he was towed behind a motorboat for about ten kilometers, flying for 24 minutes over the St. Lawrence River, but then the towing rope broke, the glider landed in the river and broke up in the recovery process.

Fig. 6.4 Larry Lesh is getting ready to fly his glider at the White City Amusement Park in June 1907. Chanute scrapbook. Manuscript Division, Library of Congress. Washington, DC

During the 1907–1908 time period, Lesh wrote several articles about his flying experiences and clarified, "The development of a full-fledged aeroplane is a difficult proposition, but much of the bitterness of disappointment may be avoided if the investigator is willing to learn the tricks step by step. As Lilienthal and Chanute pointed out, man must fly and fall and fly and fall until he can fly without falling." [16].

Born in Joliet, Illinois, **Edwin Way Teale** (1899–1980) spent much of his childhood with his grandparents at their farm in northwest Indiana. He recalled in his book *Dune Boy* how he became interested in flying [17]. "One winter night, when I was six or so, some neighbors spent the evening at my grandfather's farm in the dune country of northern Indiana. One of them told about 'The Crazy Old Man of the Dunes,' who had been building wings and sliding down the sandhills, trying to fly. Long after I went to bed I thought of that story, and when I fell asleep, I dreamed of an old man with a long beard spiraling high over the lonely dunes where the white gulls scream in the blue sky. I never forgot that fireside tale. No story I had ever heard impressed me more deeply. His first wings, it was said, had been thatched with chicken feathers. Later ones were pure white and made of pine and muslin. The hill from which Chanute had launched himself on his earliest fledgling wings lay but a few miles from Lone Oak Farm, which further increased my desire to be an aerial pioneer." Teale built himself a biplane glider in 1911, but the horse that was to pull him into the air was not interested in cooperating, so he never became airborne.

Years later Teale heard who this "crazy man" was. "When I interviewed Orville Wright, I learned that Chanute's work, so unappreciated by the dune-country dwellers, had formed a cornerstone of their research. Just before the Wright brothers flew in 1903, Chanute was a guest at their Kitty Hawk camp among those other sand dunes on the Carolina coast."

Teale earned his MA at Columbia University in 1926 and joined *Popular Science* as feature writer. Being fascinated how nature helped pilots stay airborne, one of his first articles discussed sky sailing and flying gliders, both in Europe and America. He was convinced, that no sport in the world had the same thrill as soaring like a bird. Teale's first book, *The Book of Gliders*, published in 1930, tried to catch the attention of beginners and people with a general interest in aviation. Even though gliders were special to him, Teale returned to his other great love, the outdoors and nature and writing about it.

A colorful daredevil, **Horace B. Wild** (1879–1940), called himself Chicago's first birdman and *Popular Science* published his life stories [18], the one about Chanute is interesting.

"By hopping a ride on a hayrack, I took part in a pioneer experiment that led to the airplane. One morning in 1896, I was standing in front of a Chicago fire station when an old-fashioned hayrack clattered past with a white glider stretched out lengthwise on it. I ran down the street, jumped on the back of the wagon, and rode with the driver all the way to Miller, Indiana, thirty miles away. The glider was delivered at the sand hill camp where Octave Chanute and his assistants were making their historic tests that laid the foundation for the Wright brothers. When the wagon went back that day, I stayed behind and became cook and chore boy for the experimenters. For several months we tested queer machines. Some had one wing, others two, and one craft, which we dubbed the 'Flying Staircase,' was fitted with five planes arranged one above the other. At first, we had a greased track down the side of a high dune for launching the gliders. Later, they were taken into the air by running downhill into the wind. Chanute, a friendly little man with pink cheeks and a white goatee, was as enthusiastic as a boy over the machines we built. Before he took up aerial experimenting, he had achieved world fame as a bridge builder and engineer. His bridge-building experience aided him in designing the most successful of his gliders, a biplane braced with piano wire in a bridge truss, the system still employed in modern planes. Only the other day, I flew a biplane that had its bracing wires arranged the same as that crude glider that I saw fly in the wilds of the Lake Michigan dunes.

"Octave Chanute was the first of many pioneers of aerial history that it was my good fortune to know intimately. On one of his trips to visit the Wright

Fig. 6.5 The updated Darmstadt I, named in honor of Chanute, became the highest performance sailplane in the early 1930s in the United States. Fred Loomis photo. National Soaring Museum, Elmira, New York

brothers at Dayton, Ohio, before they flew at Kitty Hawk, Chanute took me along. I remember we spent most of the time on the train debating the relative merits of airships and airplanes. I upheld the side of the gas bags. When we arrived in Dayton, we found Wilbur Wright tinkering with a bicycle in the little shop at 1927 West Third Street. At that time there wasn't an airplane in existence and more than ninety-nine percent of the inhabitants of the globe thought there never would be. When Wilbur wiped the grease from his hands and called Orville, Chanute went over their aerodynamic tables and calculations with them. He was greatly impressed and congratulated them several times. On the way home, he told me they had made remarkable progress." [18]. Some parts of Wild's story are wild and cannot be confirmed.

A few months prior to this story appearing in print, Wild had bought at an auction the damaged Darmstadt I sailplane, that Peter Hesselbach from Germany had flown in 1928 for four hours five minutes at Cape Cod. The next day, Hesselbach had crashed the glider on take-off. With the help of John K. "Jack" O'Meara, the Darmstadt sailplane was rebuilt, and Wild named it "*Chanute.*" O'Meara entered it in the 1932 National Gliding Contest in Elmira, and it was a big success; he established an American distance record of 66 miles, soaring from Elmira to Wyalusing, Pennsylvania, staying airborne for over eight hours. On the same flight he also set the national altitude record of 4,960 feet, and won the national contest. As Wild had hoped, this hawk-like motorless machine "*Chanute*" was the highest performance sailplane in America in the early/mid 1930s [19] (Fig. 6.5).

6.2   The Brothers from Dayton Push Ahead

The Wrights had submitted the claims on their flying machine with lateral, vertical and horizontal control [20] in early March 1903 without the advice of a patent lawyer. The examiner rejected numerous claims "based upon a device that is inoperative or incapable of performing its intended function. The examiner is unable to understand how the machine is supposed to operate." [21]. A year later, they hired a patent lawyer who filed amendments on behalf of the Wrights. The following year, the claims were re-written to show that their machine had actually flown.

Having heard from Wilbur periodically about their non-progress with the patent office, Chanute decided to do something in early May 1905 when he attended the 7th International Railway Congress in Washington. One of the social highlights was a garden party on the south side of the White House, hosted by the First Lady, Mrs. Edith Roosevelt. Chanute was pleased to socialize with President Theodore Roosevelt, whom he had met years earlier when the engineer and the politician were debating the pros and cons of a Rapid Transit System in New York City. With both men being interested in aeronautics, the discussion most likely included the flying activity of the Wright brothers and their patent claims, that were not accepted yet. The latest rejection by the patent examiner stated "… that the ambiguities, inaccuracies and imperfections of the specification, drawing and claims are such as to preclude intelligent action upon the merits of the claims until the defects in question have been remedied." [21]. Chanute most likely shared his belief in the eventual success of this transportation scheme and made a plea for governmental support of aeronautics, bringing the topic closer to Roosevelt's current thinking.

With the quarrel between Russia and Japan continuing in late 1904 and reading about troops being constantly shifted from site to site as part of the conflict, gave Chanute an idea. He wrote a note to Wilbur, flagging the possibility to do reconnaissance work with their new Flyer, possibly collecting $100,000 for a few months of work for Japan [22]. This suggestion might have planted a seed.

6.2.1   Contacting the American Government

Believing that their Flyer was ready to be sold as a military weapon, the Wrights asked their congressman in early January 1905 for help. Two weeks later, Wilbur followed up with a letter which was forwarded to the Secretary of War. The U.S. Board of Ordnance and Fortification responded, that they receive many requests for assistance, and "Once the machine was perfected, the Board would be pleased to hear about it." For some reason, Wilbur did not mention to Chanute that they had contacted the War Department and were turned down. In late May, Chanute happened to ask how close the brothers were to a practical machine, and Wilbur explained: "We stand ready to furnish a practical machine for use in war at once, that is, a machine capable of carrying two men and fuel for a fifty-mile trip. We are only waiting to complete arrangements with some government. The American government has apparently decided to permit foreign governments to take the lead in utilizing our invention for war purposes. We greatly regret this attitude of our own country, but see no way to remedy it; we have made a formal proposition to the British Government and expect to have a conference with one of its representatives very soon. We think the prospect favorable." [23].

6.2.2   Contacting the British Government

Lieutenant Colonel John E. Capper, Superintendent of the Balloon Factory in England, had watched the aeronautical activities in St. Louis; Chanute then suggested that he stop in Dayton to meet the Wrights, but also Langley and Manly at the Smithsonian and Zahm at the Catholic University in Washington.

Only a few days after contacting their congressman, Wilbur wrote to Capper asking if the British were still interest in their Flyer. The British War Office opened negotiations and in April, Patrick Alexander, a prominent member of the Aeronautical Society of Great Britain, who Chanute had introduced to the brothers two years earlier, came to Dayton. But then the British War Office thought it more important to possess the know-how to design different types of aeroplanes than to buy one.

6.2.3   Contacting the French

Assuming that Captain Ferdinand Ferber had leads to the French War Department, Wilbur wrote a detailed letter to the career military man on October 9, 1905. Receiving no response, he wrote a more forceful letter, first congratulating Ferber on his progress and then stating that France "may wish to avail itself of our discoveries, partly to supplement its own work, or to accurately inform itself of the state of the art as it will exist in those countries which buy the secrets of our motor machine … For when it becomes known that France is in the possession of a practical flying machine, other countries must at once avail themselves of our scientific discoveries and practical experiences. With Russia and Austria-Hungary in their present troubled condition and the German Emperor in a truculent mood, a spark may produce an explosion at any minute." [24].

Being curious about the Wrights' flying activity, Ferber asked fellow Aéro-Club member Frank S. Lahm, the European agent for the Remington Typewriter Company in Paris, privately if he knew of anyone who could verify their story. Lahm, who was also curious, wired his brother-in-law, Henry Weaver, of Mansfield, Ohio, asking him to verify the Wrights claims. After some initial back and forth, Weaver met the Wrights and some locals on December 3 and wired: "Claims fully verified, particulars by mail." Lahm translated and read the letter at the next Aéro-Club meeting, but members were not ready to believe and accept what they heard.

In the meantime, Ferber had given Wilbur's letters to Georges Besançon, editor of L'Aérophile, who translated and published them in December. Here Wilbur's words "truculent mood" read "*cherchant noise*," or seeking a quarrel or fight. Needless to say, relations between Ferber and the Wrights became tense, as they regarded the publication of their letters as "simply outrageous." Wilbur stated, "This is the worst given that he deliberately included direct references to Russia, Austria and the German Emperor, although he attacked all the embarrassing references for his 'bluff'." [25]. And Ferber explained to Chanute that "you have to make it clear to the Wrights, that the issue of authentication is of importance."

Considering himself the middleman in the possible trade negotiations, Ferber wired the Wrights on December 13, 1905: "Friend with full powers for stating terms of the contract will sail next Saturday, wire if convenient." Wilbur responded laconically: "Time convenient." Ferber's friend, Arnold Fordyce, who represented a syndicate of French capitalists, arrived in Dayton on December 28. The Wrights were at first hesitant, "Our discovery is incontestable, can render the greatest service to the army, which will adopt it; that is why we refuse to engage in a transaction intended to serve private, industrial or commercial interests." But the financial aspect was tempting, so the brothers signed a preliminary contract with Fordyce, agreeing to deliver their flying machine for 1,000,000 francs not later than August 1, 1906. It is not surprising that Wilbur shared the news with Chanute, who responded, "I am delighted to learn that you have sold a flying machine to the French for $200,000, and this without onerous restrictions as to sales to other nations. This may not make for peace as much as an exclusive sale to one nation, for now they will all have a machine, and I fancy that the French have acted so promptly because they expect early war, but it will make you world famous and eventually millionaires, if you care for the latter." [26].

Another French official was to come to Dayton to verify the Wrights' claims. Still not wanting to show their machine, the brothers needed a businessman, fluent in the French language with a worldwide reputation to vouch for their achievement. They wired Chanute on March 31, "Could you conveniently attend final conference with French at Dayton tomorrow, Monday." Always willing to help when asked, Chanute rearranged his work schedule, traveled through the night to arrive in Dayton early next morning on April 1. No agreement was reached and the contract lapsed.

6.2.4 Contacting the German Military

In early March 1905, the German patent office informed the Wrights that "the principle of twisting the wings" had been disclosed in an article by Chanute, published in Moedebeck's *Pocket Book* [27], which was sufficient to "teach anyone to fly." Actually, Chanute had mailed the article to Wilbur and asked for comments prior to sending it to Germany, but he had received no feed-back.

One month later the Wrights mailed a sales offer to the German Minister of War. But the Inspector of Transportation considered the development of dirigibles more meaningful than the purchase of a flying machine; they rejected the offer on June 15, 1906.

The German Patent Office nullified the Wrights main claim in March 1912, stating that the information had been previously published, but we have to remember that the German Patent Office in the years prior to World War I usually obstructed foreigners from taking out patents, often for nationalistic reasons.

6.2.5 Initial Struggle for Recognition

To develop a practical flying machine, the Wrights proceeded more akin to tinkering than employing a systematic study. Every time they flew, something went wrong, requiring alterations or parts to be improved. After another nasty crash in July 1905, the brothers made some radical changes. Success. On October 4, Orville flew 21 miles in 33 minutes; and the next day Wilbur flew 24 miles in 39 minutes, longer than the total duration of all the flights of 1903 and 1904. They now had an aeroplane to make turns and come back to level flight without losing control; they made longer and higher flights, and the passengers on the passing by train began to take notice. Wilbur explained to Chanute on October 19, "Some friends whom we unwisely permitted to witness some of the flights could not keep silent, and on the evening of the 5th the [Dayton] *Daily News* [28] had an article, reporting that we were making sensational flights every day. It was copied in the *Cincinnati Post* the next day. Consequently, we are doing nothing at present, but before the season closes, we wish to go out someday and make an effort to put the record above one hour. If you wish, we will try to give you notice to be present." Two weeks later, on Monday, October 30, 1905, Wilbur cabled that they were ready to make their record flight the next day, "Can you come?" Always anxious to see an aeroplane in the air, Chanute accepted Wilbur's latest invitation, but weather prevented a takeoff.

Most likely prodded by Chanute, Wilbur wrote another letter to the Secretary of War in October. The response was similar to the earlier one, and Chanute just commiserated with Wilbur, "Those fellows are a bunch of asses," but he also stated, "I still adhere to my original belief that the best disposal of your machine would be to have it go to one single government who would keep it secret and thus promote peace, while the others were meeting numerous accidents in the efforts to rival it." [29].

To obtain information about the Wright Flyer, the Aero Club in Vienna, Austria, thought of a roundabout approach. They contacted Chanute in early February 1906, suggesting to give him an honorary membership in exchange for drawings of the Flyer "to create a diploma for the Wrights." Recognizing this as an odd honor, Chanute replied: "This is the first opportunity which I had to say that I have not been what you term 'the teacher' of the Wright Brothers, in the usual meaning of the word. If the Wright Brothers have progressed so very much further than others to whom similar information was imparted, it has been entirely the result of their own perseverance and mechanical ability. As to the construction of their present machine, I regret that I can tell nothing nor can I send a sketch." [30]. A few months later, the Wrights and Chanute were elected honorary members.

Frustrated about their negotiations not going anywhere, the Wrights dismantled their Flyer and stopped flying. Wilbur wrote Chanute on November 19, "that it will be best to absolve our friends from their obligation to keep secret the results of the experiments. The strict refusing of authentic news may be harmful with widespread rumors in circulation." [31].

Now that Chanute was allowed to talk about the Wrights flying activity, one of his first letters went to Francis Wenham in England. "I think it is due to you, as the pioneer of aviation, and my valued friend, to be the first to be told, confidentially, that the Wright brothers have been in possession of a practical flying machine for the past two years and have been improving it. This season's operations being now closed I have just been permitted by Wright brothers to advise you of the results." Chanute summarized the brothers experiments of 1904–1905 and mentioned that a sale of a machine to the British or the French government was a possibility. The 81-year old Wenham was delighted to hear of the Wrights' success and added, "As you are aware, I always believed in the possibility of flight by man through the air, and this success removes me from the list of cranks who always stated I was advocating an impossibility." [32].

To make their accomplishments known, the brothers mailed reports to several European correspondents. Carl Dienstbach actually received two letters, one from each brother; his report was published in the February 1906 *Illustrierte Aeronautische Mittheilungen;* it was followed by a sarcastic editorial by the owner of the magazine, H. W. Moedebeck, reflecting the general mood in central Europe. "Show us what you can do or we do not believe you; we will judge you by what you demonstrate to us." Reading this, Chanute wrote to Moedebeck right away, "I think you are making a mistake in doubting what the Wrights have published concerning their achievements. I believe firmly that all they have asserted is entirely true." [33].

The Wright's letter to Patrick Alexander was reprinted verbatim in the *Aeronautical Journal*; it is not known if Col. Capper shared his letter with anyone.

The letter to Georges Besançon, editor of *L'Aérophile*, stirred quite a bit of attention. Due to the editorial deadline of his magazine, Besançon submitted his comments to *L'Auto*. "A device like the one the Wrights are announcing, and the possibility of reproducing it in multiple copies, for a miserable million, really, it is for nothing. But then, why don't the famous aviators address their own government? If the United States does not understand the importance of their success, why did they not find industrialists willing to pledge the required amount needed? … Among the many sportsmen and technicians, perhaps one could assemble a group to pledge the amount requested by the Wrights. If successful, subscribers will have the satisfaction of helping to shed light on a major discovery. If it is a bad joke, they will have unmasked an unworthy deception, and that without purse. Under these conditions, why not?" [34]. The December 1905 *L'Aérophile*, published a couple weeks later, showed Wilbur's letters to Ferber plus one from Chanute, as well as an editorial, in which Besançon questioned where and why the Wrights had been hiding and why journalists kept up the silence. "We were surprised by the suddenness of the latest results obtained by the Wrights. We know that science does not proceed by leaps and bounds. Great discoveries are always the result of successive progress."

The Car magazine from London also wanted confirmation of "the marvelous reports which have filtered through the press about the operations carried out by the Wright brothers during the last three years." They sent a correspondent to Chicago to interview Chanute, who stated: "It has now become evident that the first practical application of the flying machine will be in the art of war. What modifications in that art will be brought about, cannot yet be determined, but it is probable that the ultimate effect will be to diminish the frequency of wars. This may come to pass, not only because of possible additional horrors in battles, but because a fast flying machine will render the enemy's disposition of forces so easy of observation, that nations may incline more and more to universal peace." [35].

6.3   Aeronauts Organize

In early June 1905, representatives of several European aero clubs gathered in Brussels, Belgium, proposing to the Olympic Congress to form an international aeronautic federation. "This Congress, recognizing the special importance of aeronautics, expresses the desire that in each country, there be created an association for regulating the sport of flying and that thereafter there be formed a Universal Aeronautical Federation to regulate the various aviation meetings and advance the science and sport of aeronautics." The group met again four months later and the *"Fédération Aéronautique Internationale"* (FAI) was formed on October 14, 1905; representatives of Belgium, France, Germany, Great Britain, Italy, Spain, Switzerland and the USA adopted the proposed statutes. The U.S. was represented by the balloonist Cortlandt Field Bishop from New York City.

Homer W. Hedge, co-founder of the Automobile Club of America, had promoted the idea of an aero club in the New York City area early in 1905 "to show the world what America can and will do in the science of sailing in the air." The first official meeting of the Aero Club of America, the American representative to the FAI by default, was held in mid-November 1905 [36], and members listened to Charles Manly for almost two hours about flying. He told his audience that he was originally inspired by the "great patron saint of aviation, Octave Chanute" and had worked for Samuel Langley. He held his audience spellbound, telling of his thrilling escapes while experimenting with aeroplanes, that were far superior to the "gas bags." And he denied the reports that Langley's experiments were failures, he believed that the launching mechanism was defective. In closing he mentioned the secret experiments of the Wright brothers in Dayton and that they had flown more than fifty times around a small park to prove the practicality of their Flyer. The next speaker was Augustus Herring who talked about flying aeroplanes without a motor, utilizing ascending air currents as the birds do to stay airborne. The last speaker was Leo Stevens who had recently flown in a "gas bag machine;" he proclaimed that there was no sense of danger and "When I am two miles high there is not even a hint of dizziness."

Listening to the three aeronauts ignited much enthusiasm, flying appeared to be more fun and exciting than automobiling. The new club then decided to include "Aerial Technology" in the next automobile show, scheduled for January, in the hopefully finished 69th Regiment Armory. To gather material on short notice, Augustus Post and Albert Zahm contacted just about everyone in the aeronautical community, asking for models, instruments, artifacts and photos. They offered to reimburse the shipping expenses to New York, but there was no mention of returning any material.

Getting ready to move from their rented house in Streeterville on Chicago's northside, Chanute decided to help himself and the Aero Club by cleaning up. Some of his models were not in the best condition, but Post replied that the club desired all his material and that Herring would restore the models. So, he mailed $10 and joined as a charter member. Avery pulled the 1896 dome kite, the 1897 "Herring-Arnot" and the Lilienthal-type gliders from the rafters above his shop, while Chanute's daughters removed models, kites and stuffed birds from the ceiling in his library. Included in this shipment were also the 12-winged staggered wing model, Beeson's soaring device, the 1895 ladder kite and oscillating wings designed by Chanute, Brown's biplanes with a square front and one with a curved front. He also shipped Chinese warped surface kites, a Malay tailless kite, Butusov's combination of planes, a pair of dried duck wings and two birds by Pichancourt. And he suggested, "Mr. Herring is best fitted to draw up descriptions of the models, as he knows all about them" [37]. Chanute also shipped 28 large-size photos of the 1896/1897 gliding flights and some duplicate books to start the Club's library. Everything was packed into three large crates and shipped on December 18 to New York [38], with Chanute reportedly stating, "Good riddance."

6.4   Have a Fly with Me! The First Aero Club Exhibit

Early in January 1906 New York City newspapers enticed readers with stories about this new fad aeroplanes (Fig. 6.6).

The news of "France to own the first real airship of the world, an invention of the Wright Brothers of Dayton, Ohio" [39] traveled quickly. A few days later, the same papers discussed Alberto Santos-Dumont, who had abandoned the gas bag-type airship to realize Jules Verne's dreams and now worked on the same design principle in which Maxim, Langley, Chanute and the Wrights had such faith. Reading all the news in the papers ignited much curiosity: "Want to fly? Go to Aero Exhibit," was just one headline [40].

The "Exhibition of Aeronautical Apparatus" opened with great fanfare on Saturday morning, January 13, in conjunction with the 6th annual automobile show. Big ads in New York City papers attracted large crowds. Society was out in full force, as it was stylish to be seen among the latest automobiles and flying machines.

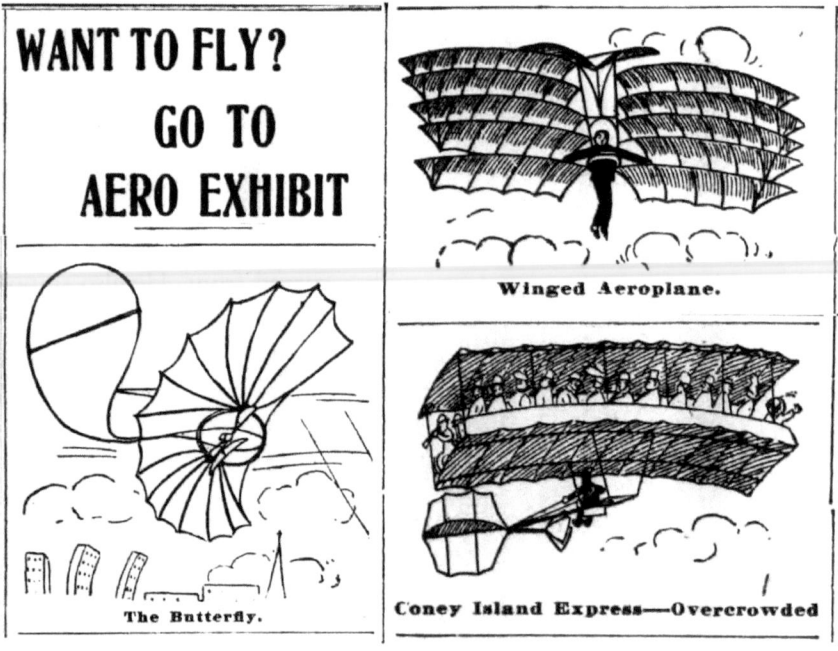

Fig. 6.6 The *Morning Telegraph*, dated January 13, 1906, was just one New York City newspaper which advertised the upcoming event, sponsored by the Aero Club of America.

The Lilienthal glider, purchased a decade earlier by the newspaper magnate William Randolph Hearst, had recently changed hands. John Brisben Walker had sold his *Cosmopolitan Magazine* to Hearst, and the rather tattered glider was part of the financial transaction. Hanging from the ceiling with its wings stretched like in flight, the glider was impressive, even though its cloth covering was torn (Fig. 6.7).

The reporter for the *Scientific American* explained: "In contrast to the great secrecy of the later aeroplane experimenters should be noted the free manner in which that first great pioneer experimenter in gliding flight gave the results of his experiments to the world. Had it not been for his untimely death in 1896, there is scarcely any doubt that he would have solved the problem of the motor-driven aeroplane some years ago; for he was not only a thorough mathematician and physicist, a clever constructor and mechanical engineer, but he was also possessed of that daring and physical dexterity which is a valuable aid to one attempting to solve such a problem." [41]. Close-by hung the cleaned up "Herring-Arnot" Chanute-type glider, an effective contrast to Israel Ludlow's aeroplane, on which Charles Hamilton had recently made a successful flight over the Hudson River.

Lawrence Hargrave, who had shipped his compressed air model of 1888 to Chicago in 1894, had agreed to transfer the model to the National Museum

Fig. 6.7 Even though the Lilienthal monoplane glider was not in the best of conditions any more, it was prominently displayed. The cleaned up "Herring-Arnot" Chanute-type glider from 1897 was hanging beside it. Alexander Graham Bell photo album, National Air & Space Museum, Washington, DC (NASM-9A18474)

in Washington, after displaying it in New York [42]. Dienstbach reported that the only sturdy part of the Hargrave model was the pipe shaped container that stored the compressed air. The ribs in the wings were like wooden matches and thin paper covered the wings. Even though every caution was used hanging the model, the wings were damaged twice [43].

Hanging beside the Hargrave was Langley's steam-driven model, which flew about half a mile over the Potomac River on May 6, 1896, and his full-size aerodrome that had made several launching attempts between late July and early December 1903. The discussion whether this aerodrome could have been the first aeroplane capable of flight continued for decades. With the approval of the Smithsonian Institution, Glenn Curtiss restored this aerodrome and William E. "Gink" Doherty made a few straight-ahead flights in 1914, trying to prove that Langley's aerodrome, powered by a radial engine designed by Balzer and Manly, could have been the first heavier-than-air manned, powered aircraft. But the fact remains, the Langley aerodrome did not fly in 1903 and the Wright Flyer did.

Other crafts on display were Emile Berliner's aeromobile, Edward Horsman's kites, W. M. Keil's "Ballo-Plane", Roy Knabenshue's airship, Wilbur Kimball's helicopter, and Leo Stevens' airship and a balloon with basket. Tom Baldwin and Carl Myers also had exhibits, and Curtiss displayed his lightweight engines for dirigibles this year, one of which Bell reportedly purchased to be used in his man-carrying kites.

Alexander Graham Bell displayed several kites, covered with red silk instead of the commonly used light-color nainsook; he spent quite a bit of time on the exhibit floor to answer questions, making it clear that "the age of the flying machine is not in the future, it is with us now." Bell was also the principal speaker at the Transportation Club Dinner and, a few days later, at the Annual Dinner of the Automobile Club of America, with about 550 members and guests attending each event [44].

Among the scientific exhibits were several highly polished wooden wing models, displayed by Albert Zahm from the Catholic University. He used these models in his wind tunnel research to measure head resistance and friction. Carl Myers considered this exhibit the most valuable to anyone interested in that part of aeronautics [45]. And the Patent Office displayed models for which patents were issued, "some of them were so ludicrous in construction as to excite a smile." The earliest model dated back to 1878 and the last from 1889 when the rule was changed to allow inventors to submit drawings instead of models.

Large photographs covered the walls of the room; to the left of the main entrance were photos from private collectors and experimenters, like Berliner, Kimball, Montgomery, Knabenshue, Whitehead and others. William J. Hammer's photograph collection showed foreign airships and balloons in flight, and George Grantham Bain showed photos of Chanute's team and of Santos-Dumont, Francois and Lebaudy of France, von Zeppelin of Germany, Spencer of England, and more. The wall to the right of the main entrance was dedicated to the development of the aeroplane in the U.S., divided under the headings: Lilienthal—Herring—Wright Bros.—Langley—Maxim— Pickering; Chanute's name was mentioned in some photo captions under Herring's contributions. When Wilbur saw this, he wrote to Chanute: "I was under the impression that I had learned somewhere that you had conducted some experiments about 1896 or 1897. Possibly my memory is at fault." [25]. Chanute was amused how Herring had arranged the photo display, but said nothing.

Dienstbach had urged the Wrights to ship their 1903 Flyer to be displayed, but having signed an initial contract with the French, the brothers only mailed photos of their gliding activity. When asked again, the brothers shipped the crankshaft and flywheel of their 1903 wind damaged Flyer, they were placed on a stool below their photo display.

As a pleasant diversion, organizers set-up a temporary theatre in the fourth floor to show movies of automobile and motorcycle races, but also of the Lebaudy and Santos-Dumont airships and Archdeacon's glider. This provided welcome entertainment and relaxed seating for the exhausted visitors.

After an informal reception in the Armory on Sunday morning, January 21, the exhibit came to an end. Most of the photos were transferred to the Aero Club and members used them in their upcoming *Navigating the Air* book. Langley's machines and photos were shipped back to Washington. As discussed with Chanute, the Hargrave model, Brown's gliders and other artifacts were shipped to the National Museum in Washington, DC, to be included in the proposed exhibit "*Man's Effort to Fly.*" Some of these artifacts are still part of the National Air and Space Museum, while others were in such dilapidated shape that they were officially burned in 1913 [46]. Upon strong urging by Bell, Walker donated the Lilienthal glider to the National Museum. Herring was asked to donate the "Herring-Arnot" glider, but he decided to sell the craft to the balloonist Dr. Julian P. Thomas, a newcomer to the Aero Club, who intended to add a French gasoline motor to the glider. The Wrights flywheel and crankshaft were probably transferred to the Aero Club, but disappeared over time. These are just a few of the artifacts that could be traced.

Chanute did not go to New York; he spent the winter with his daughters at the rice plantation of his niece, Amelia Gueydan, near New Orleans. Many years later, Amelia's daughter Pepilla recalled, that her uncle Octave was fond of watching the buzzards, and the best place to do so was the city garbage dump [47].

6.5 Reality of the Airship

Wanting confirmation of the Wrights' flying, the secretary of the Aero Club asked the brothers to submit some proof. Wilbur obliged and mailed a four-page report to Augustus Post on March 2, 1906. "Though America, through the labors of Professor Langley, Mr. Chanute, and others, had acquired not less than ten years ago the recognized leadership in that branch of aeronautics which pertains to bird-like flight, it has not heretofore been possible for American workers to present a summary of each year's experiments to a society of their own country devoted exclusively to the promotion of aeronautical studies and sports. It is with great pleasure that we now find ourselves able to make a report to such a society." The last page listed the names of thirteen Dayton residents and four non-residents who had seen them fly, but the names of Amos Root, who reportedly witnessed Wilbur's first circular flight, and of Chanute were not included.

The *Scientific American* had published an editorial in January, "*The Wright Aeroplane and its Fabled Performances*" questioning "Why particularly, as is further alleged, should the Wrights desire to sell their invention to the French government for a million francs? Surely their own is the first to which they would be likely to apply. We certainly want more light on the subject." [48]. Two months later, the still skeptical editor asked Chanute if he could verify the Wrights claims. The April 14, 1906, issue printed his letter, stating that the brothers performed two improbable feats, one was inventing a practical flying machine and the other, to keep their aircraft away from the argus-eyed American press [49].

Being concerned that the Wrights could not sell their Flyer, Chanute cautiously warned Wilbur in late summer, that they might be setting their financial goal a bit too high, as French experimenters were making good progress. Wilbur thought different, "We are convinced that no one will be able to develop a practical flyer within five years. This opinion is based upon cold calculation. It takes into consideration practical and scientific difficulties whose existence is unknown to all but ourselves. Even you, Mr. Chanute, have little idea how difficult the flying problem really is." [50]. Chanute responded a few days later. "The value of an invention is whatever it costs to reproduce it, and I am by no means sure that persistent experimenting by others, now that success has been achieved, may not produce a practical flyer within five years. The important factor is that light motors have been developed. I cheerfully acknowledge that I have little idea how difficult the flying problem really is and that its solution is beyond my powers, but are you not too cocksure that yours is the only secret worth knowing and that others may not hit upon a solution in less than 'many times five years?' It took you less than that and there are a few able inventors in the world." [51] Indeed, Aerial Experiment Association (AEA) members made the first public flight in an aeroplane less than two years later, but that is another story.

After suffering a stroke, the 72-year-old former Smithsonian Secretary moved to Aiken, South Carolina, accompanied by his brother John's daughter; Langley died on February 27, 1906. A few days later the saddened Chanute received a letter from Wilbur, stating among others, "The newspapers report the death of Prof. Langley. It is really pathetic that he should have missed the honor he cared for above all others, merely because he could not launch his machine successfully. If he could only have started it, the chances are that it would have flown sufficiently to have secured him the name he coveted, even though a complete wreck attended the landing. I cannot help feeling sorry for him. The fact that the great scientist, Prof. Langley, believed in flying machines was one thing that encouraged us to begin our studies. It was he that recommended to us Progress in Flying Machines, the Aeronautical Annuals, Experiments in Aerodynamics, Empire of the Air, and Lilienthal's articles on gliding, and started us in the right direction at the beginning" [52].

To honor their former Secretary, the Smithsonian organized a memorial meeting on Monday, December 3, in the lecture room of the National Museum. Four invited speakers discussed the different areas of Langley's interest, with Chanute focusing on aerial navigation and Langley's faith in the possibility of success. He stated that it probably would have been better for Langley's happiness and reputation, if he had terminated his experiments with the successful 1896 demonstrations. On a more optimistic note, Chanute told his audience that "We are now in the possession of a solution of a problem, which had baffled the ingenuity of man since the dawn of history," stressing that Langley's work had to be continued. Beyond that, Chanute clarified misleading statements by the press about his aerodromes, and the facts, chronicled by Chanute, became a matter of history [53].

Wanting to attend the second Aero Club exhibition at Grand Central Palace in New York, Chanute took the next train north. Since Americans, and New Yorkers in particular, were interested in "the latest thing," the combined automobile and aero club show attracted quite a crowd. Visitors had to take the elevator to the top floor to see the flying machines, aeroplanes, motors for airships, and several "gas bags," used by club members [54]. Interest in flying was also manifested by the way visitors crowded around aeronauts whenever a group discussion started. Chanute believed that the solution of navigating the air was close at hand, but "It is a most regrettable fact, that during 1907 there is every reason to expect that many experimenters with flying machines will be injured. Now that there is so much popular interest in solving the problem of aerial navigation, some men will surely break their legs, and I only hope that it will not be their necks. They should take plenty of time to become accustomed to flying in gliding machines before they trust themselves to motor-driven aeroplanes." And Wilbur explained, "The secret of our being alive today, is that we have been extremely cautious. Perhaps we could have made long flights three years ago, but it is more likely that we would have been killed. We went at the problem slowly, taking each step so gradually that by the time we had our machine in successful operation we were accustomed to meeting the emergencies that continually confront a man in the air." [55]. As a special treat, John Brisben Walker invited the visiting aeronauts to dinner at the Century Club after the exhibit closed.

For the trip back home, Chanute took the New York Central's 20th Century Limited, or the Central's new "Flyer," which had just beat the speed mark of a mile a minute, making the daily runs between New York and Chicago in 18 hours. Sitting back in his upholstered chair, which folded into a sleeping berth for the night, he probably reflected on how humanity had entered the nineteenth century, moving at about six miles an hour, or the speed of a horse-drawn coach. Seventy years later, while working for the Erie Railway, he introduced many technical improvements to speed up the rail traffic; and now, at the beginning of the twentieth century, trains were finally traveling at 60 miles in one hour.

The social schedule of the Chanute family was quite full in fall of 1906, but then came an opportunity to combine business with pleasure. Members of the Technological Society in Kansas City wanted Chanute to present a talk to his former fellow citizens. As he had to finalize the rental contract on his recently erected seven-story building in downtown Kansas City, he gladly obliged. The audience filled the Central High School auditorium to its limits, and his first sentence caught everyone's attention: "There is today a practical flying machine. I have seen it fly with my own eyes. The secret of aerial navigation has been solved." He showed lantern slides of his and the Wrights' glider flying activity and explained, "Now that we have successful airships, we are able to see their limitations. I do not believe they will be of commercial importance. They will be used in the art of war, in the explorations of swamps and for the ascent of mountains." [56].

Reading of this talk, the mayor of Chanute, Kansas, wanted the name-sake of his town to come and lecture in December. Being so late in the season, Chanute was not sure if this could be arranged, but his two daughters Nina and Alice wanted to see the town named after their father. Members of the planning committee gave the Chanutes the royal treatment; "Dee-lighted" was the Rooseveltian phrase Chanute reportedly used when asked what he thought of the city named after him. "I was not expecting to see such a large and prosperous city. It is also a very pretty town, and the streets and buildings are all well-kept. It is thirty-three years since I was here, then the city was a mushroom town of 800 inhabitants and a railroad intersection." [57]. His lecture in the Opera House was well attended, and its citizens later stated proudly that their town was the first "high flyer" designed by Chanute. Before heading back to Chicago, the family toured southern Kansas and northern Oklahoma. For Chanute it was a trip down memory lane; he was amazed how much had changed since he laid out towns and railroads so many years ago (Fig. 6.8).

Fig. 6.8 Chanute with two of his daughters traveling by train through southern Kansas and Northern Oklahoma. Courtesy Chanute Family

To satisfy the public's curiosity about the people who worked with aeroplanes, newspapers tracked down the pioneers. Chanute was one of the first to be interviewed; his comments were published in mid-May in the *New York Herald*. "Having made some experiments of my own in gliding flights in 1896, I published a full description with photos and detailed plans in the *Aeronautical Annual* of 1897, and I invited other searchers to improve upon my practice. I was written to on May 13, 1900, by Mr. Wilbur Wright who stated that he was minded to experiment, although he believed no financial profit would follow, and who asked for information and suggestions." He added that the Wrights had kept him informed on their progress, and he believed all they said about their accomplishments [58].

Half a year later, the *New York Herald* sent a reporter to Dayton to "discover the full truth about the brothers." Sherman Morse met the Wrights, talked with several locals and cabled his report to the office in Paris. In the introduction, Morse gave Wilbur's reasons for their secrecy: "We will make no exhibition test of our flying machine, nor will we permit an examination of it. For our purposes neither is necessary. Our only market must be a powerful government, and publicity would only serve to defeat our purpose to make such a sale. We do not have to have the newspapers tell us of our success, for we know, and those most important to us know, that we have accomplished all we claim to have done." [59] (Fig. 6.9).

Reading the report, the secretary of the Aéro-Club cabled at once to the *Herald*: "History is likely to attach less importance to the affidavits of people who claimed to have seen the brothers fly, than to the undisputed and undeniable flights of Santos-Dumont."

Seeing an opportunity for profit, Ulysses Eddy, an independent business agent, met the Wrights in Dayton and convinced them to meet with Charles R. Flint, who facilitated American arms sales to foreign nations. His proposal sounded like a dream come true, and Wilbur wrote to Chanute, "It seems that the favorable conditions we have been awaiting for six months have now arrived and we have some opportunities we would be glad to talk over with you, the best from a financial standpoint that we have had." [60]. After attending the Aero Club Show in early December, the brothers stopped at the Flint Company office, but the owner was out-of-town. Orville traveled to New York a few days later, and reported back to Wilbur that Flint's European agent had high-level contacts.

The Wrights did not easily trust everyone, but Flint was just as leery of starting a business with two men who refused to demonstrate their product. Having known Chanute for some time, Flint asked him to come to New York, as he wanted a confidential character reference of the brothers. There

Fig. 6.9 Some newspaper reports were not exactly what the Wrights wanted to see. With only very few images of their Flyer available, reporters frequently used the more readily available photos of Chanute's gliding activity (lower left corner). *New York Herald*, November 22, 1906

was no time to go to New York, so Chanute wrote, "I have followed their work since 1900, have seen all their machines, and have witnessed a short flight of one quarter mile in 1904. From somewhat intimate acquaintance, I can say that in addition to their great mechanical abilities I have found the Wright brothers trustworthy. They tell the truth and are conscientious, so that I credit fully any statement which they make." [61]. On the strength of Chanute's words, Flint moved forward and offered suggestions for a contract, but none satisfied both sides. As sales would not be easy, Flint's European agent requested one or both brothers to come to Europe, as many people wondered if someone just started a canard to embarrass them. An editorial, entitled "Flyers or Liars" in the Paris edition of the *New York Herald* [62] gave the general opinion: "The Wrights have flown or they have not flown. They possess a machine or they don't possess one. They are in fact either flyers or liars. It is difficult to fly ... It is easy to say, 'We have flown'." Assuming that the financial deal with the Flint Company was good, Wilbur sailed for France on May 16, 1907, and Orville shipped the Flyer overseas. Half a year later, Wilbur signed a contract with Flint & Co. to act as their sole agent, even though it did not quite meet his vision of "lots of money."

6.6   Combining Technical Knowledge with Inventiveness

Scottish by birth, **Alexander Graham Bell** (1847–1922) left his native Edinburgh at the age of 24 and became a teacher in a school for the deaf at Boston University. In the 1860s he thought of conveying the tones of the human voice over wires by means of electric currents over long distances. Bell secured a patent (#174,465) for his invention in March 1876 and had to engage in much litigation to establish his claim against others.

Interest in aeronautics started early in Bell's life. He reportedly told his partner, Thomas A. Watson, in 1877 that as soon as the telephone was put on a business basis, they should devote their attention to flying machines. Bell believed that a person should be able to lift himself into the air, but the question remained how to achieve what the birds do on a regular basis.

After receiving the book *Progress in Flying Machines* from Langley in early 1901, Bell contacted Chanute and mailed him a copy of his latest paper, the "*Tetrahedral Principle in Kite Structure*," [63] and Chanute mailed his latest article, starting an active exchange of ideas.

Having heard from Langley that Chanute was in Washington to attend the Railway Congress, Bell invited him to his weekly "Wednesday Evening" gathering on May 10, 1905. About 25 to 30 scientific friends congregated at Bell's mansion on Connecticut Avenue, and the discussion soon turned to aeronautics; Chanute spoke of several aeronauts, including Wilhelm Kress from Austria, John J. Montgomery from California, and the Wright brothers, but Bell had only one question: "Had the Wrights really flown?" Yes, he had witnessed a flight that ended in a rough landing with much damage to the Flyer, but the pilot was not hurt.

Talking privately again the next day, both men revealed that they liked working in the evening or the night, when everything is quiet. Bell reportedly told Chanute, "You can't make an owl sleep at night and I can't think in the daytime." His earliest retiring time was 4 o'clock in the morning, and if he had important work on hand, he did not go to bed for several days. Hearing this, Chanute probably chuckled, as his work routines and sleeping habits were quite similar.

In early May 1907, Bell received the honorary degree of Doctor of Science from Oxford University in recognition of his efforts to teach the deaf and dumb to speak and his invention of the telephone. Correspondents of several newspapers interviewed Bell in London about his many interests, including aeronautics: "The problem of aerial navigation is already solved and America is in advance of the rest of the world in heavier-than-air flying machines. The age of the flying machine is not one that belongs to the distant future, there is left only the question of improving the machine that has been invented." And "… to Mr. Chanute should be given a great amount of credit for what has been accomplished in this direction in America. He started the experiments with the flying machine and reproduced the gliding machine of Lilienthal and induced several Americans, among them the Wright Brothers, to experiment." [64]. The reporter from the *Evening Star* had one more question, and Bell explained with a smile why he was so interested in flight, "Aside from the scientific pleasure, it would give me great personal joy to perfect a flying machine. It has been the sad thought of my life that neither my wife nor mother were able to use or appreciate the telephone. Both were deaf. The photophone and graphophone were equally useless to my wife, but if I am able to perfect a flying machine, I will have done something that my wife will be able to appreciate and perhaps use." [65].

A decade later, in 1917, the last of the four transcontinental airmail routes was named for Bell and Chanute. The route extended from Boston, Massachusetts, across the continent to Seattle, via Albany, Syracuse, Erie, Buffalo, Detroit, Grand Rapids, Minneapolis, Bismarck and Great Falls. This is a well-deserved memorial to Octave Chanute, the pioneer aeronautical engineer, and to Alexander Graham Bell, who was a potent factor in advancing man's mastery of the air in an unspectacular way.

6.7 From Kite to Glider to Powered Flight. The AEA Gets into the Air!

The Aerial Experiment Association (AEA) was formed on October 1, 1907, under the leadership of Alexander Graham Bell and with the financial support of his wife Mabel. Joining Bell were Frederick W. "Casey" Baldwin, Douglas McCurdy, Thomas Selfridge and Glenn H. Curtiss. Members were to work with each other for one year and each member was to design, superintend and operate his machine. The goal was "to get into the air." [66].

One of the team members, **Glenn Hammond Curtiss** (1878–1930), was a motorcycle daredevil, lean and wiry, with a thick mustache and piercing eyes, who spoke little and smiled even less. He had established a bicycle and motorcycle repair business in Hammondsport, New York, in 1901 and slowly developed an interest in flying [67]. "Personal contact with many aeronauts has brought on the airship-fever, which I find quite contagious. My first opportunity to get any experience in the air was in January 1907, through Capt. Baldwin, for whom we had just completed an entirely new power plant with a new propeller design." [68]. On November 30, the Curtiss Motor Vehicle Company was incorporated with a capital stock of $600,000 to manufacture the Curtiss gasoline motor and motorcycles, Baldwin's dirigibles and a low-priced automobile [69] (Fig. 6.10).

The AEA team began their experiments with Bell's kite, the Cygnet, an imposing structure of 42.5 feet span that included over 3,000 tetrahedral cells covered with red silk. In December 1907, a motorboat towed the Cygnet, piloted by Selfridge, across Lake Bras d'Or near Baddeck, Nova Scotia, for about seven minutes, but Selfridge could not release the tow line and the craft was destroyed. McCurdy, the AEA's treasurer and later secretary, stated: "At the conclusion of experiments with the Cygnet the season was so far advanced that we decided to move our headquarters to Hammondsport and the motor-cycle works of Mr. Curtiss. Selfridge's experience in the Cygnet had fired us all and we longed to build a machine, which would glide in the air free from any attachment to Mother Earth."

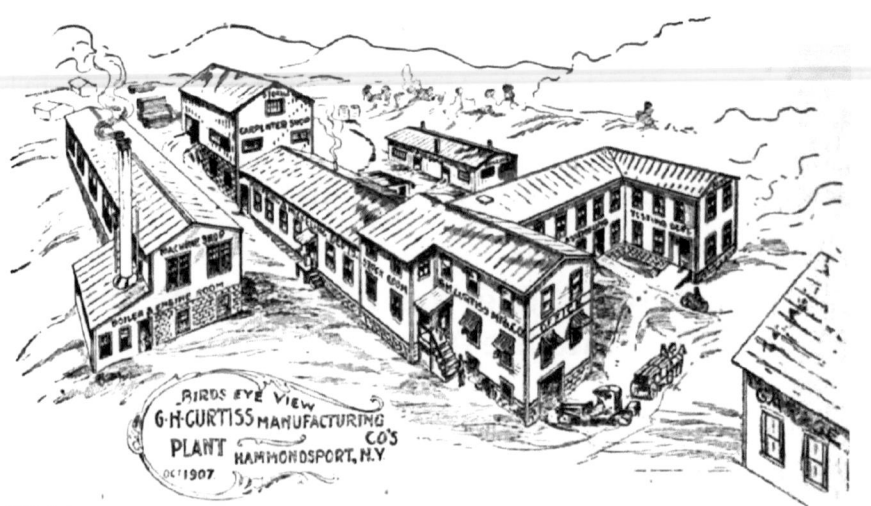

Fig. 6.10 The Curtiss factory in Hammondsport, New York, in October 1907. *Aeronautics*, March 1908

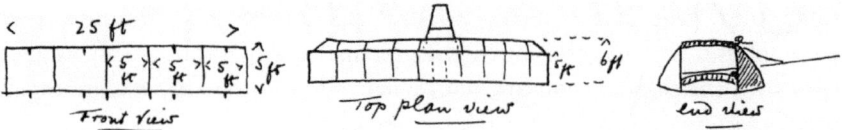

The ribs were T shaped and part of them formed the
upper and lower fore and aft chords of the panels.

Fig. 6.11 Drawings for the Chanute-type glider, built by members of the AEA team.
Alexander Graham Bell Papers, Library of Congress

To learn more about building and flying, Selfridge contacted several people
in the aeronautical community. The first letter went to Chanute, who had
mentioned to Bell some of the tricks they had learned, especially that
the uprights between the two wings needed to be shaped aerodynamically,
rounded in the front with a tail-like back to cut resistance. Tom Baldwin
and Roy Knabenshue then suggested contacting William Avery, as they had
watched him fly at the St. Louis World's Fair. Avery responded, "Transportation expenses to and from Chicago and $10/day would bring me to
Hammondsport, and I will give you the benefit of all my experience." Avery's
help was appreciated but the group went ahead without him. Writing to the
Wrights next, Orville's response did not bring anything new, but the letter
played a role in the Wright vs. Curtiss patent litigation only sixteen months
later [70].

Being anxious, Bell instructed Curtiss to start building at once a glider
using Chanute's drawings (Fig. 6.11). Two AEA members recorded the glider
flying activity, with Douglas McCurdy writing the more colorful one: "The
machine which we built is popularly known as a glider and from the information given the world by the German engineer Lilienthal, the Wright Bros. and
Dr. Chanute of Chicago, it was not very difficult to design such a machine.
The glider, we constructed, was a simple arrangement of two surfaces, one
above the other, spaced 6 feet apart. It was about 20 feet long and 6 feet
deep. At the center of the bottom surface was a hole, to allow a man's body
to pass through. A hill with a fair slope was selected from which our experiments could be performed. The glider was carried to the top of this hill and
four men were detailed to hold on to each of the four corners. You take your
position in the center of the machine with your head and shoulders coming
up through the hole on the bottom surface and hold on tightly to two firm
sticks which are secured on either side of you to the front and back edge of the
glider. We had arranged the usual signal of 1, 2, 3—GO. At the word GO,
you and the four men at each corner of the glider rushed forward to the brink
of the hill and you and the glider are thrown into the air. You half expect to be

dashed to the ground; but such is not the case. You feel for a minute as if you are on air, and you are, and much after the manner of a toboggan you gently coast down the slope. If you lose your balance for a moment, you are capable of getting quite a dump, but if the ground is covered with about two feet of snow, as it was in our case, you don't mind that very much. Each member made glides from ten feet to a hundred yards. Although not much scientific data was obtained from these experiments, we had acquired the feeling that every aviator possesses that to glide free in the air is beyond expression by words." [66] (Fig. 6.12).

Fig. 6.12 The AEA glider with members of the team and flown in mid-January from the snow-covered nearby hills. Courtesy Glenn H. Curtiss Museum, Hammondsport, New York

Building and flying the glider provided a good understanding of the underlying principles for designing their aerodromes, the *Red Wing*, followed by the *White Wing* and then the *June Bug*. And Curtiss wrote in 1909, "Much knowledge was gained with these first attempts at flight, but also much pleasure."

6.8   Spreading the Word

While not many people were tempted to try their hand at building or flying aeroplanes, the topic was of interest to the general public. Moedebeck's *Taschenbuch* of 1904 had a wealth of information on lighter- and heavier-than-air craft, but it was written in the German language. On urging by Chanute, Moedebeck agreed to have it translated and Mansergh Varley from the Emmanuel College in Cambridge, England, volunteered; Chanute helped the best he could, and Patrick Alexander made a generous financial contribution to make this literary project become reality. The translated 496-page *Pocket-Book of Aeronautics* [71] was ready for distribution in spring 1907. Thanking Moedebeck for the good-looking book, Chanute also expressed his concern that competitors were likely to overtake the Wrights. "I have repeatedly told them that there was danger of other experimenters discovering their secrets, but they have always said that they were safe for five years. It is possible that both you and myself are wrong in condemning their policy of not showing the machine until an agreement has been reached for its sale, especially in view of the fact that they say that an intelligent man can be taught in a week how to use it. The luck which experimenters will have within the next few months will determine the matter." [72].

At about the same time, members of the Aero Club of America published a 260-page book, *Navigating the Air,* in May 1907, to promote this inchoate science in America. In the preface, Cortland Bishop appealed to American patriotism to bring manflight to a position equaling that attained by French experimenters. Following the preface and an equally lengthy introduction by Dienstbach, Chanute's article discussed the progress made by the Wright brothers with their Flyer. "They made no flights in 1906, as they were negotiating with intending purchasers, who enjoined secrecy. The fame of their achievements naturally spurred flying machine inventors all over the world to emulate their success. During the year 1906 more than a score of flying machines was built and given preliminary tests, some in public, some in private." Next Chanute discussed enthusiasts who had flown with full-size aeroplanes in Europe. "They have all been distanced by M. Santos-Dumont who succeeded on November 12, 1906, in flying 220 meters in the presence of a large assemblage, thus improving his previous achievement of October 23, when he flew a distance of 60 m against the wind. He has not yet swept a circle and this he must do to win the Archdeacon and Deutsch prize of 50,000 francs. He is now hardly as far advanced as the Wright Brothers were at the close of 1903, and it is to be hoped that he will enjoy the same immunity from accident as they did." [73]. The next article was written by the Wrights, discussing the relations of weight, speed and power of aeroplanes, also providing information on the engines they used in their Flyers between 1903 and 1905. To support their statement of having flown, they attached reports from several eye-witnesses, and again Chanute's name was not mentioned.

Charles Manly offered "*Critical Remarks on Progress*," describing the work of Langley, Bell and the Wrights. And in closing, he stated, "In all of the accounts which I have lately seen of the experiments of the Wright Brothers, no mention has been made of the fact that their success has been built on the very valuable work of Mr. Chanute, who for many years carried on at his own expense work in the construction and testing of gliding machines. There is perhaps no one who has made a closer study and has a more thorough understanding of the whole subject of aerodromics than Mr. Chanute; I should like to see him given due credit for the very important work he had done."

Other authors were Bell, Zahm, Ludlow, Herring, Rotch, Lowe and Stevens, discussing kites, balloons and aeroplanes, motors and propellers, but also how to fly these craft.

Anticipating a growing interest in aeronautics, two new magazines hit the newsstands in 1907. The splashy American Magazine of *Aeronautics* edited by Ernest La Rue Jones in New York City, appeared in July 1907, and the *American Aeronaut and Aerostatist*, published by A. Kaufman and Thomas R. MacMechen in St. Louis, began its publication in October 1907.

The *Aeronautics* magazine featured Chanute's "*Conditions of success with flying machines*," [74] "Fascinated by the now acknowledged success of the Wright brothers in America, and tempted by the facilities which light motors offer for rising on the air, some 30 or 40 European aviators have built or are building motor-equipped flying machines on wheels, in the hope of speedily accomplishing mechanical flight. As of June 1907, they have made some fairly long jumps with their wheeled grasshoppers, but nothing like continuous flight has been accomplished. It is believed that these aviators are beginning at the wrong end and taking the longest path to success. Paradoxical as it may seem it is necessary to know how to use a flying machine before trusting oneself to fly with it." In closing Chanute stated, "The mode of conducting such flying experiments have been described by the various experimenters. It consists in first testing the apparatus as a kite and measuring accurately the 'lift', the 'drift', the 'head resistance' and the location of the 'center of pressure' at various angles of incidence. Then glides can be made by jumping into the wind, noting carefully the angle of descent, which should be as flat as possible with adequate stability. The sport enables the aviator to develop gradually the best shapes of surfaces and framing for his particular design; more importantly, it gives him experience and skill to manage his motor flying machine when he finally comes to testing it." [74].

Chanute's article in the *American Aeronaut*, "*Pending European Experiments in Flying*" discussed the other side of the story, "It may surprise the casual reader to learn that there are some 30 or 40 motor-equipped flying machines now being built in Europe by sane and talented men. Some of them have already been tested, with rather questionable results, and the rest will probably be experimented with later this year." European flying enthusiasts and inventors were not standing on the sideline and watching, they were marching forcefully ahead. Chanute then provided background information on each of the talented engineers currently working the problem. The last paragraph highlighted his philosophy: "One advantage which the French aviators seem to have is that they are all working in unison, publishing their plans and attending each other's experiments. They learn from one another and profit by every failure. It seems not unreasonable to expect that within a few years some European engineers will be in possession of a flying machine." [75].

6.9 The Jamestown Exposition and Its Aeronautical Congress

To celebrate their 300th anniversary in 1907, Jamestown, Virginia, exposition organizers were not interested in the hoopla of a World's Fair, but they wanted the public to come, and they wanted aeronautical activities. To add a stimulus, the *Scientific American* magazine had created a silver trophy to be awarded during the Jamestown Exposition. To earn it, the pilot had to fly the aeroplane for one kilometer (3,280 feet) in a straight line. "Aeroplanes may start by running along on wheels on the ground under their own power, but no special track or launching device will be permitted. A reasonably smooth, level roadway, or a turfed field will be provided from which to make the start. Machines need not fly more than a few feet above the ground or higher than is necessary to avoid obstacles. They should be capable of being steered both horizontally and vertically, and of alighting without being damaged." [76]. No aeroplane earned the trophy in 1907.

An Aeronautical Congress was scheduled for late October, but the public was mostly gone, so the Aero Club Board decided to move the event to New York City. Chanute, as a member of the Committee on Congress and Program, received a wire on October 27, but decided not to travel to New York.

In the morning of October 28, William Hammer, opened the meeting and explained to the audience why the Congress was moved on rather short notice. He then introduced the president of the Congress, Willis L. Moore, chief of the U.S. Weather Bureau, who discussed in his address the true value of air conquest and compared practical uses of balloons, dirigibles and aeroplanes. Moore believed that growth in aeronautics would also help increase the knowledge of meteorology, opening the way into the upper domain of the air. His address was actually ghostwritten by Chanute, and Moore acknowledged the help he had received from "that splendid man and engineer." [77].

Acting as Secretary, just as he did in 1893 and 1904, Albert Zahm read papers either in full or as abstracts, including Chanute's paper on "*Soaring Flight*," a paper by George Spratt on his pet topic "*Curvature, a Relative Term,*" Larry Lesh's paper on his experiences building and flying a Chanute-type glider and then his own, "*Wind-Tunnels for Aerodynamic Experiments, their Construction and Equipment*." On the lighter side of the program, participants of the Gordon Bennett Cup Race who had taken off in their balloons in St. Louis on October 21 and headed east toward New York, told humorous stories of their landing experiences.

The next day, Zahm read his paper on "*Principles involved in the formation of wing surfaces and the phenomenon of soaring*," and the papers by John J. Montgomery, Gustave Whitehead and Israel Lancaster. Several officers of the Signal Corps read short papers; General James Allen talked about what the army was doing to build up an aeronautic branch. Admiral Colby Mitchell Chester, United States Navy, stated, "We old fellows have seen the sailing vessels give way to the armor-clad steam warships, and I predict that the future will see the aeroplane fighting machine. From the aeroplane, high above water, the submarine can be located beneath the waves and explosives dropped upon it, which is the only effective way of fighting it. The aeroplane is the fighting machine of the future, but the navy departments are so constituted that we must follow after you have led the way." [78]. Lieutenant Colonel William Glassford from the Signal Corps was one of the last speakers, he believed that "The importance of the aircraft as a military weapon had impressed every thoughtful mind and most military men clearly saw that the mastering of the air meant the mastery for nations. The United States has this mastery within her grasp." [79]. Resolutions were adopted to call on Congress to make liberal appropriations for the development of aeronautics, and committees were appointed to call on President Roosevelt and Congress to urge action [80].

6.10 America's Native Pride Aroused by Foreign Competitors

After a short-lived contract with Archdeacon and then with Louis Blériot, **Charles** (1882–1912) and **Gabriel** (1880–1973) **Voisin** launched the aviation industry in early November 1906 in France, when they established the world's first aircraft factory, "Appareils d'Aviation Les Frères Voisin," in a Parisian suburb (Fig. 6.13).

Gabriel was a handsome man, with the eyes of a sparrow hawk and a black moustache. His equally good-looking brother Charles managed the business until 1912, when he was killed in an automobile accident.

In cooperation with Leon Delagrange, Henry Farman, and Louis Blériot they formed what would become known as the "mighty French school." The

Fig. 6.13 In the workshop of the most advanced aircraft factory, the Voisin factory in a Paris suburb. *London Illustrated News*, September 26, 1908

brothers and their manager, M. Colliex, made no secret of the fact that they based their work on pioneers like Chanute, Lilienthal, Langley and others, and they never missed an opportunity of utilizing information or data on which they could lay their hands. But they also did their own research, using an "artificial wind apparatus" with which they tested their designs on a small scale before finalizing it. And to teach their clients the art of flying prior to guiding a powered aeroplane, the Voisins supplied an improved Chanute-type glider gratis to every new customer [81] (Fig. 6.14).

Envisioning the aircraft to become a powerful weapon, Glassford consulted with Chanute, "The military leadership had recently taken interest in aerial navigation however the department had the most-crude ideas in the matter and therefore desired to enlist the cooperation of those who had studied the progress of aviation … I shall be glad to learn from you whether something can be informally said as to your views concerning an arrangement with the U.S. Government." Chanute was pleased to be asked and responded: "I think that the dirigible and the aeroplane should be taken up by our government." [82].

One month later, on August 1, 1907, the "Aeronautical Division" was established within the Signal Corps; its members were to take a leading role in all matters pertaining to military ballooning, air machines, and kindred subjects [80]. In his report to the Board on October 10, General Allen expressed his skepticism of the actual value of an aeroplane. He considered

Fig. 6.14 Charles Voisin flying their Chanute-type glider in 1907. Author's collection

the aeroplane less practical, as the "engineer" [or pilot] needed to lay prone and was exposed to a wind stream greater than 30 miles an hour, making it almost impossible to use field glasses, maps, or even draw sketches. And, "For the purpose of dropping explosives on an enemy, an aeroplane is hardly suitable, when traveling at thirty miles an hour at an altitude of 4,000 feet, to drop a projectile close to half a mile from the target."

Even though the aeroplane lacked range and load-carrying capability when compared with the dirigible, the Board of Ordnance and Fortification instructed Allen to solicit bids for the delivery of one heavier-than-air aeroplane and one lighter-than-air dirigible. In true military fashion, the Division kept their actions confidential, but Allen did consult with others on the practicality and reasonableness of their envisioned specifications. Alexander Graham Bell shared his copy with AEA members, and Curtiss simply stated that he was prepared to take contracts for building any type of flying machine using his newly designed engine. The Wrights only offered a few editorial changes. Aero Club of America members discussed the tentative specifications, with Captain Lovelace and Israel Ludlow agreeing that they would be acceptable; Herring thought that a practical machine could be had, but bids could not be drawn from any of the most advanced workers with these specifications; William R. Kimball remarked that the specifications were not impractical but probably too severe to attract the conservative element of manufacturers in view of the newness of the art; and Peter Cooper Hewitt suggested that the minimum speed of 40 miles should be raised to 50 or even 60 miles per hour [83, 84].

Wilbur, who had just returned from France, mentioned to Chanute his upcoming meeting with the Board of Ordnance. Having received a copy of the proposed specifications from Glassford, Chanute responded tersely, "… I hope that you will succeed in making an arrangement with the U.S." [85]. He also expressed his regret that the brothers did not make more positive final sales. "The papers meanwhile represent that Esnault-Pelterie and Farman are making great progress, so that it may take less than the five years for one of them to catch up with you. My feeling would be to sell, even though you do not get your original price, which I always thought too high." [86]. Wilbur was not concerned about the European competition. "I have confidence that our prediction will still stand solid after the scythe of time has reaped several fresh crops of French predictions. This judgment is based on a consideration of the net advance shown by a comparison of the design of the Santos machine of 1906 with any one of the 1907 machines." [87]. Trying to calm Wilbur down, Chanute responded, "I am glad that you consider the net advance of your competitors to be less threatening to your interests than

I feared, and that the condition of your own negotiations is satisfactory to you." [88]. Chanute then mentioned Archdeacon's remarks at the Aéro-Club dinner shortly after Farman's flight of October 26: "Today the famous Wright brothers may claim all they wish. If it is true that they were the first to fly through the air, they will not have the glory of it before history … If governments had purchased from the Wright brothers their supposed invention for a million francs, they would have thrown a million out of the window, as nothing will prevent engineers from copying existing machines which shall have functioned the best." [89].

America was contributing to the progress in aerial navigation, but the eyes of the public in Europe were focused on the failures and successes of Santos-Dumont and Vuia in 1906, and of the Voisin brothers, Blériot, Henri Farman, Pelterie and Delagrange in 1907. Their original short hops through the air became longer and longer and the attending crowds became larger and larger and the cheering became louder and louder. Aviation was here to stay and grow!

7

Internationalism, Idealism and Materialism

I have no doubt that the Wright brothers have flown with their aeroplane, but I think if they still have a great secret of aerial navigation, the French Government is not inclined to pay 1,000,000 Francs for that secret, which will become public to the world the first time that the machine will fly in France.

Alexander Graham Bell at Aldine Association Dinner, April 1908

Internationalism was, and still is, one of the striking features of the development in aviation; enthusiasts from around the globe studied stability and balance, methods of taking off and landing, best material and shape for the wing, power plant and propeller, the various types of construction using different known materials, and the relative merits of monoplane, biplane or triplane. With wide open eyes and minds, they studied what others were doing to then improve their own design. In short, it would seem that the Voisin brothers copied Chanute, Farman copied the Voisins, Roger Sommer copied Henry Farman, Maurice Farman copied Sommer, and so on.

To grow the aviation industry, the French Military Chief of Staff set part of the budget aside for the purchase of aircraft from French workshops, rekindling a certain patriotism. Gabriel Voisin made it clear that "France should not bow before so naïve a claim that America was the birthplace of aviation. None of our great men borrowed anything from the men in Dayton. The history of mechanical flight starts in fact with the research by Mouillard and Pénaud" [1]. And an editorial in *Le Matin* clarified, "Many people, including some famous aviators, believe that we can claim the biplane as the

© The Author(s), under exclusive license to Springer Nature
Switzerland AG 2023
S. Short, *Flight Not Improbable*, Springer Biographies,
https://doi.org/10.1007/978-3-031-24430-8_7

beginning of modern aviation; it finally allowed men to fly … We now have the American school, represented by the Wright brothers, and the French school, represented by the Voisin brothers. But the essential principle of the two schools lies in the invention of Octave Chanute" [2]. Being borne in France, of French parents, he was, and sometime still is, considered a French National. He was the "*père du biplan*", or the father of the biplane. Yes, he was born in France, but he was as much an American as any of the thousands of successful citizens of foreign birth, living and working in America. Again and again, Chanute had to explain: "Feel honored that my name was suggested, but deprecate publicity, especially in a connection which seems to differentiate one nationality from another, while we are all American and have been absorbed in the mass" [3]. He considered himself a true American.

7.1   Henry Farman and his Record Aeroplane Flight

The ball was dropped for the first time at Times Square in New York City on January 1, 1908, signifying the New Year. Henry Farman had special plans for this new year, he made a one kilometer flight in 1 minute 28 seconds in his Voisin-Farman I biplane at Issy-les-Moulineaux in France, and two days later, on January 13, he made the first full circle flight in Europe, winning the coveted Deutsch-Archdeacon prize of 50,000 Francs. He shared his story with *The New York Times*, explaining, "I am not an inventor. I have used the principles of Langley, Lilienthal, Chanute and then added something of my own.

"I conceived the plan of building an aeroplane after the Chanute system a long time ago. Mentioning the matter to some friends, they all declared it impossible to fly with any heavier-then-air apparatus until someone builds a motor weighing not more than two pounds for one horsepower. I felt confident that an aeroplane, built on scientific lines, weighing not more than 300 pounds, would prove successful. "In the mean time I had consulted the Voisin Brothers and took part in some experiments with gliders. These experiments were my first lessons in aviation. Traveling back to Paris I gave a contract to Mr. Voisin to build an aeroplane for me; they guaranteed that I should be able to fly at least one kilometer with it. While I was working on the plans for my machine, I consulted several prominent people in the aeronautical world, but everywhere I met with the same reply: impossible. I could find no encouragement and I was informed that an aeroplane of the size which I proposed to build could never maintain its equilibrium in the air and would be certain of coming to grief, if I attempted to turn a corner. Nevertheless, I set to work, and on May 20, 1907, the building of the apparatus patterned after the system Chanute was begun. Early in September the aeroplane was delivered to my shed at Issy-les-Moulineaux, and, after fitting an Antoinette motor, I took it out for my first trial.

"I spent many days experimenting with it, but I could not rise an inch from the ground. Then I made my first flight on October 1. Indeed, it may sound like exaggeration to call this a flight, for I only rose to a height of eighteen inches from the earth and flew a few yards. Thus, encouraged I continued almost daily until October 15, when I made my first successful ascent. The wind was blowing very strong that day, but I decided to see how I should fare against a stiff head breeze. I flew 280 meters, but no commissioner of the Aéro-Club was present, so the flight could not be considered official. I was out again three days later, October 18, and made several short flights of one, two, three and four hundred yards each. I was so satisfied with the working of my aeroplane that I asked the commission of the Aéro-Club to attend on October 20, when I won the Archdeacon Cup, which was held by Mr. Santos-Dumont. My longest flight then was 770 meters, from one end of the grounds to the other. I was satisfied that I could rely on the aeroplane to travel in a straight line, so I set to work to learn how to turn the machine in the air. Then, on October 28, I tried making a circle.

"On the first few attempts the machine responded to the rudder remarkably well. I met with an accident however in the afternoon, which resulted in the smashing of the motor and the propeller. I had a new motor of the same make and power fitted, but it did not give satisfaction. The weather at this time was very bad, and further trials were out of the question. Using the time, I altered the shape of the machine, making the tailpiece much smaller and lighter. On Saturday last, despite the cold, I took the aeroplane out in the afternoon and made a flight of nearly two kilometers in a circle and I called out the commission of the Aero Club for this morning.

"Today I was on the field early making my preparations, and when the commission declared that they were ready I got into the seat of the aeroplane to start. I must admit that I felt a little excited, but at no time throughout the flight did I feel nervous. I set the motor going and then shouted to the men holding the machine behind to let go. I soon felt the earth moving away from me. I regulated the elevating plane and rose higher still. I had to use some caution crossing the starting line. Having got safely away from the starting point, I headed for the flag on the other side of the field, which I had to fly around. My machine was working beautifully, and I was determined to win the prize whatever might happen. My confidence in the aeroplane was not

Fig. 7.1 Henry Farman's gasless airship that won him the $10,000 Prize. *The American Aeronaut,* January 1908

misplaced, for in coming around the flag, with half of the journey accomplished, the machine was flying better than ever. It was only a matter of a few seconds until I was up again in front of the starting line, which I had to cross again on the return. As I was approaching the end, I felt like showing my friends that I was going to win; I put up my hand and waved to them. It was then, that they gave a cheer; the line was crossed and I had won the much-coveted prize" [4].

Yes, Farman had flown only for one kilometer in a circular flight, but no one, except the invisible Wrights, had accomplished such a feat before. Actually, Farman's flight was nearer to 1.5 kilometers (0.9 miles), as his aeroplane did not have much lateral control. He had to skid in a wide, graceful curve around the markers with the crowd cheering their support loudly (Fig. 7.1).

7.2   The U.S. Government Steps into the Air Age

The Signal Corps advertised for aeroplane bids on December 23, 1907, and several newspapers published the news. "The aeroplane has been formally recognized as a probable instrument of warfare. Brig.-Gen. James Allen, Chief Signal Officer of the Army, today issued an advertisement and specifications asking for sealed proposals to be submitted to the War Department on or before February 1, 1908, for furnishing the Signal Corps with a flying machine of the 'heavier-than-air' kind" [5]. Bidders were to submit drawings and descriptions of all parts of their machine, prove that they held patents on all proprietary features, demonstrate that their airplane was capable of carrying two persons with a combined weight of 350 pounds and sufficient fuel for 125 miles, travel at least forty miles per hour average with and against the wind, fly at least one hour and then land the airplane so that it could immediately take off again, steer in all directions without difficulty and be under perfect control at all times, provide a simple starting device, and the operator and passenger were to be seated.

Chanute thought these requirements marked a new era in aviation. "They may bring such new elements into war as to render it far less frequent, and they may develop new uses to make the world better and happier." But he also wondered, if the Signal Corps asked for impossible things or if they did not know any better. As far as he knew there was no aeroplane in the world which could fulfill these specifications at that time.

Three weeks later, the Signal Corps advertised for bids on a dirigible, to also carry two persons and travel at least twenty miles per hour. The Corps would furnish the cover material (silk, covered with an aluminum coating, requiring no varnish) for the gas bag and required it to be inflated with hydrogen gas. At the same time, Allen also announced that bidders for the aeroplane were permitted to "preserve as confidential any features of the machine which he wishes to keep secret." The editor of *Aeronautics* commented dryly, "What is the use of submitting plans at all? The greatest value in the plan is the point, which the inventor desires to keep secret. Flying machines always fly—on paper …".

By February 1, the Signal Corps had received forty-one bids for an aeroplane [6], but only three submitters put up the required 10% of the proposed purchase price as a bond. A patriotic James F. Scott of Chicago offered to deliver his plane in 185 days and sell it for $10,000, or the estimated cost of material. Augustus Herring stated that he would deliver his machine in 180 days at a cost of $20,000. And the Wrights were to deliver their machine in 200 days at a cost of $25,000, significantly less than their original asking price of $200,000. Having received more than one bid presented a certain problem, as the government is obligated to accept the lowest bid. Secretary of War, William H. Taft, suggested to accept the three bids, and the Board of Ordnance agreed to provide funding to purchase the three machines.

One week later, the eleven bids for dirigibles were opened, and Baldwin's airship, driven by a Curtiss engine, was the winner. It was to be delivered in early August [7].

The Board of Ordnance anticipated receiving at least one aeroplane, but Allen thought, "if the Wright brothers or Herring should not succeed in carrying out their contracts with the government, it is not at all unlikely that we will bring out the old Langley machine to conduct experiments" [8].

The recently elected Smithsonian Secretary, Charles Walcott, was not keen on anyone flying Langley's aerodrome, but he did want the public to know of the Smithsonian's contribution. He invited a small party of correspondents, including Augustus Post and Albert Zahm, to inspect the fully restored aerodrome. During the viewing, Dr. Cyrus Adler, Assistant Secretary in charge of Library and Exchanges, made it clear, "It [the Langley aerodrome] has made history, it belongs to history. It is a relic and should be preserved as such along with the others produced by Prof. Langley, the pioneer of mechanical flight." One of the reporters described the "Buzzard" [9] which did not look as though it had ever been in the Potomac. "The framework and the wings, which were the most damaged in 1903, were repaired and polished wherever injured. The main frame of the machine is oblong with rounded corners, about twenty-five feet long by four feet wide. Midway of this frame are the gearings for the two propellers that work amidships. At the forward and aft ends, there are planes for attaching the great bird-like wings, of which there are two pairs. Amidships there is the engine and the seat for the operator. Roughly speaking, the main frame is of two-inch tubing that looks as heavy as steam pipe, but which in reality is but little thicker than writing paper. The bare frame as it hung in the workshop looked as though it ought to weigh 500 pounds, in reality it weighed a little over 150. The machine with all equipment and the operator weighed 830 pounds."

In mid-November, Walcott had one more question for Chanute, "Mr. Manly has suggested that he be permitted to make trial tests of the Langley machine at some future time. I write to ask whether in your judgment it would be wise to have an attempt made to fly with it." Chanute responded cautiously, "I have never seen this machine but I suppose that I understand it fairly well from descriptions. My judgment is that it would probably be broken when alightening on hard ground and possibly when alightening on water, although the operator might not be hurt in either case. If the Institution does not mind taking this risk and if suitable arrangement can be made about the expense, I believe that it would be desirable to make the test, in order to demonstrate that the Langley machine was competent to fly and might have put our government in possession of a type of flying machine, which, although inferior to that of the Wrights, might have been evolved into an effective scouting instrument" [10].

7.3   A Little Practice is Needed to Aviate

With their Signal Corps bid accepted and a contract for the formation of a French Wright company signed, Wilbur shared his thoughts about the future with Chanute. "It is not possible to know just how the financial end of our flyer will come out until the event shows, but I think the prospects are as good now as at any time since Fordyce and Letellier spoiled things for us. It would have been a great advantage if we could have organized our company before beginning business, but we will still do very well unless we are much mistaken. The governments will each spend many times $200,000 on flying machines within the next 15 years, and we think we will have patents, knowledge, and business associations sufficient to ensure a good share of it coming our way. The belief that others would succeed in attaining results equal or superior to ours has of course been one of the serious obstacles we have always had to contend with. It was based on the general principle that what one can do, others can do" [11].

To successfully demonstrate their Flyer, both in Europe and the United States, and not having flown for more than two years, the brothers needed to refresh their piloting skills. Wilbur arrived at their old flying ground at Kitty Hawk on April 9; Charles W. Furnas, who worked for the brothers as a mechanic, arrived a week later, and Orville and the crates with the Flyer arrived on April 25.

Even though the Outer Banks were at the end of the world, the free-lance reporter Bruce Salley was the first to submit a story to the wire service. This enticed other reporters to come, which resulted in an amusing struggle: the brothers were determined that no description of their machine leaked to the public, and the reporters were determined that their reports reach the public as fast as possible [12]. The Wrights then announced they would not operate their Flyer if reporters were nearby, so they simply moved into the background and watched through field glasses

One of the reporters hiding in the bushes, Arnold Kruckman, recalled 50 years later, "It was quite an experience for a young man. *The World* didn't know for sure whether it wanted to send me way down to the Carolinas just to watch a couple of daredevils flap their wings," but he received the assignment. After watching the Wrights "revve up" their little aeroplane several times, he wrote what he considered a prize-winning thriller and had no trouble talking the clerk at the Weather Bureau station into transmitting the text. It did not take long for the chief editor of *The World* to wire back: "The brandy must be pretty stiff down there" [13], Kruckman's report was not published.

The climax of the flying activity came on May 14; Wilbur made a short flight with Furnas as passenger, both in a sitting position as required in the government contract. Orville flew with Furnas next, making a complete circle of 2½ miles in 4 minutes. Then Wilbur made a flight of eight miles, but wrecked the machine on landing. Receiving only minor scratches, Wilbur explained that he made a mistake by "grasping the wrong lever of the steering gear recently installed, this being the one which directed the course downward, while the other lever directed it upward, a mistake very likely to occur with a nervous man operating a new arrangement" [14].

Reading about the Wrights latest flying in the Australian newspapers, Lawrence Hargrave wrote to Chanute, that he "simply could not see what the Wright brothers have that is new and needed to be closely guarded." He was sure, that "no corner can ever be made in aeroplane machines, the public and the makers will all reap the full benefit of free competition" [15]. Chanute may have smiled reading the letter, but responded laconically, "I quite agree with your view that no 'corners' can be made in flying machines but my friends, the Wrights, have thought differently and we agree to disagree" [16].

7.4   Aero Club of America on the Defensive

How to increase the interest in the sport was a frequently discussed topic at the Aero Club's monthly meetings. Most members were enthusiastic about ballooning, but others were more interested in heavier-than-air craft. At the February 1908 meeting the discussion centered on the actual achievements of aeroplanes. Opinions varied from doubt as to the truth of any report on the experiments by the Wrights to the belief that they had shown thus far only a small inkling of what they were able to accomplish. Albert Triaca explained that, "Octave Chanute saw them fly their apparatus twenty-four miles some years ago. They are extremely wary that their secret be discovered, and you cannot blame them."

To advance the knowledge of aeroplanes, the Aero Club Board agreed to establish the "Aviation Section," a club within the club of armchair aeronauts. Triaca, who had earned his ballooning license in France prior to coming to America, agreed to chair the new group and proposed to establish a school, as was done in France, where regular people, interested in learning more about flying, could be initiated into the mysteries of aeronautics [17]. Furthermore, James Means stressed that enthusiasts should start with gliders, "The importance of the motorless glider is two-fold. First, the skill which may be acquired by its use prepares men to encounter the dangers of powered flight with the least possible risk. Secondly, its comparatively low cost will enable many experimenters to enter the field who would otherwise be barred by lack of funds" [18].

Using the drawings from the Voisin brothers, Triaca had two Chanute-type gliders built, and one member offered the use of his estate on the northern shore of Long Island for a mini flying affair on the April 11–12, 1908, weekend. A *New York Times* reporter joined the "aviating" activity. "The investigators arrived early on Saturday morning. They carried with them material for their machine. In a few hours Daniel Blaine had put together the first glider of the club. This consisted of a framework 18 feet long and 5½ feet high, as it lay on the ground, with a width of 5 feet. Muslin covers were stretched over the upper and lower planes of the framework. In the center was space for a man to hang suspended by his arms. The top cover was over his head, while the bottom one would be on a level with his arms as the machine was in the air. The whole weighed 45 pounds. The machine was carried to the top of a hill near the place of its birth, where there was a stiff breeze blowing. A member of the party was placed in the central position, while one lifted the machine at the head and one at the foot. The operator stood on his own legs with the machine partially supported in the air by the two others. At the word go, all three began to run. Within a few moments the machine would glide through the air and the two assistants would let go at the shout of the passenger" [19]. Several members made gliding flights from thirty to forty feet in length and about ten feet above the ground. Everyone had fun but felt a bit stiff and sore the next day (Fig. 7.2).

To plan for the future, members of the "Aviation Section" had two projects in mind: (1) the necessity of substantial prizes, similar to what was offered in France, to stimulate inventors and (2) locate a large tract of land near the city, available for experiments, and where sheds and machine shops could be built. They approached Chanute, Graham Bell, James Means and Lawrence Rotch to help secure a prize fund, where 250 subscribers would each pledge $100. Bell was much in favor and stated, "We know there are a dozen types of aeroplanes now in the experimental stages, which may prove practical within a short time. A prize fund of $25,000 will encourage inventors to bring forward the products of their hands. This competitive prize might stimulate $500,000 worth of capital to get behind the inventions, which now are in the formative stage, and bring their machines to completion" [20].

Hearing that Chanute would be in town, Triaca asked him to treat members to a talk on "*Future Uses of Aerial Navigation*" at the next Aero Club meeting in April [21]. As a starter, Chanute told members that he had watched Orville fly a few years earlier, and he then offered suggestions what

Fig. 7.2 Aero Club takes wing on a glider to build up expertise prior to stepping into powered airplanes. Note that Augustus Post helped launch the glider. Glenn H. Curtiss Museum, Hammondsport, NY

to look for when selecting a field for their flying experiments. Herring interrupted the speaker several times and told the audience that grounds suitable for motor machines would not be suitable for gliders. Chanute then explained that there are several ways to launch a glider into the air when there are no hills available; he preferred using a wheeled carriage with a cable, attached to an electric motor, as Avery had used at the St. Louis Fair. But the main thing was to get into the air and practice, practice, practice.

As a follow-up at the next monthly meeting, Triaca suggested, "With Mr. Chanute being recognized as the foremost figure in the aviation world, it would seem proper that he should be represented among the honorary members of the American club" [22]. At the annual meeting on November 2, Chanute, but also Henry Farman, and Orville and Wilbur Wright, were elected honorary members.

Even though the Aero Club Board had agreed to support the work with aeroplanes, nothing happened. Lee S. Burridge explained to the press, "The governors would adopt none of the suggestions of members interested in pushing aviation to the front. Hot air is all right, but we are not going to fly with hot air alone. We want to do more" [23]. Another member submitted an opinion piece to *The Sun*, "It was my desire to see the Aero Club of America follow the splendid and glorious work of Langley, Chanute and the Wrights, but I found no support and no interest in aviation at all. We must work together. It must be a cooperation of wealth, brain, experience and work for the interest of the art and not for the ambition of the few. If we have a $10,000 prize, certainly at least ten machines will be built to win this prize, and those will be new ideas, with new and different solutions" [24]. With no support coming from the Aero Club Board, nine prominent members then formed the Aeronautic Society of New York. The "object was to advance the art of aeronautics to the fullest extent within its powers by stimulating interest therein, and to assist its members in carrying on experiments; to encourage inventors to experiment along aeronautic lines; to aid experimenters by providing the most necessary facilities to carry on their work; and last but not least, to bring together those working in the various fields of aeronautic endeavor so that each individual may have the advice and co-operation of others."

The American public needed to know that Europeans were ahead in everything related to aeronautics, so the Aeronautic Society invited Henry Farman to come and demonstrate his aeroplane and flying skills at the Brighton Beach Race Track near New York City. He arrived in late July 1908, while preparations were made at Fort Myer for the government trials. Farman, just like Chanute, believed that "It is better for others to see what you are doing and for you to see what they are doing, each improving by noticing the mistakes of the other" [25]. Asked by a reporter, if his machine was better than the Wrights' Farman responded, "Ah! I could not say. We aviators are all working in the same way for the same result. Delagrange and I have worked together for years. I do not believe in secrecy. You must remember that we foreigners owe credit to Octave Chanute, of Chicago, for the basic principles of our apparatus. Of course we have made changes, and we owe much to the Wright brothers, pioneers after Mr. Chanute" [26].

7.5   The Aerial Experiment Association and the Scientific American Trophy

After learning the basic skills of building and flying a Chanute-type glider in early 1908, members of the Aerial Experiment Association (AEA) considered themselves skillful enough to design, build and fly three different aerodromes. Army Lt. Thomas E. Selfridge, designed the first aerodrome, using the left-over red silk for the wings from Bell's *Cygnet* kite project a few months earlier. When the resulting *Red Wing* was ready to be flown, Selfridge was called away for military duty, and Casey Baldwin agreed to make the first flight. Climbing through the bamboo framework and revving up the engine, he lurched forward, skittering across the ice "like a scared rabbit," Curtiss reported [27]. *Red Wing* flew straight and true, rising roughly 20 feet above the ice-covered Keuka Lake over a distance of about 320 feet. This was the first public demonstration of an aeroplane flying under its own power in the United States. A few days later, Baldwin made a second flight and flew for about 40 yards, when one wing tip was caught on the ice. Selfridge reported that "we smashed the Red Wing to match word and rags yesterday," but neither the operator nor the engine was hurt. The self-taught aeronauts were learning their lessons the hard way (Fig. 7.3).

The AEA's next aerodrome, *White Wing*, was similar in construction, but was fitted with wheels to run over the ground to pick up speed for take-off. Having read the reports by Robert Esnault-Pelterie, members designed "horizontal rudders" (Bell's terminology), installed on the four wingtips, which entered history as "ailerons." These moveable, triangular panels at all four wingtips were to provide the roll control that *Red Wing* had lacked. On May 19, Baldwin piloted *White Wing* on its maiden flight, and Selfridge flew next. Bell was quick in assuring the public, that "Aerial Navigation was solved. Flying machines were not a dream of the future anymore." And Curtiss submitted an ad to *Aeronautics*, that his company was in the "position to accept orders for heavier-than-air flying machines of the aeroplane type, built to carry one man, start with a 200-foot run from any reasonably smooth surface of ground and alight without damage in any open field. Machine

Fig. 7.3 *Red Wing* is being readied for the first public aeroplane flight in the United States on March 9, 1908. *Aeronautics*, April 1908

to be demonstrated in a flight of one kilometer. Deliveries can be made in 60 days. Price $5,000" [28].

Having seen a photo of the *Scientific American* Trophy, valued at $2,500, the 30-year-old champion motorcyclist thought how nice it would be to bring this large silver trophy to Hammondsport. As no one earned the trophy in 1907, Charles Munn, owner of the *Scientific American*, submitted an editorial to his magazine [29]. "Despite the fact that many inventors in the United States are wrestling with the problem of aerial navigation by means of a true dynamic flying machine, no public flight has been made in this country with such a machine up to the present time. The most advanced knowledge of heavier-than-air navigation seems to be held by two young western experimenters. These men have undoubtedly made flights with their aeroplane, and these flights have been witnessed by a considerable number of people. The general appearance of their machine is known, and other experimenters are making good progress along the same lines." Interestingly, Munn's editorial was published six weeks after AEA members achieved the first public aeroplane flight, taking off from the frozen Keuka Lake, just north of Hammondsport, New York, which was widely discussed in the press [30]. And Munn now urged Orville to step forward and win the trophy; he was not interested. Maybe Orville wanted nothing to do with a trophy that showed Langley's aerodrome, or maybe it was not as simple to install wheels on their Flyer to launch it without the derrick?

This large silver trophy was spectacular, so Curtiss took responsibility for the AEA's next aerodrome; in mid-June he made good flights with the *June Bug*, the longest was 3,420 feet, or 140 feet more than 1 kilometer.

The public's perception of an engineering achievement was almost as important as the achievement itself. As the *June Bug* performed well, Curtiss suggested to Bell to make a trial flight for the trophy on Independence Day, a fitting date for an American machine to fly for a trophy offered by the oldest American mechanical journal for encouraging the development of aeronautics. Everyone agreed, and Selfridge contacted the Aero Club to request observers to come to Hammondsport to witness the first official trial for the *Scientific American* Trophy on July 4, Independence Day. Sensing political problems, Curtiss and Selfridge traveled to New York City on June 28. Baldwin and Manly joined them to discuss the proposed flight first with Munn, and then with Augustus Post of the Aero Club, neither was anxious to go along with their proposal. Selfridge recorded in his diary, "Their position absolutely untenable. Rather discouraging. Made appointment for 2:30. Had long talk during which *Scientific American* came to our way of thinking.

Went out to Empire Hotel. Dined with Post, went to Aero Club about 9:30. Completed arrangements for contests on July 4 at Hammondsport" [31].

Chanute who had chaperoned his daughters to New York for their trip to Europe, happened to be at the Aero Club, when Curtiss and Selfridge arrived; hearing of the flight on Independence Day sounded exciting. He told a reporter, "The success of the French aeronauts and the excellent results achieved by the leading members of the Aerial Experiment Association at Hammondsport have removed all doubt that the worst problems in successful navigation of the air have been overcome." He regretted that he needed to head back to Chicago, as he would have liked to witness the flight [32].

Forced to be temporarily content with the old-fashioned method of travel on a railroad train, members of the Aero Club of America and the Aeronautic Society left New York City in a special sleeper car for Hammondsport [33]. Locals from the Hammondsport area came with picnic lunches and high hopes to watch their boy make history. The Pleasant Valley Wine Company added to the festivities with a buffet lunch and samples of its products.

Waiting all day for rain showers and thunderstorms to stop and the wind to die down, the *June Bug* was rolled out of its tent in early evening of July 4 under the watchful eyes of the contest committee. "The first attempt was a very pretty flight but the machine was not on its best behavior, and landing was affected several hundred feet short of a kilometer. Upon investigation it was learned that in assembling the machine the tail had been set at not quite the right angle" [34]. On the second flight Curtiss flew over the excited spectators, vineyards and the red flag marking the end of the measured course, and then flew 600 yards further. The official 1-kilometer flight was made in 1 minute 17 seconds at a speed of 33.1 mph. This flight entitled the AEA to have its name inscribed on the trophy as the first competitor.

Chanute saw the write-up in the *Chicago Tribune* and was thrilled that Curtiss succeeded and had won the trophy, but Orville Wright was not pleased reading the report in the Dayton paper. Almost two weeks later, he wrote to Curtiss, "I learn from the Scientific American that your June Bug has movable surfaces at the tips of the wings, adjustable to different angles on the right and left sides for maintaining the lateral balance." And he reminded Curtiss that he and his brother had informed the AEA in a letter to Selfridge in January 1908 of their patent, and "We did not intend, of course, to give permission to use the patented features of our machine for exhibitions or in a commercial way" [35]. Neither Curtiss nor Bell was concerned, as the AEA's way to achieve lateral control was based on a different principal than the Wright's method of wing warping. Also, not having seen technical drawings or even a clear photo of the Wright Flyer, it could not have been a source of inspiration for AEA members.

7.6   A Whole Lot of Babbling Going on

Reading both sides of the correspondence between the Wright siblings in the summer of 1908, one can detect some tension building up; maybe the uncertainty of not being able to sell their Flyer began to wear on the brothers?

When Wilbur left for France, he had instructed Orville to keep him informed of the happenings at home and get the Flyer ready for the government trials. A few days after arriving in Paris, Wilbur wrote to his younger brother, "I am a little surprised that I have no letter from you yet. Anything mailed at Eliz. City, Norfolk, Washington, or even on your arrival at Dayton, should be here. The fact that the newspapers say nothing of a visit to Washington leads me to fear you did not stop there when returning home …" Orville responded a few days later, discussed his visit with the patent layer in Springfield, his work on the Flyer and Chanute's article in the June *Independent* magazine [14], "criticizing our business methods, and saying that we have always demanded an exorbitant price." Even Milton Wright, who normally stayed out of his sons' work, sent Wilbur extracts from the *Independent* article with a fatherly advice: "There seems to be a little meddling with your prices and judiciary process on your ideas, in the foregoing, and a little assumption of your change of view, and a little acerbity that you did not train under his advice. But age and premiership are to be considered in the case. Better no rupture with a former friend" [36].

In early July, Wilbur shared his unhappiness about Ferber, whom he had described half a year earlier as a "man France was lucky to have," but now he was "the man largely responsible for the failure of their final negotiations," double-faced and bitterly hostile. And Orville complained about Flint & Company and their European agent. "I am so completely disgusted with them that I would like to sever our connection … they have been so tricky that it keeps us busy watching them."

Acrimony turned internal as well. Orville complained of Wilbur's scanty correspondence, and Wilbur carped that Orville didn't understand the pressures he was under. Complaining to his father on July 20, "You people in Dayton seem to lack perspicacity." Trying to keep peace in the family, Katharine wrote to her brother in France that it was no fun to receive his letters "when they are all complaints from one end to the other. It may be hard for you over there, but you must remember that Orv has a terrible load on him and you both ought to be more considerate" [37].

As Orville was to write an article for *Century* magazine, Wilbur reminded him, "I have been taking pains to have the chief points of our patents well published to let the general public become accustomed to linking these ideas with us before others attempt to steal them ... I hope you bring out in your Century article the fundamental difference between our methods & those of Chanute, and call attention to the fact that we have obtained very broad patents on the general combinations as well as the particular constructions employed ...".

Receiving the latest issue of *Illustrierte Aeronautische Mitteilungen*, Orville saw another article by Chanute and reported that he "again criticizes our business methods, says we have spent 2 years in fruitless negotiations, because we have asked a ridiculously high price, but now we have gone to the other extreme in making a price to our own government. He predicts that Herring will fail, but that we will succeed, unless we meet with an accident. He seems to be endeavoring to make our business more difficult ..." [38]. Orville must have had trouble with the German language, as the text he mentioned was not part of the article. Actually, Chanute discussed briefly the Wrights' recent flying at Kitty Hawk, which unfortunately ended in a crash, and provided detailed information on the two other competitors who were to sell their aeroplane to the Signal Corps. In his next letter to Wilbur, Chanute wrote, "I had occasion to say something about you in a paper to the *Aeronautical Society of Great Britain* and to the *Illustrierte Aeronautische Mitteilungen*. Of the latter, Col. Moedebeck writes that he will thereby be able to rehabilitate in Europa the good meaning for the Wright brothers upon their merits." Life was surely not running as smoothly as hoped.

7.7   Wilbur Wright Flies in France

French pilots eagerly pushed ahead, Leon Delagrange set a duration record on May 30, staying aloft for 15 minutes 26 seconds, flying for almost eight miles near Rome. Wilbur who had just arrived in France, told a *New York Times* reporter, "We are not worried. Our confidence in our leadership rests upon the essential difference between our machine and those used in Europe" [39].

Looking for a place to work and fly, Leon Bollée offered Wilbur his facilities near Le Mans (about 130 miles southeast of Paris) for the remainder of the season. Interestingly this site is within a few miles of the very spot where Montgolfier achieved the feat of giving to the world the first balloon flight in 1783. The facilities looked good, and Wilbur moved the crates containing the Flyer, starting rail and derrick, which had been in storage at Le Havre for the past year, to Le Mans. But opening the boxes brought unexpected surprises; there was considerable damage to the aeroplane, either due to improper packing by Orville or by custom officials. It took some serious effort, but everything was finally ready on August 8; Wilbur wore his gray suit, starched white collar and green cap, turned backward, so it would not blow off in flight. The evening was calm, and so, outwardly, was he [40]. Much was riding on the success of this first flight. He had crashed the Flyer six weeks earlier, and if he did so now, the French trials would be over before they even began.

Spectators watched from the grandstand, as the Flyer shot down its rail and rose to 30 feet, higher than most French aviators had flown. Wilbur banked into a half circle and brought the plane back to where he had taken off less than two minutes earlier. Even though the flight was short, spectators ran across the field to shake his hand, including some of the French aviators who had dismissed him as a charlatan just weeks earlier, and transforming Wilbur from "*le bluffeur*" to a celebrated aviator (Fig. 7.4).

Wilbur's new fame triggered curiosity about Chanute's involvement. Paul Renard wrote to Chicago simply asking how much help the Wrights had actually received from him; Chanute responded a few days later. "As in almost all inventions, the later aviators have profited legitimately from the prior labors of their predecessors. Wenham had proposed the biplane glider; I had added

Fig. 7.4 Wilbur Wright dazzled the crowd in France. *St Louis Post-Dispatch,* September 8, 1908

to it a reinforced frame and published my designs and the account of my experiments of 1896 and 1897 in accordance with Lilienthal's method. After reading my book, *Progress in Flying Machines*, as well as my articles in the *Aeronautical Annuals* of 1895, 1896 and 1897, Wilbur Wright contacted me in May 1900, stating that he wanted to experiment without financial motive. He explained his method of experimentation from which I dissuaded him, and he also asked where he might find a favorable location for his experiments. I cannot describe how useful I was to the Wrights. Ideas are absorbed unconsciously, and the important thing is to choose the good ones." Continuing his letter, "As I have stated in public, I had improved the two-surface glider by adding a reinforced frame and had demonstrated with my experiments that there was little danger of accident. The Wrights placed the vertical control device in the front, like Maxim. I had placed it at the rear, where it is possibly less effective. They placed themselves prone on the machine; I had indicated in my writings, that this position would reduce the resistance but I had never tried it. They added the warping of the wings, which I believe was quite personal to them, although Mouillard, at my suggestion, had patented it in the United States on May 18, 1897. But the Wrights' most important merits lie in applying a motor to the biplane glider. They made many laboratory tests of lifting surfaces, and I helped calculate the resulting data. They designed a propeller with a performance superior to ordinary propellers. They designed a very effective transmission and a reliable motor. They built all this with their own hands at their own risk and peril" [41]. Renard was satisfied.

A reporter of *L'Auto* asked Wilbur basically the same question, but from a little different angle, "This is not to deny a legend, but to correct the truth regarding the role of the great American scientist Chanute in the discovery that concerns us. It has been said, that his influence on the Wright brothers' experiments had been preponderant. Better yet, the two conquerors of the air were presented as his disciples and his pupils. This is a flagrant mistake. Yes, this great apostle of aviation was interested in the Wrights' work, lavished encouragement and advice, and his influence on our final success was undeniable, there is no doubt. When Chanute came to visit them at Dayton, the two brothers had never seen him before. Better still, their experiments had been going on for two years. Funny detail should be mentioned, before the brothers confided in him, they had all sorts of reservations." In the end, the great scientist became their best friend and the wisest of the counselors. "But we often had a conflict of ideas with him, and I can tell you this: in a general way, on the principles of aviation, Chanute's and our ideas are almost totally different. We have often disagreed, but none of this could alter the deep respect that my brother and I have for his scientific glory, whose history will venerate the name as the equal of the greatest" [42]. Maybe Wilbur forgot that he had written to Chanute first and had invited him to visit them in Dayton, glowingly reporting to his brother Reuchlin about the visit of the "leading authority of the world on aeronautics" [43]. And yes, they frequently had differing ideas, but Chanute knew enough of the subject to provide information, mentoring or guiding to the best of his abilities.

Maybe feeling lonesome or obliged to stay in touch, Wilbur wrote a lengthy letter to Chanute one month after the interview. "For three months I have had scarcely a moment to myself except when I take my bicycle and ride off into the woods for a little rest. I have been received in France with a friendliness scarcely to be realized. In the reaction from former abuse they seem trying to make up for lost time. No American would think of giving so much time and trouble to assist a stranger as MM. Bollée and Pellier have given me here. I have come to believe that the French character outside of Paris is more to be admired than I had supposed. Archdeacon and Ferber make much more noise than their real importance justifies" [44].

Just as Chanute had suggested earlier, Wilbur now demonstrated the Flyer regularly; the high financial awards in France were surely attractive. And seeing the trophy, created by the Michelin brothers with a 20,000 francs cash prize, made Wilbur show his interest; hearing this, the Michelins asked the artist to rework the design on the trophy to show a Wright Flyer instead of Farman's. On Friday, 31 December 1908, Wilbur took off at 2 p.m. at Camp d'Auvours, made 56 triangular circuits (flying for 77 miles) and won the Michelin Cup and the money, closing a successful year.

7.8 Activity at Fort Myer: Dirigible Up—Baldwin Away

In early August 1908, aeronautical attention in America shifted to Fort Myer. Thomas Scott Baldwin, or Capt. Tom, and his crew arrived with the dirigible in late July for the official trial flights. While being confident of ultimate success, Baldwin did not claim that his type of machine would be the aerial craft of the future. "As in everything else, so in aeronautics, a man follows a system in working out his ideas. And no two men use the same system in endeavoring to solve the problem of aerial navigation. Let a thousand men work on the improvement of the dirigible balloon and a perfectly successful airship is not too much to expect" [45].

Baldwin's dirigible made a ten-minute flight in the afternoon of August 4 with an estimated thousand spectators watching. Capt. Tom did the piloting and steering from the rear, while Glenn Curtiss sat in the forward section of the 66-foot catwalk, handling the motor. Both men were pleased how the ship handled [46]. On August 9, Baldwin fulfilled the first requirements, and the newspapers reported, "Describing circles and figure eights, dipping and rising at will, the dirigible balloon of Capt. Thomas S. Baldwin covered itself with glory today in the army trials at Fort Myer. The balloon was up more than a half hour, and during every minute of the time was responsive to the will of its operator" [46].

One of the spectators on that Sunday was a gray-faced, gray-eyed man who hung around the Baldwin airship, making comments and offering suggestions. He could have been one of the many visitors, but it was Herring who was scheduled to deliver his airplane a few days later. He had come to request an extension as he needed more time to finish the work on his aeroplane and the engine [47].

The following week, big headlines announced that Baldwin's dirigible was accepted by the military. "Like a demon from the sky, its motor spitting fire and its long gray gas bag outlined against the sky of dusk, the Baldwin airship landed in Fort Myer tonight. For two hours and five minutes the big military dirigible flew back and forth over a concourse nearly five miles in length in the official endurance trial" [48]. Capt. Baldwin received $5,737.50 from the military, which was 15% less than what he would have received if he could have flown a speed of 20 miles. Baldwin then trained three Army officers, Selfridge, Lahm and Foulois, instead of only two as the contract had required.

7.9   Activity at Fort Myer: Orville Wright Next to Fly

Orville Wright arrived with the rebuilt Flyer at Fort Myer shortly after Baldwin's airship was accepted. Thus, a Wright aeroplane made a debut on either side of the Atlantic at about the same time (Fig. 7.5).

Orville circled the Fort Myer field for the first time on September 3, feeling good about himself and the Flyer. When asked by a reporter from the *Evening Star* about the take-off procedure, Orville explained, "There is a skeleton tower from the top of which weights are suspended. A rope runs from these weights through a block to the starting track, which is about fifty feet of single rail laid on the ground. A trigger releases the weights and in falling, they pull the rope alongside the track; this starts the aeroplane at a high rate of speed. The motor is started before the weights fall, and after the plane has run the length of the short track, it is going at a speed that carries it into the air.

AT THE HEAD OF THE PROCESSION

Fig. 7.5 Mystery of the Wright airship is fully explained to the members of the Signal Corps, showing that the Wright brothers are ahead of the procession. *The New York Herald*, May 30, 1908

There is a little wheeled car under the plane, and at the end of the track this falls off and the plane mounts into the air." Orville also stated that "a landing can be safely made at a speed of 100 miles an hour, and the operator does not know just when the machine first touches the ground" [49]. Landing the Flyer at 100 miles seems a bit high, and the sound of landing on a wooden skid is quite noticeable, but then this is a newspaper report.

Chanute arrived at Fort Myer on September 12, and reporters quickly besieged him. When asked how the Wrights got started in the aviation business, he explained, "The Wright brothers began their work in aeronautics in an unusual way. Orville was convalescing from typhoid fever and his brother Wilbur read Pettigrew's book on aeronautics to him. Soon after, the *Aeronautical Annual* was published, and the Wrights learned from its pages what had been done toward conquering the air. Their experiments started thereafter" [50].

George Spratt arrived one day later, meeting Chanute at the parade ground; they both hoped that the Wrights would be able to sell their Flyer to the military without delay. Orville saw and greeted his two friends, but did not spend much time socializing.

Among the spectators at Fort Myer were usually several distinguished visitors. Colonel John Templar of the British Army had studied Baldwin's dirigible closely and now wondered if Orville could fly from Washington to

New York, if he cared to do so. Chanute did not see a reason why such a flight could not be accomplished. "Mr. Wright only carries enough gasoline for a few hours now. I think, however, that he will have to learn more about the effect of the hills and dales and valleys and woods beneath him before he attempts any such long flights. He tells me that he can feel the difference when flying over them, even here in the parade ground" [51].

While spectators were watching and socializing, Orville worked on the Flyer, especially after having made longer flights. "Good for you, my boy," puffed Chanute, who had come racing from the other end of the drill ground. "It is splendid. How does it feel to be making history?" Brightening up, "I am not much interested in making history," said Orville. "I am a good deal more wrapped up in making speed. The long flight last night gave me a great deal of valuable practice. Never in any flight have I had better control or experienced as little trouble. After a few more flights I will be satisfied that I can live up to any reasonable requirement the Signal Corps may make" [52]. Orville then replaced the two eight-foot eight-inch propellers with nine-foot ones.

The twenty-six-year-old Thomas Selfridge, the first army officer to pilot an aeroplane (AEA's *Red Wing*) had requested permission from his superiors to be Orville's passenger "for the purpose of officially receiving instructions," but he also wanted to compare the flight characteristics of the Flyer with the two AEA aerodromes that he had flown a few months earlier. Orville was not pleased that he had to fly with him, as the brothers were convinced that AEA members stole the secrets of the Flyer for their profits.

More than 2,000 spectators watched Orville and Selfridge take off on September 17. After a few minutes in the air, there was "a light tapping" noise behind the aviators, followed by "two big thumps, which gave the machine a terrible shaking;" the Flyer then came down and crashed "with frightful force." Orville was quickly pulled from the wreckage, but it took more effort to free the unconscious, bloody Selfridge, who was pronounced dead three hours later.

By coincidence Chanute heard the Flyer take off, standing 15 feet south of the press tent and 560 feet west of the point where the machine struck, watching the accident happen. "Mr. Chanute testified that the machine was perhaps 60 feet up and circling the field to the left. He then went 40 or 50 feet to the south so as not to be behind the tents between him and the aeroplane shed. When the machine was 300 feet from him, the propeller flaked off or snapped, and the piece fluttered down to the ground; the aeroplane maintained its level for 60 or 100 feet, then oscillated and pitched down to the left side and disappeared from his view behind the bushes. He did not see it strike. When he examined the broken propeller blade, Mr. Chanute testified that the wood was brittle and over seasoned, or kiln dried. A few days later Mr. Chanute informed me [Lahm] that he thought the propeller blade had struck the upper guy wire of the rear rudder and had torn the end of the wire from its attachment to the rudder" [53] (Fig. 7.6).

A reporter of the *Evening Star* asked Chanute to comment on the accident. "It is hard to have such a thing happen at any time, but particularly so, when Wright was so near to official success. I do not think that it disproves the two-propeller theory; I think had he but one propeller and it had broken, the same effect would have been produced. Selfridge is one more name to add to the list of world's martyrs in the conquest of the air; his death will not retard the development of aeronautics any more than did the deaths of those other men who have given their lives for the cause of flying. I deeply deplore the accident, but the work will go on the same" [54].

The wrecked Flyer was crated for shipment back to Dayton, when Katharine heard of an unauthorized person examining and measuring the aeroplane. Expecting industrial espionage, she wanted Chanute's advice. "Having been at the shed at the time," he explained, "I am inclined to believe that the account, which reached you was greatly exaggerated. I met Sergeant

Fig. 7.6 While surgeons looked after Lieutenant Selfridge, members of the Signal Corps and Octave Chanute (center) discussed the broken propeller as the possible cause of the accident. *New York American*, September 19, 1908

Downey after I left you and spoke to him about the occurrence. He says that Dr. Bell went in the shed after everything was packed up, but there was no cover on the aeroplane box. That a dispute arose about the width of the wing and Dr. Bell said, 'Well, I have a tape in my pocket and we will measure it.' That this was the only measurement taken is what I gathered, but you had better see Sergeant Downey when opportunity serves and get an understanding of what occurred" (Fig. 7.7) [55].

Taking part in the accident investigation by courtesy, Chanute stated that the accident was not due to a faulty airplane design; he suggested that Katharine request a contract extension. Among the Chanute papers is a draft in his handwriting, but also a letter from Katharine to Wilbur, stating: "Major Squier said that I could sign for you and ask for an extension. Mr. Chanute will help me make out the form" [57]. And there is a letter with an unidentified signature on Wright Brothers stationary to Major George O. Squier: "In order to give Mr. Orville Wright time to recover from his late severe injuries, we beg respectfully to request an extension of time to make the official tests for the acceptance of our flying machine, now under contract to the Signal Corps, for a period of nine months or until 28 June 1909, if found possible" [58].

On September 25, Selfridge was laid to rest with an impressive military funeral at Arlington National Cemetery. The honorary pallbearers represented the Army and the Navy, the AEA and the Aero Club of America, with Chanute heading the latter.

Fig. 7.7 Several men were willing to assist Orville Wright to take up the broken strand of his aerial experiences. The first person listed in the newspaper was Prof. Chanute, the Scientist, who has been an ever-ready and valuable adviser with his experience in aeronautics. *The Sunday Star* Washington, DC [56], September 27, 1908

7.10 Activity at Fort Myer: Aviator Augustus M. Herring

There was a certain mystery over the other competitor in the heavier-than-air government contract bidding. Herring had made a 72-feet flight in 1898 in his biplane, powered by a compressed air motor (see Sect. 4.2). Reading that the Board of Ordnance had allotted $25,000 to investigate the possibilities of flying machines for reconnoitering purposes and as engines of war [59], Herring wrote to the Board of Ordnance, "… If you care to have results of my experiments, which I believe are in advance of those of any other worker in the field, I should be pleased to lay them before your representative here in New York or I could come to Washington with photographs, etc., for your personal inspection." It is not known if Herring received a response from General Greely or any member of the Signal Corps at that time.

A decade later, Herring entered his bid to supply an aeroplane to the military. We will probably never know if he intended to build an airplane or if he just wanted to underbid the Wrights. As fate went, a few days prior to his delivery date on August 13, Herring informed the Signal Corps that he had injured his finger testing his engine, and destroying it in the process. The Signal Corps granted his request for a one-month extension, but some members voiced doubts if Herring was serious; he then announced that his aeroplane was not an imaginary craft, he would fly at forty-four miles an hour or more from New York to Fort Myer to deliver it.

Two days before the next scheduled delivery date, on September 11, Herring wired the acting chief signal officer, Major Squier, that his machine was finished, but some shop work still needed to be done. He was granted another extension, which was quite agreeable with all involved, as no one wanted the Wright and the Herring aeroplanes on the Fort Myer field at the same time.

One month later, on Monday, October 12, Herring brought parts of his aeroplane in a trunk and two suitcases to Washington, making "technical delivery." During a three-hour interview he assembled sections of the framework and demonstrated the motor to the aeronautical board. They allowed another extension to November 13.

Back in Long Island, Herring reportedly made a flight on October 28. As there was rain in the forecast, he assembled the machine in a hurry and forgot to install certain parts and crashed. He requested another extension until June 1, 1909.

And the mystery about Herring continued.

7.11   Human Birds to Try Flight at Morris Park

Members of the Aeronautic Society of New York had located and then leased the Morris Park Race Track, built a clubhouse, aero garages (or hangars), set-up workshops with tools, machinery, etc., and a flying field. On rather short notice, the melancholic appearance of the old Race Track in the Bronx had an aerodynamic injection of "buzzing life" at almost every corner of its run-down site. Beneath every roof that was still rain proof was some kind of a flying machine, from small amateur models and gliders to such giants as the Beach monoplane, the Schneider biplane and others.

With nothing else needing his attention at Fort Myer, Chanute traveled to New York to wait for his daughters who had spent the summer in Paris. He took the opportunity to visit the flying field of the Aeronautic Society and thought the grounds were ideal, offering ample opportunities for wide and safe returns, not like Fort Myer (Fig. 7.8).

On Election Day, November 3, 1908, the Aeronautic Society staged the first air carnival at their new flying field. While voters would not know "which way the wind blows" in the National and State election, a crowd of aviators, kite experts and pilots of various aircraft hoped to be more fortunate. The day dawned clear and crisp, good conditions for testing and flying the various aircraft.

Ernest La Rue Jones, owner/editor of *Aeronautics*, had designed a biplane glider, which the Wittemann brothers built for him; it was displayed beside the biplane of sixteen-year-old Laurence J. Lesh, who had made sensational flights during the past summer in Montreal, Canada. His latest glider, with 120-square-foot supporting surfaces, was an improved craft that he designed

Fig. 7.8 First public race track to aid experimentation by rich and poor alike: gliders, wind wagons, catapults, sheds, tools, shops and advice free. *The New York Press*, October 4, 1908

after numerous consultations with Chanute. It was distinctly novel as the upper wing surface was placed considerably in advance of the lower one, which hopefully would improve the airflow over and under the wings. (Fig. 7.9).

Lesh's first flight, towed by an automobile, was a success, but on the second launch, he rose sharply to about 40 feet, lost his balance, and plunged to the ground, breaking his leg just above the ankle [60]. The Beach-Whitehead glider took off next and crashed; the judges then decided to make no further trial flights with gliders, either being launched by an automobile or with the Wright-style derrick. All aircraft then stood in front of the grandstand for the public to inspect.

Even though nothing novel happened in the aeronautical line, a large crowd witnessed balloon and kite ascensions, but also wind wagon races. A wind wagon is a light chassis on wheels with a motor that drives one or more propellers. One of the participants, Julian Thomas, had purchased the "Herring-Arnot" Chanute glider two years earlier, intending to install an engine and propeller to create a powered aeroplane. The powerplant of his wind wagon was to be installed into the glider, if successful. After making several high-speed rounds, Thomas crashed his wind wagon and was seriously hurt. He went back to ballooning and his medical practice, and the powered aeroplane dream became history.

Fig. 7.9   Larry Lesh on his first flight at Morris Park with his new glider. *Aeronautics,* November 1908

The staff at the box office reported that many spectators, including quite a few youngsters from the neighborhood, had slipped into the grounds through beckoning gaps in the fences and enjoyed the show without paying an entrance fee. However, an estimated 4,000 tickets were sold, which helped pay for some of the expenses of the Aeronautic Society; they considered the event a positive popular success.

To encourage old and new enthusiasts to become more involved, members of the Aeronautic Society started early to prepare for the next air carnival in May 1909. They announced a $2,000 prize for the machine that would make a successful circuit of the Morris Park course, a distance of about one and three-quarter miles. And Triaca offered a gold medal for the most efficient aerial propeller, a second one for the most efficient aviation motor, and a third one to the first lady who qualifies as an aeronautical pilot. Chanute, again, offered $50 as a first and $20 as a second prize to be awarded to the individual making the flattest glides.

7.12   Aerophobia: Flying Men Here and There and Everywhere

Each day brought further evidence of the universal interest in aerial navigation, with pilots flying faster and higher and flights becoming longer in duration and distance. An editorial summed up humorously the general mood around Paris. "Immense activity is noted among inventors of apparatus for conquest of air … there is a phenomenal amount of activity in the flying machine world. To be quite frank, there is more action than speed, but that is no fault of the inventors. It is rather the fault of gravity, for no sooner do machines get a few yards up in the air than they come down hard and go to the repair shop. However, the near future should see a perfect flotilla of airships, airplanes and dirigibles, sailing about the environs of Paris if their designers expectations are realized" [61].

Offering a forecast of man's amazing future with flying machines, the *New York Herald* published three full pages with photos and articles in their Sunday Magazine on August 30, featuring opinions and views from aeronautical experts, even flashing some futuristic visions [62]. The title on the first page tried to catch everyone's attention: "*Comet-like transit of the future in the ocean of the skies.* What will be the ultimate type of airship?" Henry Farman's write-up was highlighted with a sketch of a futuristic sleek airliner, "I am sure that within the next 10 years the flying machine will have taken the place of the automobile. It will first come into popular use as a high-speed sport vehicle. Then it will become a dangerous weapon of offensive warfare, compelling universal peace. Only a little later will it carry special delivery mail, thus establishing its commercial value. Not the scientist, but the daredevil devotees of speed will force upon the public the acceptance of wing flight." Farman continued reporting about the development in France during the past year and closed with, "The simplest device for flying in the air is the glider, a machine having rigid wings and handles to which the operator clings. Aeroplane racing will come next, it will become as popular as automobile road racing has hitherto been, and I confidently predict international contests will be held similar in every respect to the motor races within a few years."

The second page showed the bold title "*Facts and Fancies of the Conquest of the Skies*," and included a write-up by Chanute on "*Looking into the future*." As could be expected, he believed that the ultimate type of flying machine would preserve the feature of bridge trussing, because "in that way we seem to be able to attain the maximum of stiffness and strength with the minimum of weight. My interest is in the two-surface or double decker machine. It is with that type that Farman, Bell, the Wright brothers and Delagrange are getting their best results ... The ultimate heavier-than-air machine must be one, which will meet successfully its own emergencies in flight and overcome obstacles without too much intervention on the part of the operator ... I do not subscribe to the prophecy that the flying machine will remodel civilization, or do away with tariffs and frontiers."

The third page was devoted to "*Solving the Secrets of Aerial Flight*." One author, Carl Dienstbach, reviewed the current status of aeronautics and closed with a discussion on the possibilities of using the wind as an energy source for flying. And he strongly suggested that the new school of flying apprentices should take up gliding before they venture into a powered aircraft.

While enthusiasts promoted and improved aviation in the United States, there were also brilliant minds in other countries who were stimulated to work out the aeronautical problems; several of these experimenters became the first aviators in their home country to achieve flight.

Willem Hendrik Schukking (1886–1967), Second Lieutenant of the Royal Dutch Engineering Corps, had studied the gliding flights of Lilienthal in Germany, the Americans Chanute and the Wrights and Captain Ferber in France. As his first project, he built himself a Chanute-type biplane glider, using the drawings in *L'Aérophile*. On Tuesday, July 28, 1908, Schukking made the first flight in the Netherlands with a heavier-than-air craft from the highest hill near Soesterberg, located in the middle of the country. Just like members of Chanute's team, the 22-year-old Schukking launched himself

Fig. 7.10 Willem Hendrik Schukking flying his updated glider. Gedenkboek Neder-landsche Vereeniging voor Luchtvaart, September 1908

by hanging below his 30-pound glider. The longest flight covered about 15 meters (Fig. 7.10).

Next, he added a canard and used the Wright-method, laying prone in the glider. Two soldier friends lifted the glider at the wingtips and ran downhill, but he could not get airborne. Not giving up, Schukking then laid a rail on the down slope of the hill, using a wheeled carriage to stand on, attempting a few take-offs, similar to Avery's launching methods at St. Louis. These flights were not very successful either. Schukking then concluded that the hill was not steep enough for longer flights, the wind was too variable and the glider did not meet the necessary technical standards. [63] This was the end of his flying activity, but it was the beginning for other Dutch enthusiasts to be more successful.

Samuel F. Cody, born as Franklin Samuel Cowdery (1867–1913) in Davenport, Iowa, earned his money as a frontier cowboy, Wild West showman and gold prospector. He usually wore a sombrero, a long mustache and imperial with long hair, just like the famous Buffalo Bill Cody.

Sam Cody left the United States with his first wife in 1889 to set up a Wild West show in England. In 1902 he joined the British Army Balloon Factory

Fig. 7.11 Samuel Cody on the first sustained, controlled, powered flight in Great Britain, October 16, 1908. George Grantham Bain Collection, Library of Congress, Washington, DC

at Farnborough as the Chief Kite Instructor. Cody was not an engineer, but he was a good rule-of-thumb mechanic and a man of pluck and perseverance; he never tried to fly an imperfect machine. The War Office hired him as military aviation specialist and authorized him to develop heavier-than-air craft. In early summer 1908 Cody had a workable aeroplane, but kept practicing without getting airborne, and the public soon called it the "lawn-mower." His first successful flight took place in late July 1908 at Aldershot, with John Capper in charge of steering, and Cody handling the motor. His next design was a biplane with a 40-foot wingspan, a 7-foot cord and 8 feet space between upper and lower wing surfaces. On the morning of October 16, 1908, he made several circles near Farnborough at about forty feet, covering a distance of 500 yards. He then attempted to make a sharp turn to avoid hitting a clump of trees, and, struggling with a wind gust, the machine slewed around, he lost its balance and came down hard. [64] Even though this flight was not the most perfect flight, it is generally recognized as the first powered, sustained flight in Great Britain (Fig. 7.11).

Months of more experiments followed; on May 14, 1909, Cody made a flight of 1,200 yards with entire success. In August, Capper flew with Cody as passenger for over two miles, and on September 8, Cody made a world's record cross-country flight of over forty miles in sixty-six minutes, but the military gave him notice to remove the aeroplane. Cody then became a British citizen and won the support of the British people; the military reconsidered and gave him another government contract to continue the experimental work on a larger scale.

Cody and his passenger were killed in early August 1913, flying a new hydroplane design at Aldershot. The attitude of the War Office toward Cody at his death is shown in the message to his wife: "The science of aeronautics owes much to his mechanical genius and courageous perseverance. The British War Office has special reason to mourn the loss of his valuable services, both in regard to man lifting kites and his contribution to military aeronautics" [65]. In the chronology of early British experimenters, Cody is regarded as the pioneer of the twentieth century.

7.13   The Aeronautical Exhibition in Paris, France

Organizers of the 11th automobile show in France decided to include the "*Première Salon Internationale d'Aéronautique.*" This first European exhibit of flying machines was staged at the Grand Palais in the heart of Paris in December 1908, and an estimated 120,000 visitors attended the salon every day. It was a spectacle not seen before, as visitors could closely examine the monoplanes, biplanes or even triplanes, some of which had their horizontal rudders in front, while others had them astern.

Organizers assembled a comprehensive sampling of Colonel Charles Renard's collection, including the gliders of Massia Biot, Octave Chanute (imported by Jacques Balsan in 1905), Ernest Archdeacon and Ferdinand Ferber. The introduction in the exhibition catalog explains: "Before admiring appliances, airships and airplanes who proudly wear our colors in the sky, we must salute those who first achieved the glory of flight. These relics can now be admired in the first exhibition of aeronautics" [66]. The most prominently displayed aircraft was the *Avion*, designed by Clément Ader, looking like a huge yellow eagle with curved wings and two propellers. Placed on a pedestal, this bird-like structure stretched its uncanny wings in silent benediction over all who entered. The *Avion* reportedly flew a distance of almost 1,000 feet on October 14, 1897, and the exhibition catalog stated, "If this can be proved, it will establish for France the honor of recording the first flight of modern times by a mechanical apparatus, as the flights of the Wright brothers are dating from 1903" [67].

Nine months later, the Grand Palais hosted the second aeronautical salon (September 25 to October 17, 1909), focusing on the latest French development in aviation. But the event was overshadowed with tragedy, as Ferdinand

Mort tragique du Capitaine FERBER

Fig. 7.12 A few days before the second Aeronautical Salon was to open in Paris, Ferdinand Ferber was killed flying a new Voisin aeroplane at Boulogne. *L'Aérophile*, October 1, 1909

Ferber, the chief apostle of aviation in France, was killed flying a new Voisin aeroplane at Boulogne only a few days earlier (Fig. 7.12).

The papers reported, that Ferber appeared to have been foolishly reckless in his flight and had taken needless risks [68]. Reading about the death of his friend, Chanute wrote and mailed his eulogy to the editor of *Aeronautics*, "Ferber was a charming lecturer and writer, with a dash of humor. Besides articles for the press, he published several books on the progress of aviation, with more details of the inside facts of its development than any other works which have been published. To him more than to any other man is due the enthusiasm aroused and the progress made in aviation in France" [69]. And the day the Aviation Salon opened, the news came that the French airship *La Republique* had crashed, killing the four-men crew. The disaster was pushed aside, and French newspapers stated that no human sacrifice can discourage the conquest of the air.

The star of the show was the oil-stained Blériot XI monoplane, in which Louis Blériot had crossed the English Channel a few months earlier. Other highlights were Ader's *Avion* and the wicker basket from Blanchard's balloon, a replica of the Wright Flyer, and Chanute's 1904 glider, shown in the back where few visitors probably saw it [70].

After World War I, Albert Caquot, Head of Aerospace Technical Service, proposed to the Ministry of War to create a Conservatory of Aeronautics. Many artifacts from the Renard collection were transferred to the "*Depot des Collections de l'Aeronautique*" which opened in 1919 at Issy-les-Moulineaux in the abandoned factory of the Voisin brothers. When the river Seine flooded, the aeronautical collection was moved to the military ballooning center at Chalais-Meudon, southwest of Paris. The old balloon shed, where wicker baskets for captive observation balloons were manufactured during World War I, became the "*Musée de l'Aéronautique*," opening its doors on November 23, 1921 [71]. Jacques Balsan donated his authentic Chanute glider in 1923; the varnished silk covering was replaced in 1930 and the glider was again prominently displayed, hanging from the ceiling with other early aircraft [72].

When the Charles de Gaulle Airport opened in 1973, the vacated *Le Bourget Airport* buildings became the new "*Musée de l'Air et de l'Espace*." The former airport terminal became the "*Grande Galerie*," displaying the original aircraft and balloons from the early days of aviation, including the freshly restored 1904 Chanute glider, the first American aircraft to be exported (Fig. 7.13).

Fig. 7.13 The restored original 1904 Chanute glider in the new Grande Gallery, *Musée de l'Air et de l'Espace*. Courtesy Jean-Luc Claessen 'Pyperpote'

7.14   In Some Ways 1908 Was a Perfectly Good Year …

The advances made during 1908 in perfecting the art of flying aeroplanes is best illustrated by comparing what had been accomplished. In January, the winner of the Aéro-Club grand prize was to fly a circular course of one kilometer. On the last day of the year, Wilbur flew 56 triangular flights, or about 77 miles, and won the Michelin Cup. French pilots, like Louis Blériot, Henri Farman and Leon Delagrange, pushed steadily ahead to improve their aeroplanes, inspiring a new generation of "Darius Green's" to become involved. Aeroplanes were for sale in Paris for $6,000, which was no more expensive than a good automobile. The *New York Evening Telegram* announced that one could order a dirigible on Broadway, starting on December 10. Many shoppers showed interest, but there was no bargain rush [73]. Seeing the write-up, the editors of the *Los Angeles Herald* were quick to point out that their town was actually ahead of New York and the rest of the country; their hometown was the "*Flightiest City*" and the "World Center of Airship Activity." Roy Knabenshue had opened a store at 1042 South Main Street two weeks before the New York concern began its business, offering Curtiss motorcycles, airships and aeroplanes, hoping Californians would take interest in aerial travel. "No sport is equal to an aerial trip, and a short balloon journey is not only exhilarating but also perfectly safe" [74].

Members of the City Club in Chicago asked Chanute to discuss the development of aeroplanes at their November meeting. He talked about his early efforts with a smile, and then gave details of the work of others. "We cannot tell as yet what is going to be the result of an invention, which is now only five years old. We only know that the development is rapid, and that the subject has passed from the hands of cranks into the hands of skilled men who have knowledge and ability; results will be obtained within the next 10 or 20 years which will make the world better and happier" [75].

8

New People, New Designs, New Ideas, New Problems

The joy of speed and the sensation of flying—of overcoming space and distance and time—that has given to automobiling its thousands of devotees, is bound to make air-sailing the great sport of the future.
Arnold Kruckman, Country Life in America. January 1909

As the public became more interested in aviation, the initial age of the flying machine slipped into history, and the age of the practical airplane entered reality. As Hudson Maxim commented, "We of the twentieth century hardly realize the privilege which is given us to watch the revolution of the airplane."

In early January 1909, Chanute chitchatted with Ernest Jones at the Aero Club in New York when Orville Wright walked in and joined the conversation [1]. Then a reporter of *The World* arrived who directed his first question to the 38-year-old Orville, wondering if he was to finish the military trials, "I do not know. We are anxious to fill the contract; I want to finish the trials myself next summer, and probably I shall" [2].

Jones, the journalist, asked next, "Just what do you think we will have in the way of airplanes ten years from now?" The response was simple, "I am not a prophet. I can't say. I don't know." To fly from New York to Chicago was the latest award offered by *The World* and the almost 77-year-old Chanute thought, "It is possible and somebody will do it before long. Of course, there would have to be stops for gasoline to run the motor, and it might be necessary to pick out landing places, but it could be done right now." However cross-country flying was of no interest to Orville, he thought "the airplane will be developed as an instrument of war and will be taken up for sport later.

© The Author(s), under exclusive license to Springer Nature
Switzerland AG 2023
S. Short, *Flight Not Improbable*, Springer Biographies,
https://doi.org/10.1007/978-3-031-24430-8_8

We do not look for much interest from individuals, if for no other reason than that the machines are expensive. I haven't found many people who want to spend $25,000 on one." Yes, in ten years the airplane would be less expensive, but Orville did not elaborate on whether it would be materially different.

The reporter then wondered, "If you receive an order for a machine now, how long before you could deliver it?" The eyes of the elderly Chanute twinkled and Orville stated, "Well, that depends. In large quantities we can turn them out pretty fast. The best prediction I can give about flying ten years from now is that there will be a very great many people at it, and they won't be much more of a novelty than an automobile is now." Other members of the Aero Club then arrived and the discussion turned to demonstrating the Wright Flyer in Europe, where airships and airplanes were built just like the automobile.

The next day, Chanute traveled back to Chicago, Jones started on the next issue of *Aeronautics,* and Orville and Katharine sailed for France to join their brother.

Besides demonstrating the Flyer, Wilbur Wright had also agreed to train students at Camp d'Auvours. Count Charles de Lambert, who had purchased a Lilienthal glider in 1894 which he did not find useful, became the first student in October 1908. The balloonist Paul Tissandier was the second one. As the upcoming winter weather would make flight training a bit difficult, he suggested to move the operation to Pau in western France, where the town folks had offered to establish an airfield and build a "chalet-hangar." This sounded like a good idea; all equipment was now moved with Orville's help and training started again on February 15. Wilbur's first flight was with de Lambert, next he gave the Countess de Lambert a five-minute ride, and as the sun started to set, he took his sister up for a seven-minute flight [3]. She reported to her father, that it was cold but not particularly uncomfortable.

When the training was done, the Wrights left Pau, crisscrossed central Europe to show the Flyer, take politicians and royalty for rides and enjoy the publicity.

In anticipation of the Wrights returning to America, and considering himself the correspondent responsible for bringing all the news-worthy aeronautical stories to the public's attention, Arnold Kruckman submitted his thoughts to the New York Bureau of the wire service. The title was attention-catching, "Wrights come home honored, left obscure," and the *St. Louis Post Dispatch* was the first paper to publish it. "There is hardly anything more dramatically striking in the life stories of great inventors than the contrast between the return and the departure of the Wrights. A scant 24 months ago, Wilbur Wright obscurely left America, regarded by those who took the trouble to notice, as an erratic individual on a par with the cranks who try to discover perpetual motion. When Wilbur Wright left America, the majority of these zealous admirers probably would have been afraid to be seen in their company, for fear the association of names might cause undesirable effect. The Wrights return to America as men with whom kings and princes have thought to link their names. And it seems good that the Wrights should have achieved such a triumph. For, in common with almost all great inventive geniuses, their story is full of early miseries and struggles. They are not made of stuff that produces intimacy. They are self-confident, self-centered and self-possessed, and they are excessively shy." Next, Kruckman described how the Wrights became acquainted with Chanute. "With the rules they established for their own working course, they found only one man in perfect accord. That man is Octave Chanute, a distinguished engineer of Chicago. In his work on aerodynamics, they found tremendous inspiration. Entering into communication with him they found him ready to give them all the benefits of his store of knowledge. He helped them solve their problems and brought them around to his views and theories. After a fashion Chanute became their guide and their stimulus. This much they have acknowledged. But for some unaccountable reason they have never publicly given him credit for his services. In all their dealings with the rest of mankind the Wrights have been scrupulously and exactly honest. Why they should have denied Chanute his share in a triumph so great as theirs has never been understood by people who are practically on terms of intimacy with both. Chanute himself, who is an example of the finest type of cultivated and active inventive genius, does not understand it … Soon after they tacitly joined forces with Chanute, they made amazingly marvelous progress. Their gliders flew and their gliders with motors flew." Kruckman closed his article with, "The success of the Wrights in France is too recent to require recapitulation. Financially it is estimated that in France and Italy they have collected sums in excess of a half million dollars for the right to their machines alone. Outside of this they have won numerous cash prizes reaching a total of nearly a quarter of a million dollars" [4].

A few days later, Katharine, Orville and Wilbur arrived in New York, attended an Aero Club luncheon in their honor and then headed to Dayton. One week later, Wilbur shared the latest news with Chanute, who responded that he rejoiced over their triumphs and wrote probably with a twinkle in his eyes, that he "was particularly gratified with the sensible and modest way in which you accepted your honors. It encourages the hope that you will still speak to me when you become millionaires" [5].

8.1   Mathematical and Otherwise—Soaring Flight. How to Perform It

Reviewing Hiram Maxim's latest book *Artificial and Natural Flight* for *Science,* Chanute questioned the author's observations on the rising trends of winds with which soaring birds perform their feats. Maxim wrote, "We shall never be able to imitate the flight of the soaring birds." But Chanute was convinced otherwise, so he simply wrote, "It is possible that Mr. Maxim is mistaken" [6].

Having studied the possibility of soaring with a man-made machine for more than a decade, Chanute's first article on "*Soaring Flight*" was published in the *Aeronautical Annals* of 1896 and 1897, and the latest updated version was published in *Aeronautics* in April 1909. It is reprinted below.

"There is a wonderful performance daily exhibited in Southern climes, occasionally seen in northerly latitudes, which has never been thoroughly explained. It is the soaring or sailing flight of certain varieties of large birds who transport themselves on rigid unflapping wings in any desired direction; who, in winds of 6 to 20 miles per hour, circle, rise, advance, return and remain aloft for hours without a beat of wing. They appear to obtain from the wind alone all the necessary energy, even to advancing dead against that wind. This feat is so much opposed to our general ideas of physics that those who have not seen it sometimes deny its actuality, and those who have only occasionally witnessed it subsequently doubt the evidence of their own eyes. Others, who have seen the exceptional performances, speculate on various explanations, but the majority gives it up as a sort of 'negative gravity'."(Fig. 8.1).

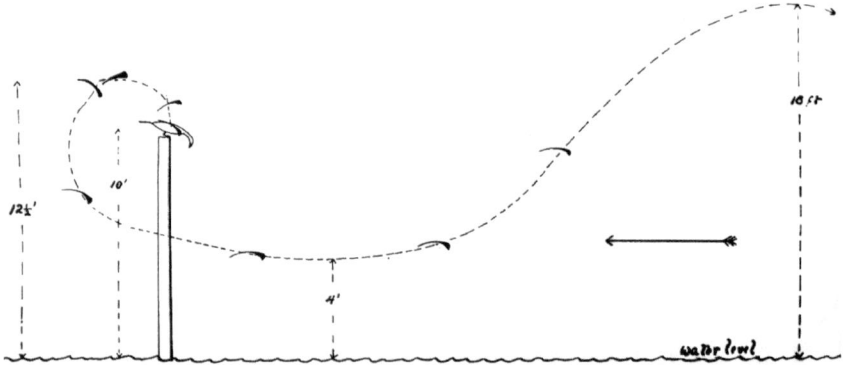

Fig. 8.1 Gull raising from a pile. Chanute Papers, Manuscript Division, Library of Congress, Washington DC

In his paper on the sailing flight of birds, Chanute gave a list of authors who had described such flight or had advanced theories for its explanation. He also gave his own observations and submitted some computations to account for the observed facts. These computations were correct as far as they went, but they were scanty. It was, for instance, shown convincingly by analysis that a gull weighing 2.188 pounds, with a total supporting surface of 2.015 square feet, a maximum body cross-section of 0.126 square feet and a maximum cross-section of wing edges of 0.098 square feet, patrolling on rigid wings (or soaring) on the weather side of a steamer and maintaining an upward angle or attitude of 5° to 7° above the horizon, in a wind blowing 12.78 miles an hour, which was deflected upward 10° to 20° by the side of the steamer (these all being carefully observed facts), was perfectly sustained at its own 'relative speed' of 17.88 miles per hour and extracted from the upward trend of the wind sufficient energy to overcome all the resistances, this energy amounting to 6.44 foot-pounds per second. It was shown that the same bird in flapping flight in calm air, with an attitude or incidence of 3° to 5° above the horizon and a speed of 20.4 miles an hour was well sustained and expended 5.88 foot-pounds per second, this being at the rate of 204 pounds sustained per horse power. It was stated also that a gull in its observed maneuvers, rising up from a pile head on unflapping wings, then plunging forward against the wind and subsequently rising higher than his starting point, must either time his ascents and descents exactly with the variations in wind velocities, or must meet a wind billow rotating on a horizontal axis and come to a poise on its crest, thus availing of an ascending trend. But the observations failed to demonstrate that the variations of the wind gusts and the movements of the bird were absolutely synchronous, and it was conjectured that the peculiar shape of the soaring wing of certain birds, as differentiated from the flapping wing, might hereafter account for the performance (Fig. 8.2).

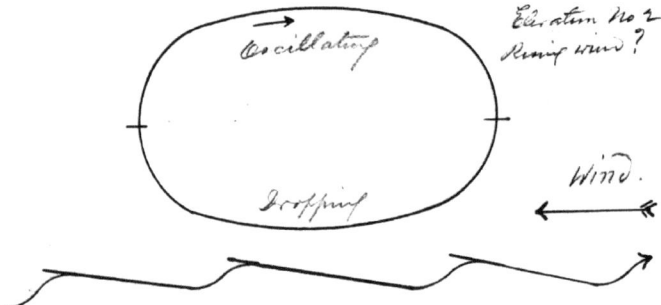

Fig. 8.2 Bird circling with unflapping wings. Chanute Papers, Manuscript Division, Library of Congress, Washington DC

These computations, however satisfactory they were for the speed of winds observed, failed to account for the observed spiral soaring of buzzards in very light winds and the writer was compelled to confess: "Now, this spiral soaring in steady breezes of 5–10 miles per hour which are apparently horizontal, and through which the bird maintains an average speed of about 20 miles an hour, is the mystery to be explained. It is not accounted for, quantitatively, by any of the theories which have been advanced, and it is the one performance which has led some observers to claim that it was done through 'aspiration,' i.e., that a bird acted upon by a current, actually drew forward into that current against its exact direction of motion."

A still greater mystery was propounded by the few observers who asserted that they had seen buzzards soaring in a dead calm, while maintaining their elevation and speed. Among these observers was Mr. E. C. Huffaker, a one-time assistant experimenter for Professor Langley. The writer believed and said that he must in some way have been mistaken, yet, to satisfy himself, he paid several visits to Mr. Huffaker in Eastern Tennessee and took along his anemometer. He saw quite a number of buzzards sailing at a height of 75–100 feet in breezes measuring 5 or 6 miles an hour at the surface of the ground, and once he saw one buzzard soaring apparently in a dead calm. The writer was fairly baffled. The bird was not simply gliding, utilizing gravity or acquired momentum; he was actually circling, horizontally in defiance of physics and mathematics. It took two years and a whole series of further observations to bring those two sciences into accord with the facts.

Curiously enough the key to the performance of circling in a light wind or a dead calm was not found through the usual way of gathering human knowledge, i.e., through observations and experiment. These had failed because I did not know what to look for. The mystery was, in fact, solved by an eclectic process of conjecture and computation, but once these computations indicated what observations should be made, the results gave at once the reasons for the circling of the birds, for their then observed attitude and for the necessity of an independent initial sustaining speed before soaring began. Both Mr. Huffaker and myself verified the data many times and I made the computations.

These observations disclosed several facts:

1st Winds blowing five to seventeen miles per hour frequently had rising trends of 10° to 15°, and that upon occasions when there seemed to be absolutely no wind, there was often nevertheless a local rising of the air estimated at a rate of four to eight miles or more per hour. This was ascertained by watching thistledown and rising fogs alongside of trees or hills of known height. Everyone will readily realize that when walking at the rate of four to eight miles an hour in a dead calm the "relative wind" is quite inappreciable to the senses and that such a rising air would not be noticed.

2nd The buzzard, sailing in an apparently dead horizontal calm, progressed at speeds of fifteen to eighteen miles per hour, as measured by his shadow on the ground. It was thought that the air [or a thermal] was then possibly rising 8.8 feet per second, or six miles per hour.

3rd When soaring in very light winds the angle of incidence of the buzzards was negative to the horizon, i.e., that when seen coming toward the eye, the afternoon light shone on the back instead of on the breast, as would have been the case had the angle been inclined above the horizon.

4th The sailing performance only occurred after the bird had acquired an initial velocity of at least fifteen or eighteen miles per hour, either by industrious flapping or by descending from a perch.

5th The whole resistance of a stuffed buzzard, at a negative angle of 3 degrees in a current of air of 15.52 miles per hour, was 0.27 pounds. This test was kindly made for the writer by Professor A. F. Zahm in the wind tunnel of the Catholic University at Washington, D.C., who, moreover, stated that the resistance of a live bird might be less, as the dried plumage could not be made to lie smooth.

This particular buzzard weighed in life 4.25 pounds, the area of his wings and body was 4.57 square feet, the maximum cross-section of his body was 0.110 square feet, and that of his wing edges when fully extended was 0.244 square feet.

With these data, it became surprisingly easy to compute the performance with the coefficients of Lilienthal for various angles of incidence and to demonstrate how this buzzard could soar horizontally in a dead horizontal calm, provided that it was net a vertical calm and that the air was rising at the rate of four or six miles per hour, the lowest observed, and quite inappreciable without actual measuring.

The most difficult case is purposely selected. For if we assume that the bird has previously acquired an initial minimum speed of seventeen miles an hour (24.93 feet per second, nearly the lowest measured), and that the air was rising vertically six miles an hour (8.80 feet per second), then we have as the trend of the "relative wind" encountered: $6/17 = 0.353$ or the tangent of 19° 26′, which brings the case into the category of rising wind effects. But the bird was observed to have a negative angle to the horizon of about 3, as near as could be guessed, so that his angle of incidence to the "relative

wind" was reduced to 16° 26'. The relative speed of its soaring was therefore:
Velocity $= \sqrt{17^2 + 6^2} = 18.03$ miles per hour. At this speed, using the
Langley coefficient recently practically confirmed by the accurate experiments
of Mr. Eiffel, the air pressure would be: $18.03^2 \times 0.00327 = 1.063$ pounds
per square foot (Fig. 8.3).

If we apply Lilienthal's coefficients for an angle of 16° 26', we have for the
force in action:

Normal: $4.57 \times 1.063 \times 0.912 = 4.42$ pounds.

Tangential: $4.57 \times 1.063 \times 0.074 = -0.359$ pounds, which latter, being
negative, is a propelling force.

Thus, we have a bird weighing 4.25 pounds not only thoroughly
supported, but impelled forward by a force of 0.359 pounds, at 17 miles per
hour, while the experiments of Professor Zahm showed that the resistance at
15.52 miles per hour was only 0.27 pounds, or $0.27 \times 17^2 / 15.52^2 = 0.324$
pounds at 17 miles per hour.

These are astonishing results from the data obtained, and they lead to the
inquiry whether the energy of the rising air is sufficient to make up the losses

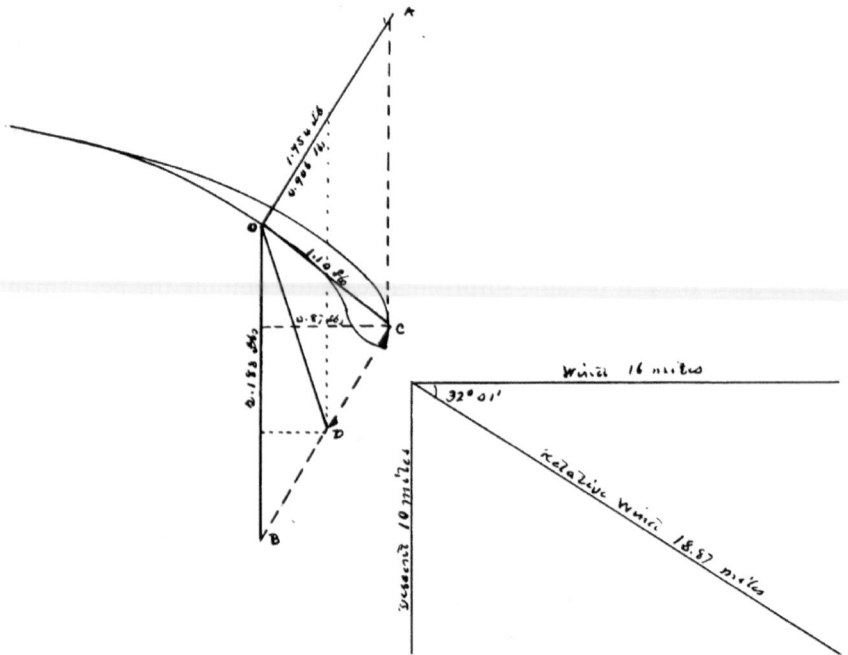

Fig. 8.3 Effect of forward motion and rising air mass produce the approximate relative wind vector shown. Chanute Papers, Manuscript Division, Library of Congress, Washington DC

which occur by reason of the resistance and friction of the bird's body and wings, which, being rounded, do not encounter air pressures in proportion to their maximum cross-section. We have no accurate data upon the coefficients to apply and estimates made by myself proved to be much smaller than the 0.27 pounds resistance measured by Professor Zahm, so that we will figure with the latter as modified.

As the speed is 17 miles per hour, or 24.93 feet per second, we have for the work done: $0.324 \times 24.93 = 8.07$ foot pounds per second. Corresponding energy of rising air is not sufficient at four miles per hour. This amounts to but 2.10 foot pounds per second, but if we assume that the air was rising at the rate of seven miles per hour (10.26 feet per second), at which the pressure with the Langley coefficient would be 0.16 pounds per square foot, we have on 4.57 ft^2 for energy of rising air: $4.57 \times 0.16 \times 10.26 = 7.50$ foot pounds per second, which is seen to be still a little too small, but well within the limits of error, in view of the hollow shape of the bird's wings, which receive greater pressure than the flat planes experimented upon by Langley.

These computations were chiefly made in January 1899 and were communicated to a few friends, who found no fallacy in them, but thought that few aviators would understand them if published. They were then submitted to Professor C. F. Marvin of the Weather Bureau, well known as a skillful physicist and mathematician. He wrote that they were, theoretically, entirely sound and quantitatively, probably, as accurate as the present state of the measurements of wind pressures permitted. The writer determined, however, to withhold publications until the feat of soaring flight had been performed by man, partly because he believed that, to ensure safety, it would be necessary that the machine should be equipped with a motor in order to supplement any deficiency in wind force.

The feat would have been attempted in 1902 by Wright Brothers, if the local circumstances had been more favorable. They were experimenting on Kill Devil Hill near Kitty Hawk, NC. This sand hill, about 100 feet high, is bordered by a smooth beach on the side whence come the sea breezes, but has marshy ground at the back. Wright Brothers were apprehensive that if they rose on the ascending current of air at the front and began to circle like the birds, they might be carried by the descending current past the back of the hill and land in the marsh. Their gliding machine offered no greater head resistance in proportion than the buzzard, and their gliding angles of descent are practically as favorable, but the birds performed higher up in the air than they.

Professor Langley said in concluding his paper upon "*The Internal Work of the Wind*": "The final application of these principles to the art of aerodromics seems, then, to be, that while it is not likely that the perfected aerodrome will ever be able to dispense altogether with the ability to rely at intervals on some internal source of power, it will not be indispensable that this aerodrome of the future shall, in order to go any distance - even to circumnavigate the globe without alighting - need to carry a weight of fuel which would enable it to perform this journey under conditions analogous to those of a steamship, but that the fuel and weight need only be such as to enable it to take care of itself in exceptional moments of calm."

Now that dynamic flying machines have been evolved and are being brought under control, it seems to be worthwhile to make these computations and the succeeding explanations known, so that some bold man will attempt the feat of soaring like a bird. The theory underlying the performance in a rising wind is not new, it has been suggested by Pénaud and others, but it has attracted little attention, because the exact data and the maneuvers required were not known and the feat had not yet been performed by a man. The puzzle has always been to account for the observed act in very light winds, and it is hoped that by the present selection of the most difficult case to explain, i.e., the soaring in a dead horizontal calm, somebody will attempt the exploit.

The following are deemed to be the requisites and maneuvers to master the secrets of soaring flight:

1st. Develop a dynamic flying machine weighing about one pound per square foot of area, with stable equilibrium and under perfect control, capable of gliding by gravity at angles of one in ten (5¾°) in still air.

2nd. Select locations where soaring birds abound and occasions where rising trends of gentle winds are frequent and to be relied on.

3rd. Obtain an initial velocity of at least 25 feet per second before attempting to soar.

4th. Locate the center of gravity that the apparatus shall assume a negative angle, fore and aft, of about 3°. Calculations show, however, that sufficient propelling force may still exist at 0°, but disappears entirely at +4°.

5th. Circle like the bird. Simultaneously with the steering, incline the apparatus to the side toward which it is desired to turn, so that the centrifugal force shall be balanced by the centripetal force. The amount of the required inclination depends upon the speed and on the radius of the circle swept over.

6th. Rise spirally like the bird. Steer with the horizontal rudder, so as to descend slightly when going with the wind and to ascend when going against the wind. The bird circles over one spot because the rising trends of wind are generally confined to small areas or local chimneys, as pointed out by Sir H. Maxim and others.

7th. Once altitude is gained, progress may be made in any direction by gliding downward by gravity (Fig. 8.4).

The bird's flying apparatus and skill are as yet infinitely superior to those of man, but there are indications that within a few years the latter may evolve more accurately proportioned apparatus and obtain control over it. It is hoped, that if there be found no radical error in the above computations, they will carry the conviction that soaring flight is not inaccessible to man, as it promises great economies of motive power in favorable localities of rising winds. The writer will be grateful to experts who may point out any mistake committed in data or calculations, and will furnish additional information to any aviator who may wish to attempt the feat of soaring" [7].

Almost twenty years later, two glider pilots and soaring pioneers, Robert Kronfeld and Wolf Hirth, succeeded in circling like a bird and climbing with their glider in a thermal to gain altitude; they now encouraged others to study the clouds and repeat their flights. Another twenty years later, the noted British glider pilot and author Alan E. Slater pointed out, "A highly intelligent gliding pioneer, Octave Chanute, described exactly how glider pilots should bank in order to circle tightly within a thermal (though he did not call it that); but even he made the mistake of advising the pilot to climb only when pointing upwind. It apparently did not occur to him that the

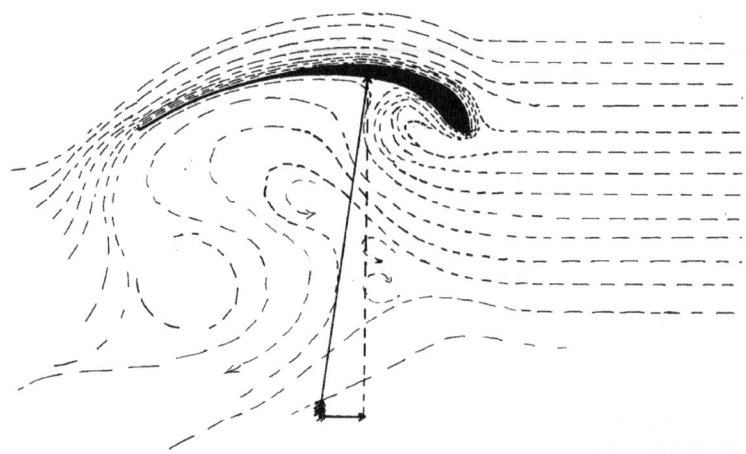

Fig. 8.4 With 3° nose down, a positive propelling force exists and is sufficient. Chanute Papers, Manuscript Division, Library of Congress, Washington DC

thermal does the pilot's climbing for him, all the way round the circle." [8] The widespread idea that birds only gain altitude on the up-wind leg of the circle is probably due to people watching birds as they drift away downwind, Slater explained. "What today's pilots owe to the observations and labors of the veteran civil engineer can only be conjectured; Chanute anticipated them by many years, making success only a question of time." Today, pilots thermal regularly to gain altitude, once in the air.

8.2   Should the Government Sponsor Aeronautical Work?

Early in 1909, Brigadier General James Allen, Major George Squier and Lieutenant Frank Lahm appeared before the sub-committee on Fortifications requesting a $500,000 appropriation to continue the aeronautical work of the Signal Corps. At first the request was looked at favorably but was dropped before being voted on. Hearing this, Jones published a short note in *Aeronautics*, "The appropriation desired for military has been ignominiously slaughtered. This is digressing" (Fig. 8.5).

With Congress not approving funding to continue work in aeronautics, maybe a national aeronautical laboratory should be established? James Means submitted his thoughts to *Aeronautics*, [9] "Members of Congress have it in their power to give to the United States an instrument of war of great importance. For three years past the appropriations made by Congress for the Naval Service and for the support of the Army have amounted to more than one hundred and seventy million dollars per annum, with future appropriations likely to be even larger … The eddies and whirls of air encountered by the aviator in windy weather bring entirely new conditions. Precise measurements cannot be obtained when machines are in free flight. I believe that the present airplane is not yet conceived on correct principles; until the correct principles are discovered by research, development will be at a standstill. Let us have the research work for a foundation" [9].

Prior to publishing Means' plea, Jones had asked a few people for comments. Albert Zahm responded, "Mr. Means' plea for a federal aeronautic laboratory seems both opportune and reasonable. Such a laboratory would do for aerial navigation what the model basin at Washington does for marine navigation. It would determine by means of accurate instruments, in the hands of trained observers, the aerodynamic data needed by inventors. There is no doubt that a liberal appropriation for military aeronautics would quickly lead to the development of such a laboratory as Mr. Means

EVERYBODY FLIES BUT UNCLE

Fig. 8.5 The hopes of the army for sufficient money to continue its experiments in aeronautics were blasted when the House reconsidered its action of last week and withdrew the increase of $500,000 then made to the army appropriation bill. *The Philadelphia Inquirer,* January 31, 1909

has suggested ... It is the wish of the Signal Corps not only to purchase aircraft, but also to stimulate inventors and manufacturers to produce the best possible. They would be furnished with physical data, would be offered an ample testing place and machinery for repairs or construction, they would be offered rewards for the best designs, and a market for aircraft of every type as soon as efficiency was demonstrated. Such an institution would be the center of aeronautical activity, where the inventor, the manufacturer, the scientist and the soldier would mutually stimulate and assist one another" [10].

Instead of sending his comment to Jones for publication, Chanute responded to Means directly, "You are quite right, there is a lot of loose talk, scolding the congressmen and criticizing officers, the authors of which deserve a rubbing. But, as the skunk said of the automobile, 'what's the use.' Such attacks defeat themselves in the minds of those whose opinion is worth having" [11].

To expand the general knowledge of aeronautics, Charles Walcott wondered if the Smithsonian should become involved. Chanute's thoughtful three-page response, dated February 22, 1910, shows his full understanding of the current aeronautical situation. "Inasmuch as the Aeronautical Institute of Koutchino has been experimenting upon the resistance of air upon screws and upon atmospheric friction, while Mr. Eiffel, after making experiments upon plane surfaces dropped by gravity in an apparatus installed in the Eiffel Tower, is now engaged in investigating curved or arched surfaces in artificial currents of air in a wind tunnel, I beg to suggest that the Smithsonian Institution might profitably engage in researches upon the one query which all aviators are now seeking an answer, i.e., the possibility of obtaining automatic stability, either by a combination of surfaces or by adding moving parts.

"Prof. Langley proposed tandem surfaces fore and aft and dihedral angles laterally, but would probably have found the latter inadequate in quartering gusts. The Wrights used to say that automatic equilibrium in the air was impossible of attainment, but the newspapers report that they have applied for a patent in England covering a device for that purpose. Hargrave exhibited a model at the Royal Society of New South Wales in November 1909, which he claims to be a complete solution of the problem. Montgomery of California experimented with an airplane, which he says righted itself even if it turns a somersault in the air, while a score of other inventors believe that they have designed adequate devices. I myself experimented with three methods for obtaining automatic stability with inconclusive results. It would be my idea to gather information upon these various proposals to investigate them and to test the more promising, and then to publish the results for the benefit of aviators who want to apply motors. The difficult thing, of course, is to find an expert competent to carry on and to analyze such delicate experiments. It now occurs to me that Dr. Zahm of Washington might be such an expert. I do not know whether he would undertake it, or whether he would be a good operator for a gliding machine. I should think, one investigator and a model maker ought to obtain valuable results in six months. I shall be happy to help" [12].

8.3   Starting an Airplane Business

Seeing the announcement of *"Another Guaranteed Flying Machine in America"* in *Aeronautics*, [13] members of the Aeronautic Society of New York wanted to buy a real airplane so that members could learn and experience flight. William Hammer traveled to Dayton to discuss a purchase with Wilbur, who was not interested; Ernest Jones then discussed the same topic with Curtiss, who was interested but not sure where the money would come from to buy the material to build an airplane. After more discussion, the Aeronautic Society purchased their airplane on March 2, 1909, from Curtiss for $5,000, to be delivered at the Morris Park Race Track. *The World* was quick in describing that the modest Curtiss with his light blue eyes "becomes the first successful American airplane maker with a machine that has actually flown to demonstrate the commercial value of the flying machine in America" [14].

Just one day after the sale was made public, Cortlandt Bishop, president of the Aero Club of America, announced that a group of the wealthiest men in New York had furnished funding to combine the interests of the Herring airplane, to be built for the Signal Corps, and the Curtiss Company in Hammondsport. And "… Mr. Herring has turned over all of his American patents to the company, for which he received a majority of the stock of the concern. Together the two men will perfect their invention, and will begin the manufacture of airplanes at once." [15] For some reason neither Bishop nor Curtiss had asked to see the patents that Herring claimed to have. Yes, he had submitted claims on his powered triplane to the Patent Office in late 1896, but they were rejected. The only patents that Herring could claim were an "Anti-drip nose for pitchers" (No. 457,715) [16], approved in 1891, and the joint British patent communication with Chanute "Improvements in or relating to means and appliances for effecting aerial navigation" (GB15,221) [17] from 1897, covering Herring's motorized triplane and Chanute's biplane; it lapsed two years later, as no one paid the required £20 for an extension of time.

To gain international prestige, Bishop really wanted an American pilot to enter and win the Gordon Bennett Trophy, so that the next international flying event could be staged in America. Curtiss agreed to travel to Rheims, France, and give it a good try. The *Grande Semaine d'Aviation* (August 22–29, 1909) was a spectacle never before witnessed, with twenty-three airplanes built by various designers from several countries filling the skies.

The competition was stiff, but Curtiss was determined to show the superiority of his airplane and won the Gordon Bennett Cup with the prize money (Fig. 8.6).

Beats All Creation, B'Gosh!

GLENN H. CURTISS, THE AMERICAN AVIATOR, WINS
GORDON BENNETT TROPHY, BEATING WORLD'S RECORD

Fig. 8.6 The international cup, the Gordon Bennett Trophy, was awarded to Glenn Curtiss for the fastest aerial journey of 20 kilometers (12.42 miles) ever accomplished by man. *The Call,* San Francisco, August 29, 1909

Wasting no time, Curtiss rushed back to the United States, attended the Aero Club luncheon in his honor, but did not spend much time in New York, even though he had signed a contract to perform at the Hudson-Fulton Celebration. He hurried to Hammondsport, where Tom Baldwin and his wife Lena were waiting to travel to their next two appointments.

Organizers of the Saint Louis centennial celebration wanted every citizen to see an airplane or a dirigible fly high in the sky. They asked several balloonists and Curtiss, and all agreed to fly. Wanting additional publicity, organizers sent a reporter from the *St. Louis Post-Dispatch* to Chicago to interview Chanute on the future of aviation in 2009, or one hundred years into the future. The civil engineer was sure that the airplane would not supplant the railway train, but Chanute also thought that the current airplane design was not perfect yet, especially because it could not carry more than five or six people. Thinking more ahead into the future, he thought the next breakthrough development could be motorless flight using updrafts "as the soaring birds, without the action of motors, save that of the wind. That will only be practical in certain regions and under certain circumstances where there is an ascending trend in the wind, for it is those ascending winds produced by irregularities in the earth's surface or by the rising of heated air, which is utilized by the birds. I think as soon as man begins to utilize the wind and air currents, this will be another great step in flying" [18] (Fig. 8.7).

Chicago politicians did not want to stand behind anything Saint Louis folks were doing, so they, too, contacted Curtiss. Yes, some Chicagoans had seen Roy Knabenshue fly his dirigible in 1905 and Larry Lesh in his glider in 1907, but no one had seen a real airplane fly above Chicago; this would be very exciting! Curtiss agreed to perform but wanted to make sure that the selected site was big enough to safely maneuver his airplane.

Fig. 8.7 What progress will aviation make in the next hundred years? What will be the state of air navigation when St. Louis celebrates its two-hundredth's birthday? *St. Louis Post-Dispatch*, October 10, 1909

The crated "Rheims Racer" and Baldwin's airship left Hammondsport by train, while Curtiss and his team travelled by car, stopping in Chicago on their way to Saint Louis [19]. Organizers showed several sites, with the Hawthorne Race Track in Stickney (now Cicero) being their preference; all looked good and Curtiss proposed, "I intend to show Chicago some figure eights, and if you had a statue of liberty loose which you could set up in the middle, I might give you an imitation of Wright entertaining the multitude at the Hudson-Fulton celebration. But you must not anticipate that I will sail into or over the downtown district and encircle the skyscrapers."

The next day the Curtiss–Baldwin troupe continued their trip to the three-day affair in St. Louis. Here, a stiff wind, mixed with rain, delayed the takeoff at the Aviation Field of Forest Park (the site of the World's Fair in 1904) until almost dusk. The initial crowd of an estimated 300,000 spectators had mostly disappeared, but Curtiss electrified the waiting rain-soaked spectators with a swooping flight of more than a mile under a brilliant rainbow that set off the golden wings of his machine [20].

After a successful event in Saint Louis, the Curtiss team traveled back to Chicago; the *Tribune* announced, "In the flights at hand Mr. Curtiss will not only try for new speed records, but will cut figure eights and do other maneuvers to show his wonderful control over this creature of wood and canvas." After the motorcycle race was completed at the Hawthorne Race Track, Curtiss leaped into the seat of his airplane, pulled the lever, and the machine rose smoothly into the air; there was an audible gasp, followed by a roar of cheers from the crowd, as the strange bird swooped close over their heads and then sank gradually back to earth. The whole flight covered a little over a quarter mile and lasted perhaps forty seconds, but the attending crowd was thrilled, including Chanute who sat with one of his daughters in the grandstand (Fig. 8.8).

Curtiss made the requested three flights every afternoon in high winds, [21] launching Chicago into the air age. At first Baldwin refused to even assemble his dirigible, but on the last day he made two flights, "successfully negotiating the difficulties to be overcome in the Windy City." Chicago lived up to its reputation or as Chanute stated, "the usual Chicago half gale prevailed" [22]. Curtiss declared Chicago to be the most unfavorable of all cities in the country in which, or rather over which, to fly as its location made it subject to the most freakish and unexpected weather conditions. It was as dangerous and treacherous aerially as its lake was to the mariner, and Lake Michigan is admittedly the most untrustworthy of all the great lakes [23].

CHICAGO SEES CURTISS FLY

Heavy Winds Limit Aviator to One Attempt at Racetrack.

SOARS 60 FEET OVER CROWD.

The first aeroplane flight ever made in Chicago was accomplished late yesterday afternoon, when Glenn H. Curtiss made a brief ascent at the Hawthorne park racetrack.

Among those greatly interested in the aeroplane were Count Mowik de Beaufort, Charles Coey, Octave Chanute, Mrs. Curtiss, and Miss Chanute.

CURTISS AND HIS AEROPLANE SHOWING LOCATION OF RADIATOR, GASOLINE TANK, MOTOR, AND PROPELLERS

Fig. 8.8 (LH) Chicago's upper regions, democratic in soot flecked breeziness, must pay homage to aerial royalty on next Friday, Saturday, and Sunday, when Curtiss will make flights over Hawthorne Race Track. The *Chicago Tribune Sunday Magazine*, Metropolitan Section, October 10, 1909. (RH) A heavy northwest wind on the first day prevented Curtiss and Baldwin to do their requested flight performances. *Chicago Daily Tribune*, October 17, 1909

Curtiss and his wife Lena spent two enjoyable evenings with the Chanute family, talking about flying but also about the control mechanism he employed on his airplane. Feeling confident, Curtiss wrote to Bell, that Chanute believed that "the Wrights have little prospect of winning" their case [24].

The day after Curtiss left Chicago, Chanute received a letter from Emerson Newell, Curtiss' defense lawyer, inquiring about some prior art of wing warping. "The bare idea of warping and twisting the wings is old, but there are several ways of accomplishing it … I obtained an American patent for Mouillard on May 18, 1897, in which he claims the warping of the rear of the wing. It will be for experts to determine what are equivalent devices. I do not remember any balancing devices such as Mr. Curtiss used. I dare say, however, that they are quite old" [25].

8.4   Summarizing Recent Progress in Aviation

Even though the 77-year-old Chanute started to show his age at times, he was still the best-informed man on the question of aeronautics. He enjoyed talking about flying machines and sharing what he knew. A good opportunity came when he was asked to present another talk to Western Society of Engineers (WSE) members.

In spite of inclement weather, about 150 members and guests attended the meeting at the Monadnock Building. WSE president Andrew Allen introduced the speaker: "It is a remarkable instance that exactly twelve years ago this evening, 20 October 1897, Mr. Chanute gave his first paper before this society on the subject of aviation, entitled 'Gliding Experiments.' The opportunity comes to few men to appear before the same body twelve years after their predictions were made and be able to point to the fulfillment of those predictions, as can be done by Mr. Chanute tonight. The airplane is a real flying machine, and Chanute had been deeply interested in its perfection for many years. He is an acknowledged authority on the subject, and he has lent his skill constantly and consistently to experiments which finally seem to have brought success. The story of the airplane experiments conducted by Chanute and other scientists in the sand dune region of northern Indiana, is probably familiar to most people living in the mid-west. They were the beginning of the end of success which now seems to be near, if it has not already come" [26].

Using lantern slides Chanute highlighted the aeronautical accomplishments of the past decade. He first described his gliding experiments with Avery and Herring, and then discussed the achievements of others in chronological order: Maxim's work of 1894, Langley's activities between 1896 and 1903, the work of the Wrights, the flight of Santos-Dumont and other French enthusiasts. In closing Chanute showed a film of the flying activity at Rheims, just received from Curtiss, who could not attend the talk, as he needed to get ready for the legal patent case back east.

A transcript of Chanute's presentation was published in the WSE *Journal*, followed by a detailed "*Chronology of Aviation*," stating that "Bewildering advance in aviation took place in 1908 and 1909. When it is remembered that the first successful manflight, landing safely, was made by Wright brothers on December 17, 1903, that it took them two years (1904–1905) to obtain entire control over their machine; that the Santos-Dumont flight of 720 feet on November 13, 1906, excited the wonder and admiration of all Europe. Now flights of over 100 miles were made and a height of 1,600 feet is said to have been attained. There are hundreds of successful experimenters in the field, and records are broken every few days. It would be quite futile to give a compendium of all the flights of 1909, as they number thousands" [26, 27]. Both papers were widely republished, confirming that aviation was indeed a topic of general interest.

Following the lecture, the WSE president announced that Chanute was elected an Honorary Member of the society, and a few months later, the originator of the "Chanute Medal" was awarded the "Chanute Medal, Mechanical Engineering" for the best-researched engineering paper in 1909.

8.5   The Monoplane Enters the Aviation Scene

Talented competitors with fresh ideas to improve the airplane's performance regularly appeared on the scene. One of these was the Romanian experimenter **Traian Vuia** (1872–1950), who broke away from the biplane design, introduced by Chanute, and improved by Farman and Delagrange in France, and by the Wrights and the Hammondsport group in the United States. Vuia discussed his "aéroplane automobile" at the Academy of Sciences in early 1903, but members rejected his design as utopia, "The problem of flight with a machine heavier than air cannot be solved." But some Aéro-Club members thought Vuia's proposal doable, and Victor Tatin offered to help design the propeller. In March 1906, Vuia's No.1 drove on four wheels on an ordinary road to pick up speed and actually was airborne for a split second. Vuia No.2 was tested in June/July 1907 [28], rising to about 16 feet and flying for some 65 feet.

Even though Vuia's design was not as successful as hoped, his idea of a tractor-propeller driven aéroplane inspired others to design swift, light and easy to handle monoplanes. The public was quick in calling the monoplane the most advanced form of a high-speed airplane, and the patriotic Ferber announced, "Today we are abandoning the solution found by the Wrights and return to the monoplane, the original French machine, and the astonished crowds of those who do not believe in aviation will cry: 'Monoplanes are unstable, they won't do, and you will never be able to use them in a high wind.' To that, I say: 'You have seen the results of my doctrine: step by step, jump by jump, and flight by flight; you will soon see even bigger results, where we shall conquer the planet from rim to rim, from town to town, from continent to continent" [29].

To help conquer the world, the flamboyant British *Daily Mail* had announced the creation of a £500 prize for the first successful crossing of the English Channel by an airplane in October 1908. As no serious attempt was made, the prize was doubled to £1,000, and the offer was extended to the end of 1909. Three aviators arrived with their crew in the Calais area in early summer, and French torpedo boats were on stand-by to guide the aviators and rescue them if needed.

Charles de Lambert came with two French-built Wright Flyers, confident that his biplanes had more stability and flying power than the monoplanes of his colleagues. After damaging both Flyers practicing take-off and landings, he withdrew.

Hubert Latham took off in his Antoinette IV monoplane from a cliff near Calais on July 19, heading for Dover, but suffered engine failure about halfway across the Channel. He landed in the sea, and the French Navy destroyer *Harpon* rescued him.

Two days later, Blériot and his team arrived with his monoplane in Les Barraques, just west of Calais. The weather was clear on Sunday morning, July 25, and he flew the 22 miles from Calais to Dover, winning the *Daily Mail* prize. Even though his flight was not a greater achievement than recent feats by other pilots, it electrified the world. The first page of the *Chicago Tribune* showed the big headline, "*Sea Gull Blériot describes flight,*" with the story continuing on page 5. Chanute was thrilled reading the lengthy report. "Dover, England. This sleepy seaport town experienced its keenest thrill for a generation when at sunrise this morning a white winged birdlike machine with loudly humming motor swept out from the haze obscuring the sea toward the distant French coast, and, circling twice above the high chalky cliffs of Dover, alighted on English soil. A calm human seagull, Louis Blériot, a portly and red mustached Frenchman, descended from the saddle, limping on a bandaged foot, which had been burned on a previous flight" [30].

Having received many clippings on Blériot's flight, Chanute forwarded "a bushel of them" with his dry comments to Glassford: "Now that Blériot has invaded England by his lonesome the disquisitions will get more and more numerous, but the British newspapers will get over their scares ..." [31]. Glassford, now Commandant of the Army Signal Corps Station at Fort Omaha, regarded Blériot's flight a step forward, "For purposes of reconnaissance, aircraft will no doubt prove useful, but as destructive engines of war and for transportation, we shall have to await for further developments before taking them into serious consideration. Blériot is to be complimented for his notable feat of crossing the English Channel in a monoplane, which is the more interesting to Americans, as it is understood that he has been a close student of our own Prof. Langley" [32].

To compare the relative merits of the monoplane and the biplane, Chanute assembled a table of "*First Steps in Aviation and Memorable Flights*" [33, 34]. He listed Jean-Marie Le Bris as the first inventor to "warp the wings" and fly more than six hundred feet. The list continued with four entries of the Wrights, followed by efforts of European pilots, but for some reason he forgot to include the flights by AEA members, made half a year earlier.

8.6   Trying Again: Activity at Fort Myer

Several nations had added dirigibles to their military fleet in the past few years, but none had purchased an airplane. Now the United States was to have two, as Herring and the Wrights were to deliver theirs during June and July 1909.

Having requested another contract extension the previous fall from the Signal Corps, Augustus Herring was scheduled to deliver his airplane in mid-June. Being a partner of the Herring-Curtiss Company, he had heard of John McCurdy, a former AEA member, making the first controlled powered flight in Canada in the *Silver Dart*; so he suggested to sell this airplane to the military instead of him building one. Neither Curtiss nor Bell agreed. Next, Herring requested to deliver his airplane at Hammondsport instead of Fort Myer; General Allen denied the request [35]. Not agreeing to any further extensions, the War Department annulled Herring's contract on August 3.

The Signal Corps had extended the Wright's contract after Orville's crash the previous fall; the brothers now arrived with their updated Flyer at Fort Myer on June 20. Encountering several mechanical problems, they too requested an extension. The Signal Corps accepted the Wright Flyer on August 2 as S.C. No.1, and Wilbur trained two officers at College Park, Maryland. Benjamin Foulois received a few hours of basic instruction prior to moving the airplane to Fort Sam Houston, Texas, to train in weather more conducive to flying. After several mishaps, S.C. No.1 was decommissioned and donated to the National Museum.

General Allen's final report to the Secretary of War makes interesting reading. "The successful outcome of the tests of the Wright Brothers' airplane at Fort Myer has fully justified the allotments made for the purchase of such devices. The recent airplane flights by the Wrights and other aviators make a short recital of the Board's efforts to develop this science of interest. At the time the Langley machine was under construction under the allotments of the Board, Messrs. Wilbur and Orville Wright were making tests of their gliding machines on the sand dunes near Kitty Hawk, N.C., and the Board was, in a general way, acquainted with their work. It was not until the latter part of 1905, that the Wright Brothers publicly asserted that they had success-fully solved the problem of mechanical flight. No definite information came to the Board until the spring of 1907, when the Board entered into corre-spondence with the Wright Brothers. The successful outcome of these tests justifies belief in the practicability of the airplane as a useful military adjunct. It is not asserted that the airplane, as furnished under the contract with the Wright Brothers, will revolutionize our present methods of warfare, but as a result of the very rapid development of aviation in the past year it has been demonstrated that, in its present form, the airplane is practicable for certain purposes of observation and reconnaissance. It is undoubtedly true that the development of the gas engine to its present high state of efficiency is hugely responsible for the practical achievements of the past year in aerodromics, but the Board, from its early and long-continued interest and support in the development of the science, feels itself in position to assert that to Samuel P. Langley and his pioneer efforts in the establishment of the basic principles of the art is due the present state of mechanical flight. History will accord to Doctor Langley his place in the development of the science, and this late expression of the Board's opinion is but a scant need of justice to a distinguished scientist who was sensitive to the ridicule and abuse heaped upon his efforts in his latter days" [36].

Wanting to continue the aeronautical work, Allen offered a few suggestions. The Board of Ordnance and Fortification allotments for aviation had provided a total of $46,000, of which only $30,000 were expended. Since Congress had made no appropriation for fiscal year 1910, he requested permission to use the balance of $16,000 for experimentation and to purchase a Curtiss machine, which was faster and easier to handle. The request was granted; in April 1911, the Army took delivery of S.C. No.2, a Curtiss 1911 Model D, and S.C. No.3, a Wright B Flyer.

Second Lieutenant Henry H. "Hap" Arnold (1886–1950) arrived for duty in June 1911 to teach at the Signal Corps aviation school at College Park. According to Arnold, who had learned to fly at the Wright School in Dayton, "No two types of controls were the same in those days, and from the student's point of view the Wright system was the most difficult." Here the pilot had to manipulate three different levers with two hands changing the shape of the entire wing, or warping. In comparison, Curtiss airplanes had one integrated control wheel and straps attached to the pilot's upper body (or shoulder harness) that controlled cables to move the ailerons, much less confusing for the pilot. A few years later, Arnold became a captain and a "Junior Military Aviator" and was ordered to Washington to assist General Squier, when the U.S. entered World War I. One of the first steps after entering the war was to begin the training of an aerial army, by organizing ground schools and establishing flying fields. The aviation field at Rantoul, Illinois, was to train and teach future military personal, and Arnold suggested that the site should be named "Chanute Field". Throughout his long military career, he pushed for improvements and safer and better performing airplanes for the military personal.

8.7   A Trigger for Aviation Enthusiasm

It was only natural for an automotive manufacturer like the German industrialist **August Euler** (1868–1957), the great-grandson of the famous mathematician Leonard Euler, to become interested in the emerging aviation industry. In October 1908 Euler negotiated a license agreement with the Voisin brothers in France, leased part of the Griesheim military field and established the first German airplane manufacturing company, the Euler-Werke, in early 1909. As part of the contract, the Voisin brothers supplied one of their Chanute-type gliders, so that Euler could learn to fly.

The *Internationale Luftschiffahrt-Ausstellung* (ILA or International Aeronautic Exposition) was staged in Frankfort in the summer of 1909. Euler rented a booth to promote and sell Voisin airplanes. While demonstrating the airplane to a potential buyer and having official witnesses, he made a flight of 4 minutes 35 seconds on September 28, 1909, earning the first pilot license in Germany.

To create entertainment for the entrance-paying public, ILA organizers arranged for gliding competitions. To launch the gliders, they built a hill near the grandstand for the public to watch [37]. Admiring the gliders, **Hermann Reichelt** (1878–1914) introduced himself to Euler, cautiously explaining that he wanted to learn everything about "*Aviatik.*" Having no immediate use for the old Chanute-type glider anymore, Euler invited Reichelt to come to Griesheim and help pull it out of storage. Back at the ILA, the two men replaced a few broken ribs and assembled the glider in one of the hangars. Euler provided some initial flying lessons and Reichelt was ready to give gliding a try.

The new owner was excited about flying his own glider and soon his glides were longer in distance than those of the other participants. Feeling confident, he entered the glider flying competition and won the Grand Prize for gliders, 3,000 Marks, which was a lot of money for someone who earned little as a free-lance photographer and artist. After the ILA closed, Reichelt returned to Dresden with his glider and became interested in powered flight. He and his passenger were killed in a powered airplane in 1914 (Fig. 8.9).

Just like the Voisins, Euler established a flying school as part of his company and supplied a Chanute-type glider to his customers, so that they could learn to fly safely before stepping into a powered airplane [38].

Having spent time in the United States, and then in England, **Hellmuth Hirth** (1886–1938) had read in the papers about Santos-Dumont and Farman and developed an interest in airplanes. When he returned to Germany, his father introduced him to Euler, and he began his flight training

Fig. 8.9 Several pilots participated with their gliders, including two Chanute-type versions, each making nice straight-ahead longer flights. *Denkschrift. Ergebnisse der ersten internationalen Luftschiffahrt-Ausstellung zu Frankfurt a/M 1909*

in a Chanute-type glider at the Griesheim flying field. "Euler, Colonel Ilse, my father and I took turns making short flights of up to 40 meters, usually in strong, steady winds. The accidents with these motorless craft were mostly light, although I have the feeling that a pure glider is more dangerous than a powered airplane" [39]. Having watched Blériot fly at the ILA, Hirth became convinced that monoplanes would have a better future than biplanes. He became a prominent racing airplane pilot, and took part in World War I. His younger brother, Wolf, became an outstanding designer and manufacturer of sailplanes as well as a competition pilot.

8.8   The War of the Aviators and Chanute's Involvement

The Wrights believed their patent [40], granted on May 22, 1906, covered all functions of controlling an airplane; they considered the airplane their intellectual property. The wording of the main claim in the patent reads, "We wish it to be understood, however, that our invention is not limited to this particular construction, since, any construction whereby the angular relations of the lateral margins of the aeroplanes may be varied in opposite directions with respect to the normal planes of said aeroplanes comes within the scope of our invention."

Hearing of Glenn Curtiss making the first commercial sale of an airplane to the Aeronautic Society in late June 1909, Wilbur filed a bill of complaint in both their names on August 18, 1909, to enjoin the Herring-Curtiss Company and Glenn H. Curtiss from manufacturing, selling, or using for exhibition purposes the Curtiss airplane. And the next day, Wilbur filed suit against the Aeronautic Society to prevent exhibition and use of their just-purchased Curtiss airplane, the *Golden Flyer,* starting what became known as the "war of the aviators." The Wrights testified that their machine was the first to accomplish successful human flight, and "we further state that the invention was made at a time when efforts at human flight were almost at a standstill throughout the world, and that the present activity and interest in matters relating to human flight, both in Europe and America, had its inspiration and origin from knowledge and information regarding the successful labors of the plaintiffs, which labors eventuated in the patent sued on" [41].

Believing that their patent (thus their secret) was fully protected, the Wrights took the next step. Three years after the Voisin brothers formed the first aircraft manufacturing company in the world, two years after the Curtiss Manufacturing Company was formed with a capital of $600,000, only eight months after the incorporation of the not very successful Herring-Curtiss Company and the more successful Euler Werke in Germany, the Wright Company was formed on November 22, 1909 in Albany, New York [42]. The new company was to deliver Wright airplanes to customers by May 1, 1910 and was to apply for injunctions against persons alleged to be infringing on Wright patents [43].

Not having corresponded with Chanute for almost half a year, Wilbur now shared the latest news. "We are all home again after a rather strenuous summer and autumn ... We have been working very hard on the organization of our business and the preparation of our case in the suit against Curtiss & Herring." Continuing his two-page handwritten letter, he gave details on their just formed company. "We receive a very satisfactory cash payment, forty percent of the stock, and are to receive a royalty on every machine built in addition. The general supervision of the business will be in our hands, though a general manager will be secured to directly have charge" [44] (Fig. 8.10).

At first, Chanute just listened to the news about the patent conflict and had good intentions to remain neutral, hoping that he would not be called to testify; but, maybe, he could just offer an opinion at times. He wrote to the editor of *Aeronautics*, Ernest Jones, "the newspaper write-ups conveyed

MILLIONS BACK TRUST

Wright Brothers Form Combination to Control Aviation.

THIS COUNTRY AND CANADA

Capital Placed at $1,000,000 and No Stock on Sale.

MONOPOLY OF AIR CURRENT

Company to Take Over Wright Patents and to Prosecute All Infringements.

"GOING SOME."

Fig. 8.10 Several newspapers published big write-ups of the Wrights forming a company with a large financial backing. *The Evening Star,* Washington, DC, November 23, 1909. And the cartoon spread the word even more. *The World*, New York, November 29, 1909

a number of false impressions. However, what is the use?" He also thought that European experimenters were infringing the Wright patents a good deal more than Curtiss [45]. Hearing that Jones wanted to publish a "legal opinion upon the claims by a patent lawyer," Chanute just responded, "… I think the Wrights have made a blunder by bringing suit at this time. Not only will this antagonize many persons, but it may also disclose some prior patents, which will invalidate their more important claims. I doubt whether a legal opinion by a patent lawyer, if published in advance of the trial, will carry much conviction. The course is to point out the important consequences to aviators and to urge that a decision shall soon be arrived at" [46].

There was certainly no general norm for or against patenting in the aviation community. Most experimenters obtained patents to document or publicize their achievement and ensure credit for priority in making an invention. Even after the Wrights began enforcing their patent with lawsuits, their opponents usually responded by attempting to undermine their suits, but did not sue the brothers for infringing earlier patents, like Chanute's trussed biplane design.

When the Wrights filed suit against the Aeronautic Society, Thomas A. Hill, one of its members, but also a member of the New York Bar, decided to look closer at the Wright's patent with its wrappers. His three-page report was published in October 1909 in *Aeronautics* [47]. "I have carefully inspected the file wrapper at Washington in the Wright Brothers' case, and their bill filed in the United States Court for the Southern District of New York in action brought against the Aeronautic Society, and am at a loss to find the motive for such a suit at this time." He described the early steps by the brothers to claim the patent and the responding statements by the examiners rejecting the claims at first. "A final amendment was filed, complying fully with the examiner's requirements, and the case was formally allowed on April 21, 1906," and patented one month later. And Hill concluded, "This cursory examination of the Wrights' patent and suit against the Aeronautic Society has not had for its object the determination of the validity of the Wright claims ... The use of supplemental surfaces appears to be indisputably a public right, and upon this apparently hinges the issues involved in the present suit. The bill, however, filed on behalf of the Wrights is not specific, and is a mere generalization of rights and grievances."

Arnold Kruckman, one of the editors for *The World*, who always had an ear for office gossip, discussed the pro's and con's of the patent fight in "*Worldwide interest in the case against Glenn Curtiss*" [48]. He wrote that Lilienthal, Le Bris and Mouillard all had ideas of changing the wing geometry and then again stated, "It was from Chanute indeed that the Wrights received the inspiration to build their airplanes ... Their persistent failure to acknowledge their monumental indebtedness to the man who gave them priceless assistance has been one of the most puzzling mysteries in their career." Kruckman finished his report with a statement by Tom Baldwin, "You can show me one hundred patents antedating every one of the features used by the Wrights. I'll admit you are right. But you can't side-step the point that they unlocked the secret of mechanical flight." Reading the newspaper report made Wilbur write a harsh rebuttal to Kruckman: "... Many of the published stories have been very embarrassing because if left uncorrected they tend to build up a legend, which takes the place of truth, while on the other hand any attempt on our part to correct inaccuracies gives us the appearance of ungratefully attempting to hurt the fame of Mr. Chanute. Rather than subject ourselves to criticism on that score we have preferred to remain silent. We, rather than Mr. Chanute, have been the sufferers from this silence so far, we see no immediate danger that he will not receive the credit to which he is justly entitled for his services to the cause of human flight" [49]. We do not know if Kruckman responded, but there was another report in *The World* one week later, "*Wrights open court fight on rival air pilots*" [50], describing the first day in the court, and closing with "The lawyers suggested that even the Wrights have not attained stability of any great value, that their fame rested largely upon their skill as 'airplane chauffeurs'."

When the suit opened in Buffalo, New York, in early December 1909, the Wrights' patent lawyer, Harry Toulmin, submitted a 160-page brief, discussing the entire history of human flight, at times extracting text from Chanute's *Progress* book and also mentioning Mouillard's patent "*Means of Aerial Flight.*" Actually, the significance of this patent appears to be quite fluid throughout the patent struggle. In the Wright Company vs Paulhan suit, Judge Hand stated that he had examined the Mouillard patent and that "… in none of the nineteen claims is there anything, which in any way even foreshadows the patent-in-suit. Indeed, the machine, which was a glider, had no tail whatever, and to depress the marginal edge of one wing would have only resulted in entirely disturbing the equilibrium which he might have attempted to restore. The depressing of one wing was meant only to turn the airplane."

One year later, another lawyer for the Wrights, Frederick P. Fish, also discussed Mouillard's patent. "He did not get so far as to see that a bird scheme was impossible and that the only thing that was really practical was Lilienthal's airplane and that only as a basis to which devices for securing equilibrium (our invention) had to be added. And in copying the wings of birds, Mouillard turned down a little bit of a corner of the wing, which can be moved separately. These corners are not connected and cannot be moved together simultaneously as our aileron surfaces and those of the defendants must be. The essential point of our construction is that the aileron extension on one side and the aileron extension on the other are coordinated so that they work together. Mouillard is clean out of this case; there are no means in his patent for maintaining lateral equilibrium. He never even thought of the subject."

To stop any arguments from the opposing side, Wilbur asked his former student Charles de Lambert to buy and mail Mouillard's book *L'Empire de l'Air* as quickly as possible. He then submitted his article "*What Mouillard Did*" [51] to the Aero Club of America *Bulletin*, refuting all claims that Mouillard conceived the use of wing warping and its application to lateral control in an airplane.

Seeing the statement that Mouillard's patented machine could not be reduced to practice, Albert Merrill decided to analyze its twelfth claim, "A soaring machine having wings adapted to move in horizontal planes, a portion of the fabric covering each wing being stiffened by flexible slats and having its rear edge free from the frame of the wing, and cords attached to said rear edge for pulling it downwards substantially as described." In his opinion, Mouillard's or Chanute's application clearly indicated their practical work early in the art along the lines pursued by the Wrights, and he simply stated in closing "I believe this system will work" [52].

8.9   The Intertwined Story of Ailerons and Wing Warping

To achieve the desired handling qualities of stability and control, an aircraft makes use of three primary control surfaces: the elevator, rudder, and ailerons. Due to the complexity of the laws of aerodynamics, Chanute explained that "science has been awaiting the great physicist, who, like Galileo or Newton, should bring order out of chaos in aerodynamics, and reduce its many anomalies to the rule of harmonious law. It is not impossible that when that law is formulated all the discrepancies and apparent anomalies which now appear, will be easily explained and accounted for by one simple general cause, which has been overlooked" [53].

Chanute had discussed in his "*Progress in Flying Machines*" articles the work of several enthusiasts who had tackled the problem of controlling their craft. Count **Phillipe D'Esterno** used pilot-activated torsion at the wing tips of his flying machine of 1864 to preserve lateral balance. Next, **Jean-Marie Le Bris** utilized a mechanism in his *Albatross* glider of 1867, which varied the angle of incidence in flight. **William Beeson** designed a machine in 1888, where the operator "might alter the angle of incidence of the mainsail, of the tail or of the rudder." Apparently, the rudder was operated by a chord running to the operator's hand. Next, Chanute described **John J. Montgomery**'s glider, built in 1884, that employed a mechanism to achieve lateral balance through pilot-operated laterally-hinged wing flaps. And a French experimenter from Marseille, **Albert Bazin**, had experimented with a "bipolar" kite in 1888, which was stable in both directions and flew steadily without a balancing tail. Bazin controlled it with two strings and a stick at each end, allowing him to vary the angle of incidence.

Louis-Pierre Mouillard and Chanute had filed their patent claims in September 1892 [54]. Here Chanute explains, "that vertical steering or equilibrium depends upon the forward and backward movement of the wings, where by the center of gravity is carried, respectively, backward and forward. The horizontal steering is affected by the downwardly movable rear portion of the fabric in the manner already described. When both sides are pulled down together, they serve as an effective brake to check the speed." Unfortunately, neither Chanute nor Mouillard considered the lifting effect, which would bank the machine in the opposite direction from what they intended.

The **Wrights** submitted their patent claims in March 1903 and Wilbur stated in the Wright vs. Paulhan suit: "We are aware that, prior to our invention, flying machines have been constructed with superposed wings in combination with horizontal and vertical rudders," and stated that none were of any value. But as Chanute had explained to Charles Renard [55], "As in almost all inventions, the later aviators have profited legitimately from the prior labors of their predecessors. Ideas are absorbed unconsciously, and the important thing is to choose the good ones."

Those who have followed the development of aviation pointed out that the mysteries of patent law are far beyond the understanding of the lay mind. But in regard to the aeroplane, it appears that a gasoline engine had been attached to the gliders of Chanute and Lilienthal and Graham Bell. "It is true that the wings of the modern aeroplane may be 'warped,' but this differs in no way from the very simple arrangement by means of which the sail of a boat can be trimmed" [56].

Historians have discussed the controversial history of wing warping and ailerons from every angle, and John Anderson gave a good summary in *Introduction to Flight* [57]: "The history of ailerons is steeped in controversy ... Ideas of warping the wings or inserting vertical surfaces at the wing tips cropped up several times during the late nineteenth century and into the first decade of the twentieth century, but always in the context of a braking surface that would slow one wing down and pivot the airplane about a vertical axis ... The Wright brothers' claim that they were the first to invent wing warping may not be historically precise, but they were the first to demonstrate its function and to obtain a legally enforced patent on its use, combined with simultaneous rudder action for total control of banking."

8.10   More Challengers for the Wrights

Reading about the Wrights forming a company, **George Spratt** wrote to Wilbur. "I'm writing to ask you if you do not think I can claim a share of your success? You will remember that the method of determining surface values you used was original with me and you have repeatedly expressed its value. I believe it has largely enabled you to reach your present first position in aviation. You can now reciprocate without compromising your future if you are in mind to do so" [58]. Wilbur apparently agreed with Spratt. "It is quite true that, before we had seriously taken up the subject of measuring the lift and drifts of surfaces, you told us your idea of balancing the lift of a surface against its shift and determining their relationship directly, instead of measuring each independently. We have not wished to deprive you of the credit for the idea, and when we give the world that part of our work, we shall certainly give you proper credit" [59] (Fig. 8.11).

Spratt decided to wait and see, but agreed to testify in the Wright patent suit in early December, stating that he shared his knowledge at a critical time when the Wrights needed advice the most. Reporting back to Chanute, he wrote that "they cannot be heroes on other men's deeds … Their attitude is a guarantee of their fall, and yet, I never met two young men that I liked better when I first met them" [60].

To document the development of their Flyer thirteen years later, Orville looked for copies of letters to refresh his memory. He also contacted Spratt who responded tersely. "I would like to call your attention to the following. The part I played in the making of your machine was not small. I pointed out the cause of the failure of your first machine, and how to correct it. The method of testing surfaces that I gave you eliminated guess work and made

Fig. 8.11 Dr. George A. Spratt from Coatesville, Pennsylvania, stated that he helped the Wrights understand the flying characteristics of their glider and then the Wrights succeeded. *Butte News*, Montana, July 31, 1909

progress positive. You sought and accepted help throughout the development period of your machine and then made the decision to work alone ... As success began to appear as a possibility, Mr. Chanute and I both noticed your growing reticence and he said, 'They are not the first young men I have helped into a fortune who have shown anxiety to forget me when they have seen it coming'." Spratt closed his letter with, "It makes little difference what you write, history writes itself. Aviation is worthy of leaders who are farther sighted and broader minded than you seem to be now" [61]. Orville's story was published as a pamphlet by the Dayton-Wright Aeroplane Co. later in 1922, where he acknowledged the young man from Pennsylvania who "had made some valuable investigations of the properties of variously curved surfaces and the travel of the center of pressure thereon" [62].

Chanute was probably responsible for arranging the contact between **John J. Montgomery** and Curtiss' lawyers [63]. Montgomery had developed a glider that utilized the principle of flexing wings and his findings were published both in *Progress in Flying Machines* and in the *Proceedings of the International Aeronautical Congress* [64] in 1893. As suggested by Chanute, Montgomery filed his patent claims on April 26, 1905, covering parabolic curvature of the wing with tail control (through combined usage of horizontal and vertical rudder and stabilizer). The patent was issued on September 18, 1906, a few months after the Wrights patent was published. Hearing of the legal problems with the Wrights' patent, Montgomery traveled in March 1910 to Washington to discuss his patent with an examiner and then met the governors of the Aero Club of America in New York. He explained to a reporter [65]: "I have certain patents that I ought to protect, but the principle of the flexing wing has been thrown open to the public, and I do not want to see it monopolized. When Prof. Chanute and I had our talk that resulted in the publication of my research, it was done with full intent and a realization of just what it meant. Chanute then said: 'This work is not for any one man; it is for the benefit of mankind.' That was the way I felt about it then, and it is the way I feel now. Any man is entitled to a reward for his labor, but he must not try to stand in the light and prevent mankind at large from progressing along any one line. Suppose Chanute and Langley and I had taken out basic patents and tried to prevent anyone or the world from profiting by it? Where would the art be today or what future would there be for it? Mankind must be allowed to profit by it."

Montgomery was killed in an unfortunate glider accident in 1911. A decade later, his heirs decided to sue both the U.S. Federal Government and the Wright-Martin Aircraft Corporation for patent infringement. The lawsuits ended in favor of the defendants; Orville Wright dismissed Montgomery's accomplishments, stating that his "work contributed nothing to the art of flying or to the science of aerodynamics."

Early in January 1910, a preliminary injunction was granted to the Wright Company restraining the Herring-Curtiss Company and Glenn H. Curtiss from manufacturing, selling, or using the Curtiss airplane for exhibition purposes. Next, the Wright Company served an injunction restraining Louis Paulhan from flying his Farman biplane and Blériot monoplane. Wilbur took up the argument and was so convincing that the Court immediately granted the injunction. But then **Israel Ludlow,** a lawyer defending Paulhan, raised the point of reasonable doubt whether the Farman biplane operates like the machine described in the Wright patent [66]. The injunction decision was reversed on the technical grounds that the affidavits presented a conflict of evidence, and Judge Hand reportedly said, "The Court is going to have a hard time understanding some of this."

To continue working the case, Ludlow contacted Chanute, who explained, "I do not think that Mr. Wilbur Wright was justified in stating in his affidavit that the method of controlling the equilibrium by warping or twisting the wings was 'unknown to' myself, although I admit that when writing those conclusions in 1893, I attached minor importance to the flexing of wings mentioned in my book on page 76, 2nd paragraph, on page 166, 1st paragraph, and page 196, top of page, which refer to bird action, further alluded to on page 258, 1st paragraph, and I failed to realize the value of the D'Esterno patent of 1864, described on pages 96 & 97, or of the patent and experiments of Le Bris, who rotated the front edge of each wing separately, thus adjusting them to different angles of incidence. Nor, obviously, could I then (1893) mention the experiment of Mouillard in 1896, nor his patent of 1897, which later provides a method of flexing the rear of the wings which may serve in controlling the lateral balance, although the object stated is to provide for the horizontal steering. I gave a copy of that patent to Mr. Wright in 1901" [67]. Ludlow then mailed a copy of Wilbur's latest affidavit and wrote that his comments were "compared to what he allowed his counsel to say before the Court" not pleasant, and he did not like hearing Chanute referred to as an "amiable old gentleman" before the judge. Chanute just responded, "I regret to say that my health is not good, and I do not feel equal to discussing the Wright affidavit you were good enough to send me. Shall I send it back?" [68].

Charles H. Lamson, another pioneer who had worked the flying machine problem longer than the Wrights, had a long talk with Curtiss at the International Air Meet in Los Angeles in early January 1910. A reporter of the *Los*

Angeles Express overheard Lamson's story of being the first person to patent the disputed "wing warping" feature and that he had built a glider for the Wrights, through the agency of Chanute. We do not know if this was a misunderstanding on the reporter's part or on Lamson's part, but we do know that the glider, built by Lamson for Chanute, was to be at the Wrights' camp in time for experimentation. Reading this report angered Wilbur who wrote Chanute that this write-up was "a sample of the class of misrepresentations connected with your name;" he wanted this corrected. The aging Chanute just wrote to Spratt that he was reluctant "to engage in a row."

In April 1910 Lamson filed a legal complaint charging that the Wrights incorporated his invention of maintaining lateral stability by warping or twisting the wings or supporting surfaces in their flying machine. Two months later, the Wright Company filed a written response stating that "the bill of complaint does not show whether the infringement was committed by the defendants jointly or severally, and that it does not allege execution of the letters patent according to law" [69]. The legal action was settled in 1912 in favor of the Wrights.

A reporter of *The World* had contacted Otto Lilienthal's brother, **Gustav Lilienthal**, in mid-February 1910 in Germany, after hearing that he had designed an airplane that could be operated by a one-third horsepower motor, driven by bicycle gearing, in which a person could launch himself like a bird. "The airplane would travel at a moderate speed, as the secret of efficient flying was to fly slowly. My attorney believes I have a good case against the Wrights. He thinks we can collect royalties on our 1894 patent, which they infringe on. But I am convinced the process would be so expensive as to be prohibitive. At any rate the Wrights have not materially progressed along the lines we developed. Anything with wings will fly if sufficient motor power is supplied. The Wrights adopted the curved wings, which was half the problem. Where they failed to progress is in failing to further pursue the problem of wing adjustment" [70]. Even though Gustav thought that he might have a good chance suing the Wrights, neither he nor his lawyer made contact with the Wrights or their lawyers.

At about the same time, Orville, working in Germany, happened to hear of Otto Lilienthal's family experiencing financial difficulties. The need was real, so the brothers mailed in December 1911 a draft for $1,000 (about $30,000 in 2022) to Lilienthal's widow Agnes as a token of their appreciation for her husband and with best wishes for the holidays. She acknowledged the check on Christmas Eve.

Continuing his legal struggles in America, Wilbur now looked for information on the Lilienthal glider which was on display in the National Museum;

he contacted Arnold Kruckman, who still worked for Randolph Hearst's newspaper. "I suspect that it is a machine which Mr. Hearst imported some 15 or 20 years ago. Can you find out for me, when, and by whom, the Hearst machine was used, and exactly what became of it?" [71] This request is curious, as Wilbur had seen the glider at the first aeronautical exhibition in January 1906; three years later he and his lawyer had examined all aircraft at the National Museum and Wilbur had told a reporter that they were checking "the tomb of airships that had failed to fly" [72].

Most likely, Wilbur was working on an article on Lilienthal, one of the last pioneers working the aeronautical problem before they became interested. In the introduction, he wrote, "Of all the men who attacked the flying problem in the nineteenth century, Otto Lilienthal was easily the most important. His greatness appeared in every phase of the problem. No one equaled him in power to draw new recruits to the cause; no one equaled him in fullness and dearness of understanding of the principles of flight; no one did so much to convince the world of the advantages of curved wing surfaces; and no one did so much to transfer the problem of human flight to the open air where it belonged. As a missionary he was wonderful." And concluding his article, he stated: "Although he [Lilienthal] experimented for six successive years, 1891–1896, with gliding machines, he was using at the end the same inadequate method of control with which he started. His rate of progress during these years makes it doubtful whether he would have achieved full success in the near future if his life had been spared, but whatever his limitations may have been, he was without question the greatest of the precursors, and the world owes him a great debt" [73].

But there was something else. Marvin McFarland explains in a footnote, "In 1909 the Wrights did two things that turned many heads in aviation against them: in August they entered suit for infringement of their patent against the Herring-Curtiss Company and Glenn H. Curtiss and against the Aeronautic Society of New York, and in November they formed the Wright Company, capitalized at $1,000,000 (Curtiss had a backing of $600,000), and turned their disputed American patent over to the new company. These actions were interpreted that the Wrights were assuming the role of active businessmen, thus making themselves fair game for anyone clever or crude enough to push them off their pedestal. When it was realized that they would insist uncompromisingly on their moral and legal rights as discoverers of the principles of controlled flight as well as inventors of the airplane, many who had hitherto appeared friendly grew cool and slipped away while others took up cudgels against them. Apart from the purely legal contest over the patent, the opposition centered in a group that liked to think of itself as 'the friends of Langley.' Of course, Langley, who had died in early 1906, had no part in its activities … Among the more prominent Langley adherents were Charles D. Walcott, his successor at the Smithsonian, Alexander Graham Bell, a close friend, and General Adolphus Greely, retired Chief Signal Officer. On the periphery was Albert Zahm, professor of mechanics at the Catholic University of America and later expert witness for Curtiss. Chanute, though he was a friend of Langley, was not identified with this circle. For one thing, he had differed with Langley as to the proper approach to the problem of flight, and Chicago is a long way from Washington" [74].

8.11   Differing Philosophies

The many letters between Chanute and Wilbur Wright shed light on their personalities, but also how harmony waned slowly. In the beginning, Wilbur usually confided in the elder engineer whenever he ran into problems or needed input or advice; in later years, he used him to distribute select information to the press. Chanute usually did as he was asked, but he started to notice a growing reticence to share information (Fig. 8.12).

During his missionary trip to Europe in spring 1903, Chanute enthusiastically talked about how much fun it was to fly, about the aeronautical activities by his team in 1896/1897 and the Wrights glider flying in 1901/1902. With a curious public, enterprising reporters used the readily available photos of Chanute's gliders or Langley's aerodromes to highlight their stories on the Wrights. Naturally, the brothers did not like seeing their name in the photo caption of another man's plane, but the public, or even the reporters, barely knew the difference.

Besides being disturbed by what Chanute thought were plain signs that the Wrights were pursuing financial reward over aeronautical progress, he also worried about them losing their flying skills. He believed that he always tried to give them sound, objective advice and was deeply concerned that they, complacent in their original success and involved in pointless (in Chanute's mind) litigation, were losing the race of aviation development, a recurring theme in their correspondence. "I still differ with you as to the possibility of your being caught up, if you rest on your oars. It is practice, practice, practice which tells and the other fellows are getting it," Chanute wrote in November 1906 [75]. Wilbur responded by return mail: "Fear that others will produce a machine capable of practical service in less than several years does not worry us" [76]. He knew they would make their fortune in due time.

Fig. 8.12 Octave Chanute in late 1908, George Grantham Bain Collection, Library of Congress, Washington, DC. LC-USZ62-106,858

The brothers were proud of their accomplishments, and they did not appreciate reading that their knowledge came from that "old man." Graham Bell gave what appeared to be the public's opinion in a talk at the Washington Academy of Sciences in March 1907: "In all of the accounts which I have seen of the experiments of the brothers, no mention has been made of the fact that the success of the Wrights has been built on the valuable work of Mr. Chanute, who for years carried on work in construction and testing of gliding machines, and who I understand furnished the Wright brothers with the design for their first gliding machine. There is perhaps no one who has made a closer study and has a more thorough understanding of the whole subject of aerodromics than Mr. Chanute, and I should like very much to see him given due credit for the very important work he has done" [77].

8.12   A Good Friendship Sours …

As part of the AAAS, Section D, the 77-year-old Chanute lectured on December 31, 1909, on his favorite topic, "*The present status of aerial navigation.*" Josiah B. Millet, a member of the Boston Aeronautical Society who had known Chanute since the mid-1890s, convinced fellow Bostonians to arrange a special dinner on January 12 to honor the elder statesman. About 130 Bostonians attended, including Willis Moore from the Weather Bureau, Lawrence Rotch of the Blue Hill Observatory and Orville and Wilbur Wright, who talked briefly on the future of flying and about their former mentor. Chanute's unhappiness with Wilbur peaked at this event.

Was it coincidence that Kruckman requested an interview with Chanute a few days later? He submitted the transcript to the wire service and his relatively short write-up in *The World* stated that "*Dr. Chanute denies Wright flying claim*" [78]. He quoted Chanute as saying, "I admire the Wrights. I feel friendly toward them for the marvels they have achieved; but you can easily gauge how I feel concerning their attitude at present by the remark I made recently. I told him that I was sorry to see them suing other experimenters and abstaining from entering the contests and competitions in which other men are brilliantly winning laurels. I told him that in my opinion they are wasting valuable time over law suits which they ought to concentrate in their work … There is no question that the fundamental principle was well known before the Wrights incorporated it in their machine."

As could be expected, several papers picked up Kruckman's interview, at times rewriting the text to fit their readership. The Philadelphia *Ledger* [79] published the write-up the same day as *The World*. Here the text starts, "Pioneer Aviator praises Wrights. Octave Chanute, the pioneer in the world of successful heavier-than-air aviation, said today that because Wilbur and Orville Wright have brought to fruition the incomplete efforts of others in the field of aeronautics, they were entitled to their full credit and monetary reward. Mr. Chanute said further that the work of the Ohio inventors was very similar to that of Bell with the telephone and Morse with the telegraph, in that the Wrights had reduced to practice the more or less unfinished conceptions of others. As to the wing-warping patent under which they demand damages from others for alleged infringement, this is not an absolutely original idea with them. On the contrary, many inventors have worked on it from the time of Leonardo da Vinci. Two or three have actually accomplished short glides with the basic warping idea embodied in their machines." And in closing, "… the aeroists who have no commercial interests are likewise hoping that the matter will be settled one way or another soon, so that an obstacle to progress in aviation may be removed" [80]. The *New York Times* gave its lengthy write-up a short title: "Chanute deplores Wright Patent Suits" [81]. And the Hammondsport *Herald* [82] published its version a week later, "The Wrights' suits for infringement. Chanute, pioneer glider, who gave Wrights their start, is in favor of public flights. Wing warping idea not wholly original with Wright brothers." The Boston *Evening Transcript* [83] discussed the controversy objectively, "The connection of Octave Chanute with the Wright brothers has been so frequently the subject of discussion of late that it may not be out of place to quote the words of Mr. Chanute himself, thus showing exactly how he has stood in relation to them."

It almost appears like the rising tide of hostility in the aviation community caught the Wrights off guard. Maybe Thomas Hill's write-up in *Aeronautics* or Kruckman's various newspaper posts might have been passed off as steeped in self-interest, but Chanute's criticisms could surely not be viewed in the same light [84].

Reading the write-up in *The World* made Wilbur compile an angry letter on the typewriter and mail it with a copy of the clipping to Chanute's old address, from where it was forwarded. Reading the letter, Chanute first thought about not replying, but he then pulled the wooden box with Wilbur's letters and his letterpress books from the shelves and reread the many early letters to refresh his memory. Chanute's reactions to Wilbur's letters can be sensed in his letter to Spratt, who had mentioned the write-up in the Philadelphia paper. "I have made no statement at all to the Phila. Ledger. That which you saw may have been a reproduction of an interview in the N.Y. World of Jan. 17, concerning which Wilbur Wright took me to task, thus giving me an opportunity to partly free my mind concerning the mistakes which I believe he is making. He has greatly changed his attitude within the last three years" [85].

Chanute then wrote probably the harshest letter he had ever written in his life. "I did tell you in 1901 that the mechanism by which your surfaces were warped was original with yourselves. This I adhere to, but it does not follow that it covers the general principle of warping or twisting wings; the proposal for doing this being ancient ... If the courts will decide that you were the first to conceive the twisting of the wings, so much the better for you, but my judgment is that you will be restricted to the particular method by which you do it" [86].

Being unhappy with Wilbur, who he thought should show a little more gratitude for the help he had received, Chanute continued: "I am afraid, my friend, that your usual sound judgment has been warped by the desire for great wealth. If my opinions form a grievance in your mind, I am sorry, but this brings me to say that I also have a little grievance against you. In your speech at the Boston dinner on 12 January you began by saying that I 'turned up' at your shop in Dayton in 1901 and that you then invited me to your camp. This conveyed the impression that I thrust myself upon you at that time and it omitted to state that you were the first to write to me in 1900, asking for information which was gladly furnished, that many letters passed between us, and that both in 1900 and 1901 you had written me to invite me to visit you, before I 'turned up' in 1901. This has grated upon me ever since that dinner and I hope that in the future, you will not give out the impression that I was the first to seek your acquaintance, or pay me left handed compliments" [87].

Writing his three-page letter to Wilbur made Chanute feel a little better, but reading the letter did not help Wilbur's inner attitude. He believed that the world owed them as inventors; they did not obtain their first experiences on a Chanute machine, even though their first glider design was based on Chanute's patented trussed biplane; they did not consider themselves his pupils, even though the working relationship between Wilbur and Chanute in the early stages in their development could be interpreted as one of student and mentor. Wilbur vented his frustration with another lengthy letter in late January. "I cannot understand your objection to what I said at the Boston dinner about your visit to Dayton in 1901. I certainly never had a thought of intimating either that you had or had not been the first to seek an acquaintance between us. You also object to my expressing an appreciation of the influence, which your friendship had on our work and lives. One of The World articles stated that you had felt hurt because we had been silent regarding our indebtedness to you. I confess that I have found it most difficult to formulate a precise statement of what you contributed to our success … We on our part have been much hurt by your apparent backwardness in correcting mistaken impressions, but we have assumed that you too have found it difficult to substitute for the erroneous reports a really satisfactory precise statement of the truth. If such a statement could be prepared it would relieve a situation very painful to you and us … I have written with great frankness because I feel that such frankness is more healthful to friendship than the secretly nursed bitterness which has been allowed to grow for so long … We have no wish to quarrel with a man toward whom we ought to preserve a feeling of gratitude" [87]. The aging engineer decided not to respond.

Not hearing back from Chanute, Wilbur composed a kinder letter three months later: "My brother and I do not form many intimate friendships and do not lightly give them up. I believe that unless we could understand exactly how you felt, and you could understand how we felt, our friendship would tend to grow weaker instead of stronger … It is our wish that anything, which might cause bitterness, should be eradicated as soon as possible" [88]. The increasingly frail Chanute was gratified to receive the letter and responded: "I hope, upon my return from Europe, that we will be able to resume our former relations." But his clock was running out; he died half a year later.

The ten-year friendship ended during the patent war, when Chanute critically and publicly challenged the Wright's story. The correspondence in the early part of 1910 marked the most poignant quarrel in aviation history up to that point. In their philosophical disagreements, they both had vented their pent-up feelings. This so-called controversy, which had built up over time, was part of human relations, upbringing, age difference, ambitions and egos. Neither Chanute nor Wilbur should be faulted, these developments are part of the turbulent history of aviation, human characters and a less-than-glorious chapter in the development of the airplane.

9

When Will We All Fly?

> The promoters of aviation whose palms itch for the gate receipts may point out that the great death rate among aviators is essential to the development of the science of aerial navigation. It may be true now, but it will not be the case five or ten years from now.
> **Editorial, Syracuse Herald, 18 June 1911**

The year 1909 showed tremendous progress in aviation, and the newspapers eagerly devoted space to this new science. And to make the general public more aware of the advances in transportation, the almost two-week long Hudson-Fulton Celebration (September 25–October 9, 1909) celebrated the 300th anniversary of Henry Hudson's discovery of the Hudson River and the 100th anniversary of the first successful application of steam to river navigation by Robert Fulton, when he and his team traveled in the *Clermont* from New York to Albany in 36 hours. Forty some years later, the Hudson River Railroad covered the same distance in less than four hours, and the budding civil engineer Octave Chanute was part of this achievement. Another fifty years later, *The World* offered a $10,000 prize to the first person to travel by air from New York to Albany. No one claimed this prize in 1909; it was re-advertised and Glenn Curtiss thought such a flight could be done. On Sunday, May 29, 1910, he took off at 7 a.m. from Albany, New York, flew along the Hudson, made two stops and landed on Governor's Island at noon with thousands of spectators screaming and every boat on the river blasting its whistle. He won the prize for the first true cross-country flight in the United States [1].

© The Author(s), under exclusive license to Springer Nature Switzerland AG 2023
S. Short, *Flight Not Improbable*, Springer Biographies, https://doi.org/10.1007/978-3-031-24430-8_9

Fig. 9.1 The sign on the cartoon says it all: "Sensational acrobatic aerial performances for the amusement of the public. All the most dangerous feats of the air men." *New York Herald*, January 4, 1911

For Curtiss there was more joy in flying above rivers and hills and meadows, than circling over race courses to delight the crowd. But the general public thought different, they wanted to see airmen dart above, below and around in the sky and close to the ground. The appetite for air shows was insatiable, and the ticket box holders were keenly aware of the airplane's entertainment value, thus, one of the first "practical uses" of the airplane was to provide thrilling displays to an entrance-paying crowd. Every death attracted larger crowds, as the cheering crowds drove aviators to outperform within their personal safety limits, and bold headlines in late 1910 and early 1911 told the story: "*The silent spectator*," standing in an open grave, or "*It is going to get us all*" (Fig. 9.1).

On October 15, 1911, the *New York Times* was just one newspaper publishing the news that "*Aviation Victims now Number 100.*" An editorial in the *Boston Evening Transcript* summarized the situation. "The sober second thought regarding aviation is becoming very sober indeed. The mounting death toll, rising more rapidly than ever since the one hundred mark was reached, reveals the luring of the unfit into the field" [2].

The causes of these accidents, mostly with Wright airplanes, were not completely understood, but it was generally believed that stability and control were partially responsible. Ludwig Prandtl had discussed the topic during the ILA in September 1909, and commented in closing "… how difficult it is to balance an airplane with three levers at the same time. The Wrights' students have to practice with the completely tried-and-tested apparatus for a long time, until they can take over its guidance independently. Steering would be easier to learn, if an automatic device would be available as the guide to balance and stay on course in turbulent air, leaving him only with the actual 'steering.' Perhaps the next few years will bring such an invention" [3]. And yes, Prandtl, who used the *Progress* book writing his thesis, was familiar with Chanute and his vision of automatic control of an airplane.

With the death of each new pilot, the discussion flared up whether the French theory of "stable planes," as originated by Chanute, was better than the Wright's "unstable planes." Believing that the "unstable plane" was a better choice, Wilbur explained to William Glassford from the Signal Corps why they designed their Flyer as they did [4]. "My understanding of the fundamental difference in the ideals and practice of Lilienthal, Chanute, and the Wright brothers has been that (1) Lilienthal sought equilibrium by the conscious effort of the man through the medium of shifting the center of gravity; (2) Chanute sought to make equilibrium independent of the man and employed surfaces which, if movable at all, were actuated only by the air; while (3) the Wrights, believing Lilienthal's method inadequate and Chanute's unattainable, sought control solely by conscious effort of the man through the medium of operable planes and rudders. In regard to wing structure, we considered Chanute's double-deck truss superior to that used in Lilienthal's double-surface machine and succeeded in adapting it to our own ideas and principles of control, but we were not believers in his main idea, i.e., automatic control." Inventors eventually moved toward longitudinally stable "self-flying airplanes," which they believed to be safer and easier to fly.

9.1 Before Becoming a "Bird-Man" You Must Learn Gliders

More than once Chanute had told his listeners that flying with wings was fun and when a person cannot afford to fly a powered airplane, he or she could probably afford to fly without power at first. To him the glider was not only a method of learning to fly, but also the medium to teach anyone the meaning of loving to fly.

Many newspapers and magazines shared his opinion and published drawings with instructions how to build and then fly the Chanute-type biplane. The write-up in the *Brooklyn Citizen* is typical for the time. "Think of putting a few sticks, some rubber bands, a couple of hairpins and some tissue paper, mixing them with brains, and having in the end a machine what will whirr away across the lot, sink to the ground when its stored-up power is exhausted, and run along on its wheels until its momentum is gone. Yes, you can make a flying machine and you can make one that suits your fancy to a T. And you are going to get a whole lot of satisfaction over it when you see your own creation whizzing through the air and realize that you are one of the 'flying men' in the country. And this machine you are to build is not a toy. You will solve new problems in aviation for yourself, and you may win a little fame for yourself. You and your friends will get a whole lot of fun out of racing these different machines. Which fly the longest? Which are the swiftest? And while you have been having all this fun, does the science of flying seem a little more real to you? a little more intelligible? Your main problem will be to get balance and little 'head resistance', and you may make good in a small way that will be a big way some day. Remember that the science of flying is yet in its infancy, and there is plenty of room for you to speculate and to experiment" [5].

History tells us that almost every famous, and not so famous, aviator who started flying in the early twentieth century, had the first taste of being airborne in a homebuilt glider, acquiring a practical knowledge of equilibrium in the air and experience gravity (Fig. 9.2).

The British enthusiast **Alec Ogilvie** (1882–1962) had built a quadruplane glider in 1908, which showed some resemblance to Chanute's *Katydid* of

Fig. 9.2 Cartoon, published in conjunction with the St. Louis Centennial Celebration, where organizers wanted every citizen to see an airplane or dirigible in the air. The *Cairo* (Illinois) *Bulletin*, October 5, 1909

1896. Towed behind an automobile, it reached an altitude of about 30 feet; but then, the towing rope broke and the builder's brother, who piloted the craft, fell through the wings. Not giving up, Ogilvie went to France and introduced himself to Wilbur Wright who mentioned that the Short Brothers were building the Wright Flyer in England. Re-energized, Ogilvie ordered a Flyer, which however could not be delivered due to a shortage of engines. Wanting to fly, Ogilvie sketched a glider and asked Wilbur for comments. He and his partner Ted Searight then hired T.W.K. Clarke to build an updated Wright 1902 glider, making the design similar to the big Flyer they had ordered [6]. The glider, known as the Clarke-Wright glider, became a useful training tool; Ogilvie made good glides with coordinated turns in late 1909 after removing the front elevator and fitting an "elevator tail" [7].

With the knowledge gained building the Wright-type glider, **T.W.K. (Thomas Wigston Kinglake) Clarke** (1873–1941) set up business in Kingston-on-Thames, near the site where the Sopwith factory would be producing Pups and Camels. They offered four gliders, either completely built, as a kit or only the drawings.

(1) **The Popular**. A Biplane or Chanute-type, span 20 ft. × 4 ft. This machine has been designed to supply the demand for a simple, easily constructed, and tough machine, that will be within the reach of those of moderate means. We can supply all materials for this machine from £4.15.

(2) **The Chanute**. Biplane. Span 22 ft. × 3 ft. 9 in. A beautifully made, double surfaced little glider, rounded ends, and main spars of ash with our type of small skids on the tips. Complete £34 (Fig. 9.3).

(3) **The Aero**. Biplane. Elevator in front. Span 24 ft. × 4 ft. 3 in. as described in the "*Aero*" of Dec. 7th and 14th, 1910. A very tough well-made machine, and suitable for home building. Complete £50.

(4) **Wright Type.** Biplane. Span 32 ft. × 5 ft., with Wright-type controls, working ailerons and elevator. A magnificent machine, almost identical with the one we supplied to Mr. Ogilvie in 1909 and which held world's records, making glides of over 1,200 ft. Complete with transport wheels, rail trolley, 100 feet of iron-topped rail, 200 feet of cotton rope; extra high portable derrick, pulleys, weights, and starting gear £100.

To make his company better known, Clarke submitted a two-part article to the new *Flight* [7] magazine in Great Britain. "Gliding is a side of flight that is a little apt to be neglected in the present rush to achieve the higher art; but it is nevertheless useful, and the present is a particularly appropriate

Fig. 9.3 T.W.K. Clarke advertisement in *Aeronautics,* August 1911

time to learn the mastery of the motorless flyer, seeing there is some difficulty about obtaining a proper supply of engines in this country." The next article included a "*Word of Warning*" in the final paragraph, stating that the Searight-Ogilvie glider was built with the permission of the Wright brothers. Another article was published two years later in the American *Aeronautics* magazine, "*Gliding as a sport and as an aid to flight*" [8]. In later years, Clarke went into the model airplane and glider business, selling them either as kits or completely built.

While the T.W.K. Clarke Company promoted their products in England, the Voisin Brothers built up their business in France. Stressing that any newcomer should fly a glider prior to stepping into a powered airplane, they submitted a detailed article to *La Revue de L'Aviation* which was translated and re-published as "*The Practice of Aviation*" in the *Aeronautics* supplement of the British *Knowledge & Illustrated Scientific News* [9]. Here the authors explained, "We sincerely trust that the readers of *Knowledge* may be interested in our words, and may follow our advice. Aviation is a wonderful sport, and we are glad to see that we are no longer looked upon as lunatics. Man will yet have wings!" The first article highlighted the benefit of learning to fly in gliders, and, in their opinion, their redesigned Chanute-type glider was the best and safest to start with. "Gliding flight is one of the most entrancing forms of sport imaginable. Besides the physical development resulting from continued practice, the aviator acquires from his pursuit rapidity of decision, accuracy in movement, and, above all, an almost instinctive response to emergencies that cannot fail to stand him in good stead at some later period when steering one of the large powered machines. Usually there is no danger in gliding, as long as the novice begins cautiously, and his aircraft is of sufficiently strong construction to withstand the wear and tear of the first few days." The next article provided instructions for building a biplane glider, including the dimensions of the wood to be used, the best method of attaching the pieces, and the estimated cost of material. The last article discussed gliding in general and how to control the glider with the help of moving the legs or body. The final statement was, "And now, if you have had the patience to construct your glider and have experienced the delights of gliding flight, we shall be glad to advise and assist you in any difficulties you may have encountered. At any rate you will heartily agree with Mr. Chanute that infinitely more is learned from a week of practical experience than from two years of fruitless discussion" [9]. The simple Voisin Chanute-type glider conquered the world in the next decades, and no one knows how many thousands of these primitive gliders were built and flown.

Juan de la Cierva Codorníu (1895–1936) from Spain described in his biography *Wings of Tomorrow* [10] his early steps in aeronautics. "Mine is not a long story, as years go, but neither is the airplane a very venerable affair. At the age of an American high-school freshman, I had studied as thoroughly as I could the work of Langley, Maxim, Ader, Chanute and particularly the designs of Otto Lilienthal. With the supreme confidence of my fourteen years, I was not deterred from aerial experiments by my knowledge of Lilienthal's fate. We did what we could with kites, and then dared further." Two years later the teenager collaborated with his friends José Barcala and Pablo Díaz to build and then fly their Chanute-type biplane. "Diaz pulled the glider into the air, when a gust of wind lifted the glider with Juan to a considerable height and dropped them as precipitately. On the way down Juan let go of the glider and landed with a formidable and painful thump, while the glider blew over the hills far away. That was the end of our glider."

Cierva first studied civil engineering in Madrid, Spain, but then focused on aerodynamics. Speed plus economy plus safety was Cierva's goal in designing the perfect airplane. His resulting autogiro, or the "flying windmill," was not as perfect and efficient as the other airplanes developed at the same time, but it was a step in the overall development, and it was fun to fly.

To encourage aeronautics in the Antipodes, the Australian Government offered a prize for the first Australian designed and built aircraft that would meet the requirements for military purposes, as derived from the British

Fig. 9.4 Taylor's glider under construction. It was flown by George Taylor on December 5, 1909. National Museum of Australia. Courtesy Alan Patching

regulations. The prize money was tempting, and Lawrence Hargrave encouraged his chief assistant **George Augustine Taylor** (1872–1928) to design and build a glider. Both men thought how gratifying it would be if an Australian could invent the real airplane (Fig. 9.4).

The date for the first flight of the imported Wright airplane was December 4, 1909; it was to be flown at the Victoria Park Race Course, a southern suburb of Sydney, but for some reason the plane could not get airborne. The next day, on December 5, Taylor was ready with his biplane glider, based on Hargrave's box kites; the pilot flew in the prone position, with both feet laying on the well-padded rear spar. Taylor made several good flights at Narrabeen, [11] and his wife Florence became the first woman in Australia to fly a heavier-than-air craft, her husband's glider. But to win the prize money, as advertised by the Australian government, the glider needed to "poise" (or fly slow circles) to make observations from the air; unfortunately, Taylor could not achieve this with his craft, thus could not earn the prize money.

Carl Sterling Bates (1884–1956) from Cedar Lake, Iowa, was probably the most successful aerial tinkerer in Chicago in the first decade of powered flight. He had designed and built a Lilienthal-type glider in 1898, which did not perform as hoped. The following year, despite objections from his parents, he built a Chanute-type biplane and flew it with some success. In 1903 he moved to Chicago to study mechanical engineering at the Armour Institute of Technology and became interested in flying again after meeting Chanute. Attending night classes allowed Bates to operate a small shop during the day; the sign over the entrance said it all: "*Aeroplanes Built to Order.*" In 1908 he built a smaller copy of a Curtiss airplane with ailerons and a front horizontal and rear vertical rudder, powered by a ten-horsepower engine. Bates made some short hops, but the battery proved insufficient to keep the engine running. Chanute heard of his plight and bought him a magneto, allowing Bates to make flights of about 460 yards in length [12].

Bates next project was a biplane glider, based on the Voisin drawings; he submitted his write-up to the new splashy large-size monthly *Fly* magazine from Philadelphia; they published it in March 1909 as "*How to build a gliding machine*".

To give the article wider publicity, Chanute suggested contacting *Popular Mechanics*, based in Chicago, with a subscription base of more than 200,000 subscribers. Their April 1909 issue featured Bates' article on three pages with a scale drawing and instructions how to build the glider [13] (Fig. 9.5). "A gliding machine is a motorless aeroplane, or flying-machine, propelled by gravity and designed to carry a passenger through the air from a high point to a lower point some distance away. Flying in a glider is simply coasting downhill on the air, and is the most interesting and exciting sport imaginable." Bates continued writing, "To make a glide, take the glider to the top of a hill, get in between the arm sticks and lift the machine up until the arm sticks are under the arms, run a few steps against the wind and leap from the ground. You will find that the machine has a surprising amount of lift, and if the weight of the body is in the right place you will go shooting down the

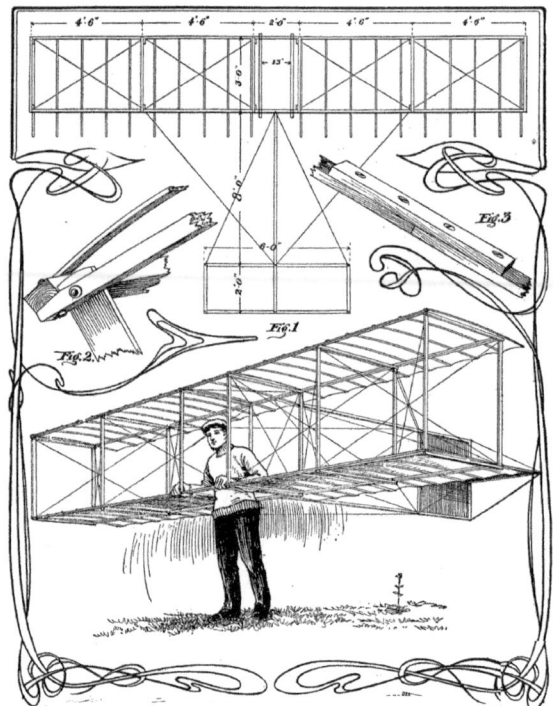

Fig. 9.5 Illustration for Carl Bates' article on how to make a glider. *Popular Mechanics*, April 1909

hillside in free flight. The landing is made by pushing the weight of the operator's body backwards. This will cause the glider to tip up in front, slacken speed and settle. The operator can then land safely and gently on his feet. Of course, the beginner should learn by taking short jumps, gradually increasing the distance as he gains skill and experience in balancing and landing." This how-to-article in *Popular Mechanics* was read by hundreds of adventurous teenagers, and we can only guestimate how many of them used Bates' write-up to build their glider at home and then experience flight, gliding straight down the hills.

Bates sold his mini airplane company in 1912 to Edward Heath who formed the Heath Aerial Vehicle Company.

Waldo Dean Waterman (1892–1976) from San Diego, CA, read all that was available on flying, especially by Curtiss and the Wrights. Seeing the 1909 *Popular Mechanics* article by Bates encouraged the 15-year-old Waterman to construct and then fly his glider down the slopes of a canyon near San Diego, California. Wanting to learn more about flying, Waterman became acquainted with Curtiss and advanced to powered airplanes. In 1912 he enrolled at the University of California at Berkeley to study aeronautical engineering and earned his degree. As World War I began, he became head of "Theory of Flight" at the U.S. Army Signal Corps School of Military Aeronautics. In the 1930s Waterman was lured back to aircraft design by the U.S. Department of Commerce proposal for a simple, easy to fly, low-cost airplane. He built and flew the first tricycle-landing geared flying wing, and then designed and built six "Arrowmobiles," or aeroplanes, that could be converted to automobiles. When the hang glider movement started in the 1960s, Waterman built an updated version of his 1909 Chanute-type hang glider [14] and helped enthusiasts experience the affordable sensation of flight.

Waterman, being amazingly vigorous and active, held the oldest commercial flying license in the United States. He continued flying and teaching and contributing to the sport of aviation until his death in 1976.

Another enthusiast who used Chanute's glider design as a starting point, was **Matthew Bacon Sellers** (1869–1932). He too had built a Lilienthal-type glider in 1902/3, but was not satisfied with its performance. Seeing drawings of Chanute's *Katydid* inspired him to design and build a quadruplane. Hearing that Gustave Eiffel had discovered in his research that one could increase the performance of an airplane by staggering the wings, Sellers contacted Chanute. He wanted his opinion about how far to project the leading edge of the top wing forward, and how far the two lower planes should be set back, one step at a time (Fig. 9.6).

After making a few good flights, Sellers installed a powerplant with a four-bladed propeller, and wheels for take-off and then made the first powered

Fig. 9.6 Matthew Sellers' "Step Glider." *The World*, April 17, 1910

flight in Kentucky on December 28, 1908. His goal was to develop an airplane which could sail against the wind without any artificial energy, using the motor and propeller only for the initial take-off and to continue flying after finding an atmospheric area of relatively dead calm [15]. Sellers stated that his motorglider, weighing 110 pounds with an eight-horsepower engine, was the smallest, lightest and safest airplane built, it was the ideal craft for enthusiasts on a lean budget. While his system of control was different to the Wright's in detail, it did have a certain similarity in principle, but Sellers considered it a step up in performance. The reporter from *The World*, Arnold Kruckman, commented, "But under Judge Hand's opinion in the Louis Paulhan case, however, the Wrights have such latitude that it is impossible to accurately define the boundaries of their property" [16, 17]. Sellers then joined the Aeronautical Society and became their technical advisor, was appointed in 1912 to the Aerodynamic Laboratory Commission, assisted Ernest Jones with his *Aeronautics* magazine as technical editor, joined the Navy Consulting Board in 1915 and continued to contribute in the aviation industry.

A local carpenter and friend of William Avery in Chicago, **John Zimmerer**, had become interested in gliders. When Smithsonian Secretary Walcott asked Chanute for scale models of his gliders to be displayed at the National Museum, Zimmerer agreed to build two sets of ¼-scale models: [18]: the 1896 Katydid, the 1897 biplane and the 1902 oscillating wing glider. One set was shipped to Washington in spring 1909, while Chanute kept the second set. The family donated the models to the Chicago Academy of Sciences after their father died. Not knowing what to do with the models, the box was moved to the attic. The relics were rediscovered in 1927 by H. H. Slawson, a member of the Chicago City News Bureau, who photographed them and published several articles. The models were then transferred to the Museum of Science and Industry in Chicago and are displayed periodically [19].

With the knowledge gained building the models, Zimmerer and another Chicagoan, Howard Wixon, built a full-size biplane glider, with a wing area of 150 square feet and weighing almost 50 pounds. As could be expected the aging engineer provided input, technical and otherwise. Always concerned about accidents, Chanute warned them to be careful until "the tricks of the wind are learned; do not fasten yourself into the machine in any way, you may want to get out quickly. Do not start from a wagon, but select a hill. You will probably make some mistakes at first and break the machine, but be careful not to break your bones" [20]. In late summer of 1909, Zimmerer and Wixon invited Chanute to come along to observe, take pictures and time their glider flying in the dunes along the southern shore of Lake Michigan. Watching the two men fly on that hot summer day brought back many memories. Being 77-years young, Chanute came home thoroughly tired after walking many miles up and down the sandy dunes [21].

The essential thing in the sport is to use a safe glider. **Charles Wittemann** (1884–1967) and his brother Adolph, formed the C. & A. Wittemann Company on Staten Island, New York, selling at first a Chanute glider, followed by the simpler Voisin-design. Their glider sold for $75, and for a small extra fee, they offered instructions in learning the tricks of gliding. To advertise their low-cost craft for lean pocketbooks, they ran small ads in newspapers and magazines and encouraged their buyers to submit reports on their flying experiences to the press, publicizing the sport and their company.

Fig. 9.7 To speed up the learning process, the Wittemann Company introduced a glider that could carry the student and the instructor. *Aeronautics,* January 1910

For more efficient training, the Wittemann brothers redesigned the biplane glider to carry two people, the instructor and the student, both hanging in the upright position (Fig. 9.7).

Starting in a small way, the Wittemann brothers built up a surprising business; a decade later they moved their shop to what became Teterboro Airport, offering their expertise to enthusiasts who wanted an airplane built.

William Aitken (1887–1964) had soloed in a Wittemann glider in June 1909 and submitted his report to *Aeronautics* [22]. He then joined the Wittemann brothers as instructor and could arguably be called the first commercial glider instructor in the United States; this "glider expert" knew how to save the bruises and laughable smash-ups. He guaranteed to turn out a proficient pupil in one day for $10, introducing many newcomers to the sport [23]. Aitken worked first as an instructor in gliders, then in airplanes, but returned to his carpentry trade when he got married three years later.

After graduating thousands of chauffeurs from its automobile schools, the Young Men's Christian Association (Y.M.C.A.) in New York City decided to offer evening classes in aeronautics at their West Side YMCA [24]. The classroom talks, offered one night a week, were designed to give teenage boys and girls a good working knowledge of building a glider. Students were also encouraged to compete for silver cups, like the one offered by Chanute for the longest flight, or glide, of their glider. Educators hoped that their students would become the "aerial chauffeurs of the future" and succeed in the new aeronautical industry.

Fig. 9.8 The 16-year-old Ralph Barnaby learning to fly in a Wittemann glider. Author's collection

One of the teenagers who took advantage of these classes was the 16-year-old **Ralph Stanton Barnaby** (1893–1986). He built his first model glider in 1909 and won the 3rd place in the Octave Chanute Silver Cup Race for the longest flight (114 feet). Next, he built himself a small full-size Lilienthal-type monoplane, which was not a success. His parents then encouraged him to take flying lessons with the Wittemann Brothers in their Chanute-type glider (Fig. 9.8).

Barnaby joined the Navy in 1917, starting a lustrous aeronautical career. He learned to fly powered planes, but his love remained with motorless flight. When the glider movement began in the United States in the late 1920s, the Navy ensign signed up at the glider school at Cape Cod, Massachusetts, in August 1929, earned his "C" badge and established an American soaring duration record in a Prüfling secondary glider (15 minutes 6 seconds), breaking Orville Wright's record flight of 9 minutes 45 seconds, set in October 1911 at Kitty Hawk, North Carolina [25].

The Cape Cod school folded a year later, and Barnaby convinced the Navy to purchase the Prüfling glider he had flown, as he wanted fellow Navy men to become familiar with motorless flight to become better power pilots. Prior to the Prüfling entering service in the Navy, Barnaby agreed to completely overhaul the glider; each step was documented with photos which he shared with the aviation community. [26] Barnaby used this freshly overhauled glider to be launched from the belly of the U.S. Navy dirigible *Los Angeles* over Lakehurst, New Jersey, on January 29, 1930, proving to the military that an aircraft could take off from a large dirigible airship and be flown to a conventional safe landing.

Fig. 9.9 The first glider pilot license by the National Aeronautic Association of the United States, was issued to Ralph S. Barnaby in 1930, signed by Orville Wright. Ralph Barnaby papers, National Soaring Museum, Elmira, New York

As America's first "official" glider pilot, Barnaby received the National Aeronautic Association License No.1, signed by his friend Orville Wright on May 26, 1930 (Fig. 9.9).

The competition pilot became a director of the National Glider Association, and when the group folded, he joined the just-created Soaring Society of America. After retiring from the Navy, Barnaby joined the Franklin Institute in Philadelphia as Aviation Curator in the late 1940s to assist cataloging the papers from Orville Wright's estate. He became a Board member of the National Soaring Museum when it was formed in 1969, and remained active in the gliding community until his death in 1986.

Barnaby was part of the "first generation of flight;" he was ten years old when the Wrights flew at Kitty Hawk in 1903 and seventy-six when Neil Armstrong set foot on the moon in 1969.

Another teenager started his aeronautical career at about the same time. Living with his family in Washington, DC, **Paul Edward Garber** (1899–1992) learned the tricks of flying kites higher and better than any other youngster in the neighborhood under the tutelage of his neighbor, Alexander Graham Bell.

In 1913 Garber and schoolmates formed the Capital Model Aeroplane Club to build and fly model aircraft and kites. He studied Chanute's biplane glider model at the National Museum, measured carefully and then increased the overall dimensions about five times beyond that of the model (Figs. 9.10 and 9.11).

Garber used barrel staves, sawed into thirds as ribs for his 20-foot wingspan glider and covered the wings in red chintz fabric, as suggested by Graham Bell. The teenagers carried the glider to an area called the Red Lot. About eight or ten boys ran with the rope and the glider started to rise, which caught the boys off-guard; they stopped running and the glider came down in a hurry. Young Garber repaired his glider and tried again. This time the boys ran hard and kept pulling until the glider with its pilot was about forty or fifty feet into the air; he then coasted ahead of the boys over trees, the road, and a fence, and landed in a rose bed, scratched but elated.

In 1920, Garber began his career of nurturing, procuring, preserving, protecting, and sharing all things aeronautical. He was originally hired for a three-months assignment to build models and prepare exhibits for the National Museum. He stayed for 72 years, devoting his life to the preservation of the nation's aeronautical heritage and helping create one of the best airplane collections in the world.

Fig. 9.10 Chanute's biplane glider, ¼ size, at the National Museum, Washington, DC. *Handbook of the National Aircraft Collection*, 1932 (NASM A-23603)

Fig. 9.11 The Chanute biplane glider is reinstalled to be displayed in the National Air and Space Museum's Early Flight Gallery, opened in October 2022 (Flickr nasm 2021-05,180)

One of Garber's lesser-known projects was the Frieze in the Rotunda of the United States Capitol, a painted panorama depicting significant events in American history. When Constantino Brumidi finished his work in 1889, there was a gap of about 31 feet due to some miscalculations about the height and width of the overall frieze. Some attempts were made in 1917 to fill the remaining space, but nothing was finalized. After World War II, Garber promoted the idea to finish the freeze, suggesting that the remaining section was to commemorate the Civil War, the Spanish-American war, and the Birth of Aviation. Allyn Cox accepted the commission and submitted his proposal to Congress for approval. "… Passing to the right, one comes to a group of the precursors to the invention of practical flying. I have put in Leonardo da Vinci, and two figures [Samuel Langley and Octave Chanute], representing American precursors to the Wrights. Each has in his hand a miniature model of his flying machine. The space is filled with a symbolical representation of the first flight at Kitty Hawk, on December 17, 1903, with Orville Wright laying in the plane and Wilbur running along to steady the wing. To localize this great achievement as an American flight, an eagle flies to meet them with a laurel branch in its claws. The eagle has been carefully designed to fit closely on to the group next to the right, that of Liberty and her attendants, so that the frieze ends as it begins, with the favorite personifications of the American spirit" [27]. It is interesting that Cox showed Chanute holding his not so well-known *Katydid* multiplane glider, which indeed helped pave the way to the very successful biplane (Fig. 9.12).

Fig. 9.12 The "Birth of Aviation" frieze in the dome of the United States Capitol in Washington, DC, shows three precursors, including Chanute and Langley. The photo of the complete "Birth of Aviation" painting is shown on page vii in the Preface. Courtesy Architect of the Capitol Archive

Only a person like Garber could have thought about this detail. For him, aviation and the profusion of knowledge were deeply ingrained in his personality and in a way, Garber picked up where Chanute left off, as an eminent historian of flight.

James H. "Jimmy" Doolittle (1896–1993), who had graduated from the Massachusetts Institute of Technology, discussed building and flying his glider in his autobiography [28]. "I had found an article in an old 1909 issue of *Popular Mechanics* by Carl Bates. Fancying myself now very handy with tools, I followed the drawings carefully and built a bi-winged glider similar to what today we would call a hang glider. The pilot was supposed to stand in the middle and hold on to it with his hands; his legs and feet would serve as the landing gear. My mother supported my efforts by giving me a little spare change for materials like muslin, wood, and piano wire, and I earned the rest by doing odd jobs in the neighborhood and selling newspapers. I did all the work except that my mother sewed the unbleached muslin for the wings. Bates had made flying a glider sound very simple. Proud of the finished job, I placed it on a wagon and towed it to a nearby bluff about 15 feet high. I carefully strapped it on and ran fast toward the edge, jumped off, and tried to glide down. Unfortunately, the tail of the glider hit the edge of the bluff and I came straight down. The glider fell on top of me in pieces and banged me up pretty badly. I dragged what was left home and assessed the damage as well as my ineptitude at getting it airborne. Obviously, I was not able to run fast enough." More speed was needed. "One of my friends had access to his father's car. So, after I made repairs, we got a rope and tied it to the bumper, and I donned the glider. He started off and I ran to keep up. I ran faster and faster and soon couldn't run any faster. I leaped into the air, put the tail down and planned to ease upward into full flight. But there was no lift and I was dragged quite a few feet while the glider splintered around me. After I had begun repairs, a storm came up one night, snatched my 'baby' and blew it over the back fence, scattering it in small bits over several neighboring yards. I was out of money, out of materials, and out of enthusiasm. The more I think about it, the more I realize how lucky I was that I didn't succeed" [28].

Doolittle joined the Army Air Service during World War I and enjoyed an exceptional military career. He is probably best known for being the lead pilot in the carrier launched raid on Tokyo and other targets on the Japanese mainland in 1942.

At the tenth anniversary of the delivery of the first airplane to the United States Military, Major General **Charles Thomas Menoher** (1862–1930), the recently appointed first Chief of the United States Army Air Service, told again how he, a 35-year-old infantry man, became interested in aviation [29, 30].

"Twenty years ago, as long ago as 1897, I was taken by Octave Chanute into the attic of his house in Superior Street in Chicago. There he had samples of every kind of kite known to men and boys of the world. He was studying the biplane kites of China and the monoplane kites of other countries. In addition to these toys, if they may be so called, he had mounted specimen of several of the long-flighted species of birds of the world, the gull, the albatross and the man-of-war bird. He was studying their structure, their wingspread and relating these things as well as he could to their ability to sustain themselves in the air for long periods without visible motions of the wings. Later on, I went to the sand dunes of northern Indiana, on the shores of Lake Michigan to witness some of the early flight attempts of men under the direction of Octave Chanute. The experimenter and his assistants fitted themselves by means of harness into great reaching wing-like planes similar to those, which an airplane uses today. Then they would start down an incline and launch themselves into the air from the hillside against the wind. In this way, and without any motive power, it was possible to determine the sustaining strength of the wings and many other things, which it was necessary to determine before the problem came of attaching motor power. At that time the present type gasoline engine had not been invented, and proper motive power was one of the chief problems.

"Octave Chanute was a pioneer in research, Langley was a pioneer in preliminary accomplishment, and the Wright brothers were the accomplishers. The entire development of the airplane has occupied only about a score of years. In Europe others were experimenting, their work followed the lines of Chanute, who was in correspondence with his European colleagues in the experimental field. It is something for a man to know that he witnessed the first attempts of Octave Chanute to make progress through the air with planes as wings, and 20 years later to see the perfected combat planes of the army go into the fight over the German lines. There was skepticism on the sand dunes of Indiana and full belief at Hattonchâtel, with 20 years intervening between the twain" [30].

At the end of the first World War, Menoher stated strongly in his annual report: "Unless the government aids the airplane industry it cannot hope to depend upon the availability of suitable commercial aircraft and facilities for their employment nor upon the existence of manufacturing plants and supplies of materials necessary for the rapid production of aircraft in time of war."

9.2   Medals for the Wrights

Alexander Graham Bell had suggested late in 1908 that America should honor the Wrights, just like Europeans were doing. On March 4, 1909, Congress awarded the Congressional Medal "in recognition of the great service of Orville and Wilbur Wright, of Ohio, rendered the science of aerial navigation in the invention of the Wright aeroplane." And the ever-critical Arnold Kruckman commented, "It is hoped to show the Wrights the magnitude of the thing they have accomplished and the esteem in which they are held by their fellow countrymen. This will be very gratifying to the reticent gentlemen from Ohio; but just how they will contrast it with the amused incredulity with which they were regarded a year ago, would be interesting to know" [31].

The Aero Club of America asked President Taft to present their medals at the White House. On June 10, 1909, Herbert W. Parson, the New York Congressman, introduced the brothers, and the good-natured Taft stated with a certain humor, "I don't like to think, and I decline to think, that these instrumentalities that you have invented for human use are to be confined in their utility to war" [32]. Chanute did not attend, even though he was part of the committee.

The Smithsonian Institution took a different approach. They established the "Samuel P. Langley Medal for Aerodromics" to be awarded for meritorious investigations with the science of aerodromics and its application to aviation. Walcott asked Chanute to chair the awards committee, with Graham Bell, James Means, George Squier and John Brashear joining him. What at first must have looked like a simple task of selection became stalled on the question of whether the medal should be conferred for achievements in pure science or in the applied science of flying. As Chanute explained to Walcott, "The results obtained by Wright Brothers are far superior to any others, but the resolution mentions 'meritorious investigations,' not achievements, and this implies that those investigations should be given to the world, a thing which the Wrights have hitherto declined to do" [33]. Committee members overrode Chanute's concern and nominated the Wrights to receive the first Langley Medal "for advancing the science of aerodromics in its application to aviation by their successful investigations and demonstrations of the practicability of mechanical flight by man."

As Chanute could not attend the ceremony on February 10, 1910, Graham Bell spoke for the committee [34]. "Who are the pioneers responsible for the great developments in aerodromics of the last few years? Not simply the men of the present, but also the men of the past. To one man especially is honor due, our own Dr. S. P. Langley, late secretary of the Smithsonian Institution. We have honored his name by the establishment of the Langley medal; and it may not be out of place on this first occasion for the presentation of the medal to say a few words concerning Langley's work." In closing, Bell read parts of a letter from Wilbur to Chanute, "The knowledge that the head of the most prominent scientific institution of America believed in the possibility of human flight was one of the influences that led us to undertake the preliminary investigations that preceded our active work. He recommended to us the books, which enabled us to form sane ideas at the outset. He was a helping hand at a critical time, and we shall always be grateful" [35]. Little did Bell or Chanute realize that Wilbur's letter, written after hearing of Langley's death in 1906, would be the beginning of the decades long squabble between the Wrights and the Smithsonian.

9.3 The World's Inventive Genius Centers on Propulsion

Gliding with wings like the birds was just one way to get airborne, but one could also use a screw propeller, jet propulsion or some kind of propelling device not discovered yet. Chanute had studied the various means of propulsion for "aerial machines without aerostats" (or gas-bags) in the 1860s and thought that "the rocket is proof of the power of powder to carry vessels through the air. It is the crudest form of flying machine, and when the genius of man is fully directed to economizing and guiding the great power which is the cause of the rocket's flight, we will have a speedy, practical, and safe flying vehicle which will astonish the world by its simplicity and tardy discovery" [36].

Twenty some years later, Chanute gave a brief review of ideas on achieving flight in the introduction to his *Progress in Flying Machines* articles. Discussing the interaction between the flow of solids and fluids, he stated, "Indeed, it may be said with respect to curved surfaces and solids, that a glimpse has been caught of a still more mysterious phenomenon. It is known that certain shapes, when exposed to currents of air under certain ill-understood circumstances, actually move toward that current instead of away from it." Or, the force acting on the fluid through viscosity (action) generates a force on the curved surface (reaction). And he concluded, "Thus it is seen that in such complicated matters, theory cannot progress in advance of experiment, and the extreme importance of those experiments hitherto tried will in part be appreciated."

Believing that progress of aviation would not stop and that man would seek to go faster and higher, Ferdinand Ferber offered this prophecy: "To go higher, and man will want to go higher, it will be necessary to adopt a different principle. The principle of the rocket is indicated, and the jet engine is deduced therefrom. The pilot will be enclosed in an enclosure where the breathable air will be made. To tell the truth, he will no longer mount a flying machine, but rather a dirigible projectile. The realization of this idea is not impossible for human thought and power" [37].

To use a jet engine was intriguing. René Lorin from Paris, France, theorized at the same time as Ferber that the exhaust from two gas engines could push a flying machine forward and could assist in taking off vertically. He submitted his claims of a reciprocating engine with an exhaust jet, where each cylinder allowed the combustion gases to exhaust through a nozzle to the atmosphere. The French military was not interested, so Lorin submitted a report to L'Aérophile. This write-up caught Chanute's attention, as he had mulled over a direct reaction explosion motor for some time [38].

Knowing that Chanute had helped others to advance aeronautics, Lorin mailed drawings of his envisioned aircraft and engine to Chicago. As this design could possibly drive a flying machine at higher speeds, Chanute forwarded the information to Spratt, who was looking for a power plant for his future airplane (Fig. 9.13) and suggested, "I have just received a French patent for a method of obtaining propulsion by exploding gasoline and air. The man proposes to compress the mixture in a cylinder, then to explode it, so that it will open a valve and issue forth through an adjutage of taper to be determined by experiment. I am willing to defray the expense of experiments in this direction, but I think they will be costly. Are you in position to undertake them?" [39].

At that time Spratt was trying to develop a glider, with a stable control, or automatic equilibrium, and to eventually install an engine. Chanute's proposal sounded interesting and Spratt began experimenting, he tried different sizes and shapes of nozzles on a stationary engine to achieve compression and exhausts. He worked closely with Chanute and discussed his results in lengthy letters and sketches, [40] but he then stopped, as he could not see using such an engine for his airplane.

Looking for a second opinion, Chanute shared Spratt's finding with Victor Lougheed, who thought that Lorin's basic proposal was ingenious, but he did not think it would work in a practical application at the current time.

Travelling to Europe in May 1910, Chanute met Lorin in Paris and gave him copies of his and Spratt's letters, explaining their findings. He advised

Fig. 9.13 George Spratt was working on designing an aeroplane that would not nose dive, tail spin or side slip, that would right itself and float to a graceful landing, and that it could be flown with or without a tail. *Butte News*, Montana, July 31, 1909

Lorin to increase the initial take-off speed of the aircraft and continue experimenting. With these fresh ideas, Lorin continued his work, [41] but could not accelerate his aerial torpedo to a higher speed for his "air-breathing jet engine" to operate, thus could not demonstrate a practical use for propulsion. With some disappointment he summed up his work in his biography. "In the twelve years that the author has been studying the question, he has gathered some documentation about it. The few scattered efforts do not represent anything interesting, including the experiments made in America at the investigation of Chanute. The creator of the biplane came to see the author in Paris in 1910, bringing him reports of his experiments. A new program of more precise research was drawn up by mutual agreement, but Chanute's death occurred a few months later" [42].

Lorin's basic idea could be considered the precursor of today's rocket or jet propulsion systems. In the 1930s several researchers proved that reactive force is an effective propelling medium for high-speed aircraft and developed the design further.

9.4   Academics Approach to the Airplane Dream

The Morrill Land Grant of 1862 moved universities from the traditional institution to a career-oriented training. With the Industrial Revolution pushing from an agrarian economy to using machines, engineering educators, like Robert Thurston of the **College of Mechanical Engineering at Sibley College** in Ithaca, New York, pioneered a balance between engineering laboratory experience and mechanical skills on one side and the science of mathematics and physics on the other. To broaden the horizon of his students, and increase the status of the school, Thurston introduced lecture series by nonresident lecturers, experts in their respective fields. For the 1890 winter term, he invited several well-known researchers to discuss their aeronautical work. The fifty-eight-year-old Chanute, an engineer of the old school who had acquired his knowledge through on-the-job study and self-improvement, accepted the invitation and went to Ithaca on May 2, 1890. Thurston announced that Chanute had turned his attention from railroad work to aeronautics and believed that young adults needed to become involved to help solve the underlying principles and perplexing problems of aerial flight. Going public was a breakthrough for Chanute; it is generally believed that formal instruction in aeronautical engineering started with this lecture at Sibley College.

To carry on investigations along various lines of engineering, but also to provide a broader education for engineering students, "engineering experiment stations" were introduced in several universities in the late 1890s. As extensions of the experiment station, aero clubs were formed a decade later at universities and technical schools, where students received hands-on training in designing, building, and flying gliders or airplanes to prepare them for a job in the budding aviation industry.

At **Columbia University**, Dean William Hallock pointed out in the fall of 1908 that the flying machine was no longer a fantastic dream, "but those who are carrying on the experiments are paying little or no attention to science. The ingenious inventors, having spent years of time and bushels of money in setting up an airplane, discover at last that someone constructed precisely the same machine ten or twenty years ago, and had failed just as his successor had failed. They shorten the tails of their machines and add wings and planes without any regard for natural or scientific laws. The last thing is paying attention to indisputable facts" [43].

A few weeks later, on November 2, 1908, the **Columbia University Aero Club** was formed; their first project was to build a Chanute-type glider, learn to fly and then proceed to study and improve a Farman-type airplane. One of the active members was **Grover Cleveland Loening** (1888–1976), who had watched the flying activity at Morris Park, New York, in November 1908 and had become enthusiastic about aviation. He now studied aeronautics from every angle, consulted the available literature, including Chanute's book, and was the first to earn a M.A. (was this a Master of Arts or a Master in Aerial Engineering?) degree ever conferred by any university in the United States. His 188-page thesis focused on automatic stability of airplanes, and *Scientific*

American published it in six installments in their *Supplement* as "*An investigation of the Practice and Theory of Aviation.*" Loening expanded the text and published *Monoplanes and Biplanes* in 1911, acknowledging the work of the pioneers. "Langley, Lilienthal, and Chanute have contributed so large and so well to the progress of aviation, that practical airplane designers of the present owe them a debt of gratitude that can hardly be repaid."

The **Cornell University Aero Club** was formed in late October 1909. Professor McDermott made practical suggestions for the study of aerial navigation and discussed the history and principles of both heavier-than-air and lighter-than-air machines. He described the Zeppelin dirigible, which he considered the most practicable form of lighter-than-air craft thus far produced. Taking up the subject of heavier-than-air machines, he described the biplane and the monoplane and explained by blackboard diagrams the action of air on the wings. "Today," he said, "stability is the great problem of air navigation, and which kind of machine will solve it best is yet to be determined. There is need for experiment, and such is the purpose of this club. We need further knowledge in the field of applied mechanics, and upon that knowledge depends further progress in the field of aeronautics" [44]. Six months later Sibley College introduced an aerial engineering course in the first term of 1910, covering not only the fundamental principles of aviation but also physics, mechanics, and higher mathematics. An editorial in the *Sibley Journal* stated that students who combined their theoretical learning with modern practical knowledge would have good chances for advancement and financial rewards [45] (Fig. 9.14).

The **University of Pennsylvania Aero Club** was formed on November 1, 1909, with an initial membership of fourteen. The now 17-year-old Laurence

Fig. 9.14 The Cornell Aero Club was doing much to stimulate interest in aeronautics at the University. A Chanute-type glider was built as part of the course of instruction. *Sibley Journal*, December 1910

Lesh offered his services, which the Club gladly accepted. Besides designing a Curtiss-type airplane, they ordered a Wittemann biplane glider and built the "*Philadelphia 1*" biplane. Here the lateral control followed an invention by Lesh, who had worked out the design with input from Chanute prior to the Wrights' starting their patent suit. The wings had a 36-foot spread and the frame measured 35 feet from the front edge to the tail in the rear. It was equipped with two rear propellers, placed horizontally across the lower plane. "This airplane will weigh about 1,500 pounds, will carry fuel for a flight of 150 miles, and it is expected to attain a speed of at least 45 miles an hour" [46]. The most unique feature of the "*Philadelphia 1*" was the arrangement for the student, who sat behind the instructor with an extra steering wheel, working in tandem with the front one.

Almost two weeks later, on November 11, 1909, the **Harvard Aeronautical Society** was formed; about 300 students and faculty members attended the first meeting. To help students increase their knowledge in aeronautics, a library was established in Gore Hall and some fifty books and current periodicals on aviation were purchased. A full-size Chanute-type glider was to be built and flown by its members first. In addition, the group purchased small working models of the Blériot monoplane and a Wright-type biplane to illustrate the principles upon which those machines operated. Clearly, Harvard also had caught the aviation bug [47].

Only one day later, on November 12, 1909, the **Massachusetts Institute of Technology (M.I.T)** formed the **Technology Aero Club**, about 600 undergraduate students joined and several faculty members provided support, [48] including Gaetano Lanza, Chair of the mechanical engineering department. He had installed a three-foot square "wind tunnel" in 1896, which gave good results to several students for their thesis, but it was not adequate for present day problems. Believing that the greatest need was experimental research, Lanza suggested that M.I.T. should arrange for academic instruction in aeronautics. "By so doing we can also aid in preventing loss of life due to accidents that could be avoided were there more exact knowledge of the underlying principles and their application" [49]. Lanza then asked Albert Merrill to lecture at the aero club. The classes were well attended and laid the ground work for future teaching at M.I.T.

In April 1912, Earle Ovington discussed the "Modern Aeroplane" at M.I.T and stated in closing: "Believing as firmly as I do in the future of the airplane, I cannot do better than to conclude my remarks by emphasizing the fact that there is a great future for the aeronautic engineer. Here is a new science waiting for competent men and women to grapple with the problems it holds forth. Personally, I am of the opinion that the young college graduate of today, and the graduate of the technical school, could do nothing better than adopt as his life's work the profession of aeronautical engineer. The time has arrived in the development of the heavier-than-air flying machine when the services of the expert engineer are required in order to carry the development to its logical conclusion. You young folks who are casting about in an effort to decide which profession you will adopt, will do well if you give weighty consideration to the new profession of aeronautical engineering. And the work is wonderfully fascinating" [50].

The **University of Chicago Aero Club** was formed in early 1910 to first build and then learn to fly a Chanute-type glider. Members combined their efforts with the Chicago Athletic Association, renting space at Beloit, Wisconsin. William Avery, Chanute's co-worker of the past two decades, intended to build an airplane that would not interfere with the Wrights' patents. And to avoid a potential conflict of interest, Chanute sold Avery his "Soaring Machine" patent for one dollar. Avery drew up plans of his big wind craft, a "warped wing" biplane called the "Cherry Circle," and worked with a Chicago syndicate of business men for funding. The most radical feature was a peculiar curvature of the wing surfaces, resembling a theoretically ideal sail on a boat. The airplane was to have a 55-foot wingspan and 20-foot over-all length, a space of seven feet between the surfaces, and weighing about 700 to 800 pounds with the engine aboard. Avery hoped to have the airplane ready to fly by the end of the summer 1910, [51] stating that "This machine I hope will be a monument to his [Chanute] memory and skill, as it was his opinion that the construction and principles were far superior to anything yet designed" [52]. For various reasons, the "*Cherry Circle*" was not completed.

The **Institute Aérotechnique** was created in July 1909 at **Saint-Cyr** in France, when Henri Deutsch de la Meurthe gave 500,000 francs to the **University of Paris**. The goal was to do theoretical and practical research to improve aerial locomotion. A few months later, the council of the University received 700,000 francs from M. Basil Zakaroff to fund a chair of aviation. To fill this position, some Parisians suggested to hire Wilbur Wright, who refused to even consider the idea, saying that his business was flying and not teaching. "Moreover," said he, "I would have to learn French first" [53]. Next they asked Chanute, but he too was not interested, giving his age and business as the main reason [54]. The committee then hired Lucien Marchis, who had taught mechanical engineering at the University of Bordeaux and who was familiar with Chanute's civil engineering work. Reading about his lecture at the Aéro-Club in Paris in April 1903, Marchis developed a curriculum and taught aeronautics starting in 1904. [55] Being curious if the future airplane was really only for reconnaissance or war, he wrote to Chanute in mid 1906, who mailed several articles for further study [56].

These were just a few of the early academic aero clubs, many more were formed at about the same time around the world.

9.5   New Books to Aspire Aviators

In early April 1910, the Aero Club of America, with Bishop still president, reached an agreement with the Wright Company, wherein the club recognized the Wrights' ownership of certain airplane patents and agreed not to sanction any event at which airplanes would operate without first compensating the Wright Company [57, 58]. The contract also implied that any pilot, licensed by the Aero Club, who competed or performed for compensation without prior approval from the Wrights, would be subject to license revocation. Many members could not afford to fly without some financial support from the public, they filed a protest with the FAI.

Two months later, on Tuesday, June 14, 1910, the United States Circuit Court of Appeals, declared the injunction between the Wrights and Curtiss dissolved due to conflicting affidavits. An editorial in the *Scientific American* explained, "It is far from our intention to censure the Wright brothers for the attitude which they have taken. They are in every way justified in seeking to uphold their patents and in defending rights which, unestablished as they are as yet from a legal point of view, are nevertheless the fruit of painstaking experiments extending over years. Probably the counsel of the Wright brothers themselves were as much astonished by the willingness of Judges Hand and Hazel to grant injunctions against Curtiss and Paulhan as was the patent profession at large. Curtiss was a successful aviator before the Wright brothers decided to cast aside all secrecy. He used hinged wing tips in his earlier machines, with which he made public flights antedating the open flights of the Wrights. It is astonishing that the lower court should have failed to find in these facts a sufficient conflict of evidence to deny the granting of an injunction. With the reversal of the decision of the lower court by the Circuit Court of Appeals, the development of aviation in this country is now unhampered. American inventors will have the opportunity of improving existing machines provided with wing-warping devices, without fear of incurring a fine for contempt of court. Much as we should like to see justice done to the Wrights, we cannot but feel that the reversal of an injunction granted contrary to established precedents will be viewed with satisfaction by every aeronaut" [59] (Fig. 9.15).

Fig. 9.15 The remarkable success of clever boys in mastering the first steps in the new science of aviation, with the enthusiastic and admiring crowd watching. *The World*, November 28, 1909

With the legal hurdle out of the way, a crop of new literature appeared to teach the mechanically inclined person how to build a glider or an airplane, and have fun flying. And of course, the smash hit "Come Josephine in My Flying Machine" helped spread the word about the excitement of flight.

William Jackman, who had interviewed Chanute several times in the last decade, was now working on a how-to-book on *Flying Machines, Construction & Operation* [46]. The author stopped at Chanute's home in late April to show him proof pages of some chapters. But the 78-year-old engineer now wanted to see the complete manuscript, so he walked to the publisher's office on Wabash Avenue, started to read and offered suggestions.

Jackman's book was to guide the novice seeking practical information on the theory, construction and operation of the modern flying machine. Chanute thought that some propositions might be better stated in technical terms, but this would have defeated the purpose of a practical guide for the beginner. "Anyone with an ordinary intelligence can read this book and obtain a clear, comprehensive knowledge of flying machine construction and operation. He will learn, not only how to build, equip, and manipulate an airplane in actual flight, but will also gain a thorough understanding of the principle upon which the suspension in the air of an object much heavier than the air is made possible. This latter feature should make the book of interest even to those who have no intention of constructing or operating a flying machine. It will enable them to better understand and appreciate the performances of the daring men like the Wright brothers, Curtiss, Blériot, Farman, Paulhan, Latham, and others, whose bold experiments have made aviation an actuality." Chapter 1, the introductory chapter in Jackman's book, republished Chanute's "*Evolution of the Two-Surface Flying Machine*" and "*Soaring Flight*" was Chapter 20. The 220-page book, bound in red leather, sold for $1.50.

When the Chicago Aeroplane Manufacturing Company formed the Chicago Aviation Institute in early 1911, they hired Jackman to teach the correspondence courses. Needing a text book, he updated and expanded his book, acknowledging in this 2nd edition, "All this was 'a labor of love' on

Mr. Chanute's part. He gave of his time and talents freely because he was enthusiastic in the cause of aviation, and because he knew the authors of this book and desired to give them material aid in the preparation of the work—a favor that was most sincerely appreciated" [46].

Victor Lougheed, Aeronautical Editor of the *Chicago Tribune*, had published several articles in the Sunday Magazine of the *Tribune* on how to become a flyer at comparatively little cost and experience the sensation of flight (Fig. 9.16).

His 480-page book *Vehicles of the Air* with the subtitle "Popular exposition of modern aeronautics with working drawings" discussed in fifteen chapters almost everything known about aeronautics, thus giving the interested person the knowledge to build and then fly an airplane.

Chanute was happy to review the book for the *Chicago Tribune*. "Mr. Lougheed is evidently an optimist. In his introduction he argues that the popularly supposed danger of accident has been much exaggerated, and he anticipates considerable commercial uses for flying machines. It may seem to the casual reader a venturesome thing to predict any similarly extensive development of aerial vehicles. Yet it is to be remembered that even the most accustomed forms of modern transportation, the railway, the steam vessel, the bicycle, the automobile, etc., all had their inception within the lifetimes of people now living, while without exception their development from the experimental stage to the status of utility has covered much shorter periods." Lougheed acknowledged the work of Hiram Maxim, and J. J. Montgomery, and duly mentioned the Wrights and Voisins. One chapter was devoted to the

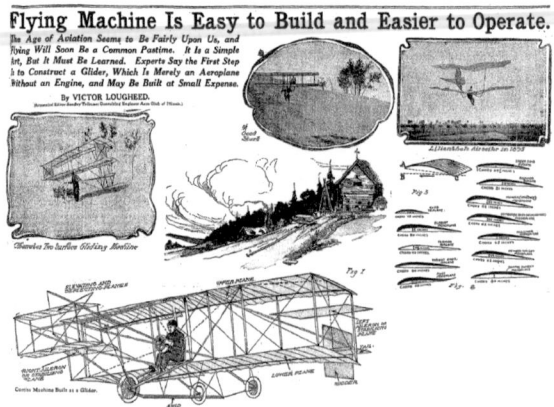

Fig. 9.16 The age of aviation seems to be upon us, and flying will soon be a common pastime. It is a simple art, but it must be learned. The first step is to construct a glider, which is merely an aeroplane without an engine, and may be built at small expense. *The Chicago Sunday Tribune*, June 19, 1910

construction of gliders and airplanes, followed by a chapter on power plants. In closing Chanute mentioned that there were a number of minor mistakes scattered throughout the book, but they were not a distraction [60].

As powered flight gained popularity, interest in motorless flight started to fade; **James Means** became concerned that men like Lilienthal and Chanute would be forgotten, even though their experiments laid the groundwork for solving many of the problems that current aeronauts were still working on. To highlight the aspirations and experiments of the men who were instrumental in guiding others towards the conquest of the air, he compiled the *Epitome of the Aeronautical Annual* which appeared in May 1910. Means wrote in the introduction, "This compilation has for its primary object the encouragement of those who are just beginning the study of aviation. In the effort to reach a good understanding of the achievements of today the student may do well to learn of the work of the pathfinders, who laid the foundation of the science." Means republished articles of several pioneers to open the topic for further research and wrote in his "*Editorial,*" "Aviation as a sport is here and it is here to stay. Military experts seem to agree that in the unfortunate event of war the dynamic flying machine in its present state of development will be of great importance for scouting purposes. These experts are far from being in agreement concerning the utility of the flying machine for offensive purposes in warfare. It may be said that those who have given the most study to the subject of aviation seem to agree in thinking that at no distant day a useful machine will be developed … There are conservatives and there are visionaries. Looking back a few centuries to the time when man began to conquer distance and looking at his present powers, we see that the visionaries have come out ahead. The great utilitarian task, which is now before us, is to navigate the air through darkness, fog, and storm. Can the obstacles be overcome?" [61].

To document the "initial steps in fostering the scientific study of aeronautics," the Smithsonian Secretary Charles Walcott wanted a record of all past aeronautical publications, books, pamphlets and articles. **Paul Brockett**, assistant librarian of the Smithsonian, agreed to assemble a *Bibliography of Aeronautics*. Langley's library was extensive, but Chanute's holding was more complete in other areas, and he was happy to assist. Much to everyone's surprise almost 1,000 pages were required to present the 13,500 entries, which were arranged alphabetically by authors, subjects and titles. While a classification of aeronautical literature was not attempted, cross-references were made to important phases of the subject. In his introduction Brockett mentioned the institution's long association with aeronautics, and that "the greatest era in the development of aeronautical science in connection with the Institution began when Dr. Samuel Pierpont Langley became Secretary in 1887." Updated versions of this listing were published after World War 1 on a yearly basis until 1937, and digital copies are available to current researchers on early aviation.

Another Smithsonian literary project was the *Langley Memoir on Mechanical Flight*, assembled and edited by **Charles M. Manly**, who had joined the Smithsonian in 1898. Secretary Walcott thought that the story of Langley's aeronautical work could serve as an inspiration to newcomers. The first part, written by Langley, discussed his preliminary work and experiments up to the successful flight of his model in 1896. The second part was written by Manly, who acted as pilot of the not so successful flights in 1903. The 540-page book with many drawings and photos was published in late summer 1911, and Manly mailed review copies to several influential newspapers, which had been critical of Langley's work. Manly's personal note to the *Herald* stated, "In recognition of the most important encouragement which the New York Herald in recent years has given to the science of aviation, which suffered so greatly from the criticism and ridicule of the newspaper press during the epoch-making experiments of the late Professor S. P. Langley, the Smithsonian Institution is sending at my request a copy of the just released 'Langley Memoir on Mechanical Flight'" [62].

All these books, and others not listed here, popularized aviation and were widely available to anyone wanting to build and then fly their own glider or airplane; they marked the true beginning of America's grass roots aviation movement.

9.6   Chanute's Last Trip to Europe

After a thorough medical check-up, Dr. Walter H. Allport, the Chanute family doctor, thought there should be no problems for his 78-year-old patient to take a lengthy trip to Europe, but he prescribed a new heart medicine, just in case. Chanute and his three daughters left Chicago on May 14 and boarded the most luxurious and fastest ocean liner, the *Kaiser Wilhelm der Grosse*. [63] Arriving in Dover, England, he heard that Jacques de Lesseps, whose father he had met decades earlier, had made the second successful crossing of the English Channel in a Blériot monoplane just two days earlier. How he wished he could have been at the landing site!

Arriving in Paris, the French aeronautic community gave Chanute a hearty welcome. In the next few weeks, he and his daughters visited family members, the Voisin airplane factory, attended banquets in his honor, and in mid' June, he accepted the honorary membership in the *Société des Ingénieurs Civils de France*. In addition, he was to receive an honorary membership of the American Railway Engineering and Maintenance of Way Association at the eighth International Railway Congress in Berne, Switzerland. There was so much happening!

Chanute granted a few interviews, but avoided any discussion about the Wright's patent struggle. A correspondent for the *Chicago Daily News* asked why Chicagoans are so far behind everyone else in regard to aviation. "Chicago men are practical when a question of dollars and cents is involved. I am convinced that for the present, aviation is unavailable for passengers and still less for freight, as the technical difficulties are almost insurmountable. The only uses for airplanes are the exploration of inaccessible places on desert mountain tops, reconnoitering in time of war, which I hope will be the rarest of opportunities, and in sport, which nowadays shows the greatest development. I believe that as soon as Chicagoans realize the value of aeronautics as a sport, they will contribute prizes for its encouragement. And I wonder who will be the first to fly across Lake Michigan" [64] (Fig. 9.17).

Fig. 9.17 Snapshot taken at one of several interviews in Paris. *La Vie Automobile*, August 8, 1910

Having heard that the works of Lilienthal and Chanute had inspired the two American inventors from Dayton, a *Le Petit Journal* reporter asked for an interview. "In spite of his seventy-eight years, Octave Chanute is a handsome old man with a ruddy complexion, with large blue eyes, a barely wrinkled forehead and a goatee. As he suggested during our conversation, Mr. Chanute thinks in English but speaks very well in French." After providing a short recap of his civil engineering career, Chanute explained that he became interested in aeronautics when he saw reports of the Lilienthal experiments. "I had noticed that Lilienthal and Pilcher were forced to restore balance in their gliders by quick movements of their bodies. It was these movements that needed to be overcome. And we did it, making more than two thousand flights."

Next, the reporter asked, "Did you have to use mathematical formulas to arrive at the result?" Chanute reportedly stated, "We do not use much mathematics. Sometimes we put so many things in it that we do not know how to get away with it anymore. I am not saying that I did not use mathematics, but I do not think that it is necessary. What is needed is a great deal of judgment, a great deal of observation and patience. Look, the Wright brothers are not mathematicians, they are good mechanics, a skill they inherited from their mother who was a real genius in the field. All the modifications they made in the construction of the biplane, they made without calculation, but by deduction."

Asking what Chanute considered to be his main contribution in aviation, he simply responded, "What I discovered is that using two wings, one above the other, with a rigid space in between, works perfectly and keeps the airplane stable. This is my main discovery. The idea was very simple, but someone had to think of it."

Wondering about aviation back home, "In America we are very practical. It was quickly realized that present aviation is of no commercial interest; there is no advantage unless it is for wartime. You see, as long as we do not find a way to keep a craft in the air without the use of the explosion motor, we will not be able to use the airplanes practically." For Chanute this was the future, "Let's accept it" [65].

In late June the family travelled to Carlsbad, a social spa town in the western Bohemia region of what is now the Czech Republic. Unexpectedly, Chanute's health deteriorated rapidly. As none of the doctors spoke English or French, the family decided to return by train to Paris where communication with the medical staff would allow them to figure out how to proceed. A waiting ambulance rushed Chanute to the American Hospital, where doctors diagnosed him with double pneumonia, a serious concern due to his age.

Chanute had agreed to present a talk to members of the Aeronautical Society of Great Britain in September; the Council then voted unanimously to confer their Gold Medal in recognition of his distinguished services to the advancement of aeronautics. Arrangements were made for a fitting public presentation of the highest award in the world of aeronautical science to the distinguished engineer. But then reading in the press, that Chanute was transferred to a hospital in Paris, the secretary of the Aeronautical Society decided to mail their gold medal to Paris, hoping that the recipient would still enjoy receiving it.

Sitting in his chair outside the Paris hospital, pilots flew overhead to do him honor, which made the older pioneer feel good. By the end of September Chanute felt strong enough to travel back to Chicago. A large crowd gathered at the train station in Paris to say good-bye to the elder aeronautical pioneer, whose writings and lectures had paved the way for the eventual triumph of the heavier-than-air principle. The Chanute family boarded the French liner *La Savoie* at Le Havre on October 1, and he quickly became acquainted with the three aeronauts who were traveling on the same boat to America to compete at the aviation meet at Belmont Park. [66] One of the aeronauts was John Moisant, originally from Chicago, who talked about his biplane glider, built years earlier, and the Blériot monoplane, that he and his mechanic had just flown from Paris to London. Chanute knew that he had laid the groundwork for their accomplishments and after ages of yearning, effort, and ridicule, "we have at last learned to fly" was his reported comment.

Arriving back in New York on Saturday, October 8, Charley Chanute and his wife Emily waited at the dock. The family stayed for a few days at the Stratford House to relax prior to returning to Chicago.

9.7   A Life of Achievements Comes to an End

Back home in Chicago, Chanute's health improved a little, but he did not feel like writing letters or socializing with the many acquaintances who came to see him. Avery was a regular visitor, talking of old times, Jackman dropped off his book *Flying Machines*, and Means shipped his *Epitome*; all these pleased the aging engineer. With a certain inner satisfaction, Chanute also admired the Gold Medal from the Aeronautical Society of Great Britain that finally reached him.

Chanute's shining light went out the day before Thanksgiving, on Wednesday, November 23, 1910. The long career of the seventy-eight-year-old civil engineer ended in the sleep that knows no awakening. Charley reported for the family that his father suffered no pain and was conscious a few minutes before dissolution took place. "His poor body was so worn out that when he fell asleep the machinery simply ran down, and he left us." [67] The family shared the sad, but expected, news by submitting a report to the wire service, also stating that a memorial service would be held at the Chanute home on North Dearborn Street on Friday, November 25. A number of prominent engineers and associates from around the country rushed to Chicago to pay their last respects. Bion Arnold came from Michigan and William Glassford took the next train from Fort Omaha. Wilbur Wright, who had heard the news from the Dayton newspaper office, also travelled to Chicago. These men, as well as William Avery, were Chanute's closest aeronautical friends in attendance.

Octave Chanute was buried beside his wife Annie on December 2 in the James and Chanute family plot at Springdale Cemetery in Peoria, Illinois, with Avery acting as the ceremonial pallbearer.

10

Some Final Words …

The age of aviation seems to be fairly upon us, and flying will soon be a common pastime. It is a simple art, but it must be learned. Experts say the first step is to construct a glider, which is merely an aeroplane without an engine, and may be built at small expense.

Victor Lougheed, *Chicago Tribune,* **June 1910**

Tackling the frontier of flight at the end of the nineteenth and the early twentieth century inspired a new generation of innovative thinkers. One of these men was Octave Chanute, a civil engineer who had vision and dreamed dreams, usually making them come true. His brilliant engineering triumphs included the laying of railroad tracks through the undeveloped western frontier, allowing settlers to move in and prosper, and at the same time, he improved the iron rail designs allowing smoother train travel; he initiated standardization of hardware like nuts and bolts, designed and constructed bridges across broad rivers and deep valleys, using first wood, then reinforced concrete and steel construction; he master-minded plans for the Chicago and Kansas City stockyards, participated in the feasibility study to introduce the rapid transit system in New York City, soon to be introduced in other big cities. A lesser-known fact is his contribution to the growth of the economy in the United States that faced an ecological collapse in the 1870s, when the expanding railroads required an estimated 100 million railroad ties every year. With the available wood decomposing in less than one year, Chanute joined Joseph Card in 1884 to treat wooden ties, and he convinced railroad

© The Author(s), under exclusive license to Springer Nature Switzerland AG 2023
S. Short, *Flight Not Improbable*, Springer Biographies,
https://doi.org/10.1007/978-3-031-24430-8_10

managers that the increased initial cost would save money in the long run; today treated railroad ties usually remain in service for 20–40 years [1].

However, Chanute's most far-reaching achievement was his contribution to aviation. At an age when successful men retire and step back from active live, he turned his attention to the subject that had always fascinated him, the study of manflight. Between the mid-1880s and 1910, Chanute played his role of information networker, believing that men can always be led more successfully than they can be driven. He collaborated with practically every earnest student of aviation, providing guidance, extending counsel and friendly sympathy [2]. Collaboration is a way of being, an approach to how we work that is steeped in values, beliefs and a desire to achieve the best possible outcome for all involved. He believed that a free exchange of ideas and results, good or bad, would make success of the flying machine possible. Considering himself the "intelligencer," he rekindled the "Republic of Letters" concept, fostering communication among intellectuals and fellow researchers by writing letters. Chanute's letters exemplified gracious writing, always in an encouraging tone. Looking through his letterpress copy books, there are many hundreds of letters addressed to experimenters around the world, including over 220 letters written to the Wright brothers alone. His open approach facilitated his substantive role as an information broker, he knew what needed to be known in this nascent field of aeronautical engineering.

The reader may wonder, if this eminent engineer is best described as a bridgebuilder, a railroad executive, a city planner, or America's first aeronautical engineer? There is no clear answer, because he contributed significantly in each field. And as we know, with determination and willingness to work, there is nothing that can stop progress.

10.1 Remembering Chanute

Hearing of Chanute's death and having known and worked with him long before he started his *Aeronautics* magazine in 1907, Ernest La Rue Jones wanted the world to know about the man who was rightfully called the "Father of Aviation." He asked several people to write an eulogy to be published in his January 1911 *Aeronautics* issue. The Aero Club of America then decided that they should honor Chanute at their meeting in early January, but with all their internal problems, Jones was unsure what would happen. He buried a brief editorial toward the end of the January issue: "It says somewhere in the Bible that the lion and the lamb shall lie down together. The Aero Club of America is to have a meeting on January 6th in memory of Mr. Chanute. The president of the Aeronautical Society, Hudson Maxim, is to be asked to address the meeting and the members of the Society are to be invited" [3]. Close to the front of the same issue, Jones published his personal eulogy with the simple title, "*The life and work of Octave Chanute*" [4]. He explained that the majority of people associate Chanute's name with aeronautics, but there was more to this gentleman. "Mr. Chanute was one of the foremost railroad builders of the country. The Kansas City Bridge was declared impossible of construction. Mr. Chanute accomplished it." And Jones reprinted an article from *Engineering News*, published twenty years earlier, when Chanute was elected president of the American Society of Civil Engineers [5].

The next contribution came from James Means: "… For many years Mr. Chanute was in regular communication, personally and by correspondence, with all the men he could find who were intelligently trying to do research work in aviation. The help, which he gave to such men can never be fully known or measured. The counsel and encouragement that he gave to Wilbur and Orville Wright have been gratefully and gracefully acknowledged by them. It came to them at the time when it was most needed, when they were at the foot of the steepest part of the unblazed trail. It gave them the courage and confidence, which were essential to enable them to keep on alone and to emerge at last at the summit triumphant … Those who knew Chanute will always remember his loveable character and will think of the oft-repeated saying: He was more willing to give credit to others than to claim any for himself. We may well believe that whenever in the future the history of aviation shall be reviewed, the name Chanute will stand forth as one of the great founders" [6].

The third testimonial was written by Wilbur Wright, who had heard of Chanute's death from the Dayton newspaper office. Wilbur reached Chicago in time to attend the memorial service at the Chanute family home; it is safe to assume that he spent much of that trip thinking about his complex relationship with Chanute, which had begun with a single letter more than ten years before. He may have also thought back to the events of their 1901 camp and his pivotal first talk to the Western Society of Engineers in Chicago. He mailed his comments to Jones after returning to Dayton. "By the death of Mr. O. Chanute, the world has lost one whose labors had to an unusual degree influenced the course of human progress. If he had not lived the entire history of progress in flying would have been other than it has been, for he encouraged not only the Wright brothers to persevere in their experiments, but it was due to his missionary trip to France in 1903 that the Voisins, Blériot,

Farman, Delagrange, and Archdeacon were led to undertake a revival of aviation studies in that country, after the failure of the efforts of Ader and the French government in 1897 had left everyone in idle despair … His labors had vast influence in bringing about the era of human flight." In closing Wilbur stated, "Although his experiments in automatic stability did not yield results which the world has yet been able to utilize, his labors had vast influence in bringing about the era of human flight. His double-deck modification of the old Wenham and Stringfellow machines will influence flying machine design so long as flying machines are made. His writings were so lucid as to provide an intelligent understanding of the nature of the problems of flight to a vast number of persons who would probably never have given the matter study otherwise, and not only by published articles, but by personal correspondence and visitation, he inspired and encouraged to the limits of his ability all who were devoted to the work. His private correspondence with experimenters in all parts of the world was of great volume. No one was too humble to receive a share of his time. In patience and goodness of heart he has rarely been surpassed. Few men were more universally respected and loved" [7].

After talking with Lieut. Colonel Glassford from the Signal Corps at the memorial service, Wilbur explained in a follow-up letter that Chanute's "books and correspondence have inspired and encouraged others to action, especially in France and America, to such an extent that I think none of the older workers can justly claim to have influenced progress in the art more powerfully than he" [8]. Without doubt, Chanute was the greatest aid and possibly the wisest advisor the Wrights had in their initial stages of developing the airplane.

10.2 Acknowledging Chanute's Pivotal Role in Aviation

Invitations to the general membership meeting of the Aero Club of America on January 6, 1911, were mailed not only to their own members, but also to members of the Automobile Club of America, the Aeronautical Society of New York, the American Society of Civil Engineers, of Mechanical Engineers, of Electrical Engineers, and the New York Electrical Society. The program included an address by the incoming President, Allan Ryan, and a tribute to the pioneering work of the late Octave Chanute by Hudson Maxim, president of the Aeronautical Society.

The auditorium of the Engineering Society Building at 39th Street in New York City was fairly well filled. Many distinguished men and women, including Orville Wright, were in attendance. After short introductions by president Ryan and past-president Bishop, Hudson Maxim stepped forward to the podium to deliver the principal address of the evening. He wore his black frock coat with the air of one who is kept on the ground yet aspired to higher things. His silver-grey hair and beard only served to heighten the bright enthusiasm of his face. Maxim's eloquent tribute is reprinted below:

"When men unite to do honor to greatness, they but perform a just duty and meet a just obligation, both to the man they honor and to themselves. Great inventors, discovers, scientists, philosophers, are men who stand a little in advance of the world and help pull the world after them. The merit of what they do is seldom appreciated or recognized until the world has caught up.

"Octave Chanute was the veritable father of aviation, and he was always, in all things, a vedette of progress. He was one of those whom duty sent far to the fore, where, unaided and lonesome, they make their landmarks, beckon their fellows to follow, and move further on; and, when the world comes up, then, and not until then, are their landmarks seen to be true and their labor found to be worth-while.

"Chanute was the Chief Engineer of aviation. He was inventor, mechanician and mathematician. He had the courage of his convictions, for they were born of scientific knowledge and experience. Long before he began to pave the uncharted skyway with mathematical equations, Chanute was a master spirit of rapid transit and practical transportation. He was one of the vulcans who hewed the hills down for the roaded thunders of locomotion.

"Chanute was one of those rare intellectual giants big enough and generous enough to endow other inventors and workers with his knowledge and to lend a hand to help them utilize it, and all without jealousy or envy. Aviation was a thing dearer to his heart than any self-greatening. It is gratifying to know that before his lamp of life went out he had the satisfaction of seeing fairly accomplished that master achievement which he had so generously patronized and for which he had so long labored and prayed, the conquest of the air. It is easy enough now, when we review what has been done, to see how it was done, but it took a genius to foreknow the advent of aviation and in imagination to foresee the actual aeroplane soaring in the coming sky, and it took courage to face the ridicule of unbelief and the sneers of ignorance. Chanute foresaw it all. He knew that mechanical flight was surely to be accomplished. He was one of the biggest and bravest of those whose labors finally launched the airship, and raised the eyes of doubt to behold accomplishment hung in the cloud, turned ridicule to wonder, the sneer to the loud bravo. Now, when it looks so easy, it is hard to realize that, but a few years ago, any serious talk about aviation was a thing to be whispered in secret. The sure fulfillment of recurrent prophecy by repeated disaster made the immortal Darius Green of Trowbridge a symbol and a type of the flying machine inventor.

"As the great human throng goes parading by in the avenue of life, it is a strange procession. First come the pioneers: the discoverers, inventors, scientists, philosophers, and a scattered few who clear the way. At their heels, and dogging them, are the standard bearers of ignorance, and the drum-beaters of prejudice, with their dancing, gibing, jeering, knaves, clowns and fools, making huge sport of their uncomprehended leaders who are clearing the way for them. These scoffers take nothing seriously which they do not understand. They do not understand their pioneers, and so they treat them as a joke. In the middle of the procession, we see men with books in their hands. They are finding out the meaning of what their great men did when they passed along. Later, we see men waving banners, playing music, singing songs, and cheering in honor of their great pioneers who have preceded them. There are but few, even among these, who actually understand the merit they are cheering. They know only that it has become a creditable thing to praise and honor now, instead of to scoff.

"From the viewpoint of present accomplishment, aviation is an amazing triumph of human ingenuity and perseverance; but still more amazing is the almost inconceivable bravery of our aviators. There have never been in any war, even where heroism made a nation's glory, braver men than are our flying heroes of today, who are so nobly fighting to conquer and subordinate the sky to man's use. One brave fellow after another loses his life. Heroic Johnstone plunges to death, and his friendly competitor, Hoxsey, all undaunted, climbs a screaming hurricane to a height of more than two miles. Such heroism makes us readjust and raise our respect for human nature. The passion for flight has in every age burned in the human heart. Paleolithic man looked with envious eyes upon the eagle's flight, and wondered at the disfavor of his Maker in having foot-tied him to earth, while to the birds had been given the winged freedom of the air; and man began then to think upon the problem of aerial navigation.

"In spite of his prowess and his cunning, man was obliged to make his miles with slow-paced weariness, while the dove could wing its way as freely as the flight of thought. When man beheld the mystery of life forsake the tired clay at the earth-journey's end, love and hope made him follow in his imagination the flight of that departed life into a spirit world, which he substantialized and visualized. He there gratified his passion for flight by giving wings to the immortal spirits of his dead, and he honored his great heroes by making them winged gods. He who first plumed himself for flight and essayed the navigation of the air had a god for his model.

"All of the old-time aeronautical inventors had to combat two very strong forces, the antagonism of contemporary prejudice and gravitation. To the minds of all the bone-headed wise-acres of the past, human flight was palpably impossible, not only impossible it was a wicked thing for man to try to invade the empire of the birds. Had God intended man to fly, he would have given him wings. Just as every innovator who has found one sphere of action too circumscribed for him, and broken out of it into broader fields of endeavor, has been obliged to face prophesied disaster, and the old admonition, 'Shoemaker, stick to your last!' so the first aviators were advised to stick to their earth.

"There has always been proof enough that the conquest of the air was an utter impossibility. Had it not been for the few big, progressive spirits in every generation who have undertaken the palpably impossible, and continually accomplished it, had it not been for the few, the courage of whose convictions was great enough and the sense of whose duty was dominant enough to face the sneers of envy, the ridicule of prejudice and the opposition of ignorance, our dwelling would still be a cave in the hill, our electric light a pitch torch, our library a few rude pictures and hieroglyphics scratched on the ledge wall; and the piece de resistance of our banquet some tough old patriarch slaughtered and put to his last use.

"Always it has been a devoted few, who have stood in the vanguard and fought the hard fight of progress. One of such few was Octave Chanute. He belonged to the true nobility of brains. He was a man to make the constellated eyes of heaven look our way and honor us and the gods to boast of kinship" [9].

10.3 The National Aeronautical Research Laboratory

Formal aeronautical engineering classes were introduced at several universities in 1908, as the new industry needed the services of the scientist and theorist to calculate how an airplane ought to be built and how the material used in its construction should be distributed to give the greatest possible amount of strength and efficiency [10].

After leaving the Catholic University, Albert Zahm had joined a group of aeronautically minded men who shared Chanute's opinion that the airplane was a world invention, the result of a slow growth of gathered information over a period of time. This system of acquiring and distributing knowledge needed to continue, as the airplane was still far from being commercially practical. Zahm also believed that a national institution for the advancement of aeronautics was essential, not only to the military but to society in general. America should not trail behind Europe. France, Russia, Germany, Italy, and Great Britain all had active aeronautical laboratories with varying degrees of government support. After discussing the basic idea with Secretary Walcott, Zahm started to push for at least one broadly planned aeronautical laboratory with ample endowment and equipment, a wise and devoted directorate, and an able and highly trained technical staff. Lawrence Rotch, a meteorologist from the Boston area, supported Zahm's views and noted that "the establishment of aerodynamical laboratories marks the entrance of aeronautics into the domain of engineering," where theoretical knowledge based on experiments would be the foundation of progress.

Struggling through the labyrinth of bureaucratic intrigue and congressional politics, Smithsonian Secretary Walcott emerged to guide the aeronautical laboratory movement. He began his campaign early in 1913 by reopening the Langley Aerodynamical Laboratory within the Smithsonian Institution; for this he needed only the approval of the Board of Regents. Granted, one of his reasons was to honor Langley and his work, but he also thought that the Smithsonian ought to look into the future. While the laboratory would conduct such research "as may serve to increase the safety and effectiveness of aerial locomotion for the purposes of commerce, national defense, and the welfare of man," it was in no way to "promote patented devices, furnish capital to inventors, or manufacture commercially, or give regular courses of instruction for aeronautical pilots or engineers." It was to "exercise its function for the military and civil departments of the Government of the United States, and for any individual, firm, association, or corporation within the United States provided that they defray the cost of all material used and of all services of persons employed in the exercise of such functions." In sum, the laboratory would use and compliment the resources of the federal government for the advancement of aviation in general, but avoid favoritism to special interest groups.

To learn how Europeans approached aeronautical sciences, the Smithsonian dispatched Zahm on a fact-finding tour of European aeronautical laboratories. Jerome C. Hunsaker joined him as he was developing a curriculum in aeronautical engineering at the Massachusetts Institute of Technology. Zahm's report, published in 1914, emphasized the galling disparity between European progress in aeronautics and American inertia.

But then, the opposition pointed out that government employees could not work at an aeronautical laboratory without consent of Congress. Legislative action was needed, but not given, so the Langley Laboratory was deactivated again.

10.4   Forming the National Advisory Committee of Aeronautics and the Manufacturers Aircraft Association

The few people who understood the importance of aeronautics were acutely aware of the rapid progress being made in Europe. Chanute's outspoken fears had materialized. The sorry record of past attempts to establish a national aeronautical laboratory led Walcott to conclude that it would be best to propose the formation of a modest committee, authorized by Congress. It should be independent of the Smithsonian Institution and members, drawn from private life, should not outnumber government members. The armed services should endorse the proposal in draft, and it should then be submitted through friendly congressmen to equally friendly congressional committees.

A mixture of national pride and a sense of urgency led to a rider, attached to the Naval Appropriations Act, signed by President Woodrow Wilson. On March 4, 1915 Congress authorized the creation of an Advisory Committee on Aeronautics, which quickly added the prefix "National," creating the acronym NACA. The President then appointed twelve members who served without pay: "two officers from the Aviation Section in the Army, two from the office in charge of aeronautics in the Navy; a representative each from the Smithsonian Institution, the U.S. Weather Bureau and the Bureau of Standards; and five additional persons who shall be acquainted with the needs of aeronautical science, either civil or military, or skilled in aeronautical engineering or its allied sciences." Zahm was appointed Recorder of the Committee.

NACA's purpose, approved by Congress, was given as: "That it shall be the duty of the Advisory Committee for Aeronautics to supervise and direct the scientific study of the problems of flight, with a view to their practical solution, and to determine the problems which should be experimentally attacked, and to discuss their solution and their application to practical questions. In the event of a laboratory or laboratories, either in whole or in part, being placed under the direction of the committee; the committee may direct and conduct research and experiment in aeronautics in such laboratory or laboratories."

Anticipating the ability to design and construct airplanes as required for the war effort, one of the first steps taken was to begin the training of an aerial army, by organizing ground schools and establishing flying fields. One of the first such fields became known as "Chanute Field" at Rantoul, Illinois, a technical training campus, both for flying as well as maintaining airplanes. It fittingly recognized the value of Octave Chanute's work in the interest of aviation.

The United States entered World War I on April 6, 1917, and the lack of preparedness in the aircraft industry was exacerbated by the requirements of the war. Then the U.S. Government commanded a truce to resolve the patent dispute. Benton Crisp, the lawyer who had helped Henry Ford win his long patent fight with George B. Selden, developed a plan to bring all concerned parties together in a new organization, the Manufacturers Aircraft Association, to implement and administer a cross-licensing agreement. The U.S. War Department reportedly told Curtiss and Orville Wright to come together. The settlement took over the Wright and Curtiss patents and administered a cross-licensing agreement that applied to all members of the association.

The agreement was a long step forward toward progress and expansion. The NACA, which was instrumental in helping the adjustment of difficulties among patentees, did a remarkable piece of work very quietly.

The effects of this settlement also bound the aircraft manufacturers closer together, as the arrangement involved financial considerations which would eventually run into the millions. Now, any patentee with an improvement of value to the industry could have his patent used and received compensation, which would be just and equitable [11]. Chanute's philosophy of researching, mentoring and sharing results had become reality, again.

10.5   In Closing …

Half a year after Chanute's death an editorial discussed the Chicago Aviation Meet (August 12–20, 1911), stating, "There probably will not be a contestant who will not turn his thoughts back to the days when Octave Chanute was conducting aeroplane experiments on the shore of Lake Michigan, only a few miles from Grant Park. Every man who guides a machine in its flight will be willing to acknowledge that to Chanute more than to any other man belongs the credit of making flying possible. The debt to the Chicago experimenter had been acknowledged by most successful aviators. In every machine which will start from Grant Park there doubtless can be found substantial evidence of the constructive genius of Octave Chanute. He lived long enough to see the practical application of principles that he had formulated" [12].

Good airplane design requires a knowledge of earlier designs, or a knowledge of history. Chanute's trussed biplane design was as much evolutionary as it was revolutionary, as he drew from a prior century of aeronautical work by others, just like most new airplane designs depend on knowledge gained with previously designed airplanes.

With the rapid development of the airplane during World War I, motorless flight was mostly forgotten. Augustus Post envisioned in 1921 that "Soaring flight is conceivable, and it may revolutionize flying as we know it today. Instead of a powerful motor pulling an airplane through the air regardless of wind or air currents, a successful development of soaring machines may bring about airplanes either motorless, or with comparatively weak little motors, soaring the high altitudes, taking advantage of every whiff of breeze and every upward current just as the buzzard does" [13]. But the American public showed little interest in flying, either with or without a motor.

Charles A. Lindbergh successfully crossed the Atlantic on May 21, 1927, in his *Spirit of St. Louis*, which caught the attention of many Americans, stimulating interest in flying. One year later German glider pilots came to the U.S. to show Americans how motorless flying is done, and then Frank Hawks was aerotowed with his glider across the whole United States, displaying the motorless craft to the public at mini airshows along the way.

Now, gliding, which is older than flying a powered airplane, came into its own again. Quite apart from recreational use and in competition, gliders were deployed during World War II to carry troops and goods more stealthily than by any powered airplane. And today, gliders, or sailplanes, are still used as research tools to study air currents without the influence of an engine, and they are enjoyed as a challenging sport.

The next time you are outside on a clear day, look up. Most likely you will see evidence of some airplane; you might see a long wing modern glider, a sailplane, circling high in the sky, gaining altitude to reach its next destination; or there could be a small, private aircraft hanging low in the sky, making its way to a nearby airport; or maybe you can see a distinct white contrail high in the blue sky produced by a fast jet airliner on its way from one end of the continent to the other. Most of us take airplanes for granted, as they are part of our everyday life, but it did not just happen, the airplane evolved over the past hundred (+) years, with many people contributing.

Chanute stated more than once that aeronautics was just a "hobby of his old age," but by applying his trained engineering mind to his "hobby," he paved the way for today's airplane. Maybe, he dreamed that the tiny acorns he planted would bear such giant oaks, for this new science of aeronautics has revolutionized our world today [14].

THE END.

Notes

Information for any scholarly book comes from many different sources and this book is no different. The foremost source of original information for this book comes from two sources: (1) the University of Chicago library system owns the Octave Chanute Collection, donated by the Chanute estate to the original John Crerar Library in 1911. These books, pamphlets, and scrapbooks are available for research, and are part of the Special Collections Research Center, University of Chicago Library, and can be found at either the Crerar or Regenstein Libraries. They are referenced as **Chanute Papers, UoC**. And (2) the Manuscript Division, Library of Congress in Washington, D.C. All letters from and to Chanute, including 24 of his letterpress books, more scrapbooks, and other personal papers were donated by the family in late 1932; they are part of the "Octave Chanute Papers" (https://lccn.loc.gov/mm78015560) and are references as **Chanute Papers, LoC**. Leonard Bruno had been extremely helpful when I started my research in the mid 1990s, and when he retired (at about the same time as "*Locomotive to Aeromotive*" was published), Lewis Wyman became my new in-between-person. Their knowledge of the material available at the Manuscript Division was very helpful. Thanks to both gentlemen!

Personal information on Octave Chanute came largely from the family and is noted as **Chanute Family Papers**. In the 1950s, Pearl I. Young, an employee of the National Advisory Committee for Aeronautics (now NASA), worked closely with Elaine Chanute Hodges, Chanute's granddaughter. When Ms. Young died in 1969, her research material was shipped to Ms. Hodges, who subsequently donated all to the Denver Public

© The Editor(s) (if applicable) and The Author(s), under exclusive
license to Springer Nature Switzerland AG 2023
S. Short, *Flight Not Improbable*, Springer Biographies,
https://doi.org/10.1007/978-3-031-24430-8

Library (https://catalog.denverlibrary.org/). Ms. Hodges also arranged to have Chanute's fragile letterpress books copied in the 1960s to microfilm; they are still available through interlibrary loan from the Library of Congress . I used these microfilms in my research.

Incoming/outgoing letters have been transcribed by Chanute's daughters in 1910–11, others by Albert Zahm and his team in the 1930s, by members of the Institute of the Aeronautical Sciences in the 1940s, and more were transcribed by Pearl Young and her team in the 1960s. Bound copies (13 volumes total) are available at the Library of Congress and the National Air and Space Museum, some have been digitized. In the 1990s, the family of George A. Spratt donated his correspondence to the Library of Congress, they are part of the Chanute paper holdings.

In 2003, the digitized correspondence of the Wright brothers and various family members became available as the "Wilbur and Orville Wright Papers" (http://lcweb2.loc.gov/ammem/wrighthtml/wrighthome.html). They are referenced as **Wright Papers, LoC**. These original letters make a good addition to Marvin W. McFarlands two volumes of *The Papers of Wilbur and Orville Wright, Including the Chanute-Wright letters*.

The National Air and Space Museum owns more information on Octave Chanute, including material from the William Avery estate, donated in the 1940s. Finding aids are available and some photos and other material have been digitized and are available on (https://sova.si.edu/).

In the past two (+) decades, the internet, with its variety of search engines, became a tremendous help. Information was retrieved from the Library of Congress Chronicling America: Historic American Newspapers (http://chroniclingamerica.loc.gov/), the google digitized book site (http://books.google.com/), Internet Archive (https://archive.org/), the HathiTrust Digital Library (https://www.hathitrust.org/), the Bibliothèque nationale de France (BnF) Gallica (https://gallica.bnf.fr), American Patents (https://patents.google.com/), various local newspapers, and my favorite newspaper archive site, created by Tom Tryniski, the Old Fulton Post Card Site (https://fultonhistory.com/). Each site had a different approach on what material to reproduce, they were all very helpful.

On a more personal note, two books gave me a start on researching Octave Chanute, Marvin W. McFarland's *The Papers of Wilbur and Orville Wright*, including the Chanute-Wright Letters and Tom Crouch's *The Dream of Wings*. McFarland noted in his diary, that the Wrights "had but one true predecessor, Lilienthal, and only one tutor, actually mentor, Chanute," and Tom Crouch stated, that his book describes the story that culminated "in the Wright brothers taking the great leap that brought men into the skies, but only because they were launched from the shoulders of giants." Chanute was clearly one of these giants.

References

Chapter 1 (Page 1–16)

1 Schelbert, L.a.H.R., Alles ist ganz anders hier. Auswandererschicksale in Briefen aus zwei Jahrhunderten. 1977, Olten, Switzerland: Walter-Verlag AG.

2 Chanute, E., My Father, in Collection of papers and photos at the home of Joseph & Jean Hodges. 1912: Denver, CO.

3 Chanut, J., letter to Emilie E. Fourchy, dated February 1860, in Collection of papers and photos at the home of Joseph & Jean Hodges. Denver, CO.

4 Niehaus, E.F., Jefferson College, The Early Years. The Louisiana Historical Quarterly, 1955. 38(4): p: 63–89.

5 Williams, W., The Traveler's and Tourist's Guide through the United States of America, Canada, etc. 1851, Philadelphia: Lippincott, Grambo & Co.

6 Editor, "Wire Suspension Bridge over the Monongahela at Pittsburg." American Railroad Journal, 1846, 13 June, Vol 2 no 24, p: 376–379.

7 Roberts, S.W., Account of the Portage Rail Road over the Allegheny Mountain in Pennsylvania. 1836, Philadelphia: Nathan Kite.

8 Smith, J.C., Illustrated hand-book: a new guide for travelers through the United States of America. 1847, New York: Sherman & Smith.

9 Editorial, "Railroad System of the United States." The Merchants' Magazine, 1848, January, Vol 18 no 1, p: 98–99.

10 Boyd, A.C., Some Memories of my Father, in Collection of papers and photos at the home of Joseph & Jean Hodges. A later manuscript contains an account of his trip to New Orleans. 1915, Denver, CO.

© The Editor(s) (if applicable) and The Author(s), under exclusive
license to Springer Nature Switzerland AG 2023
S. Short, *Flight Not Improbable*, Springer Biographies,
https://doi.org/10.1007/978-3-031-24430-8

11 Chanut, Octave. Good-Bye letter to Joseph Chanut, his father, dated 7 December 1850, in Collection of papers and photos at the home of Joseph & Jean Hodges. 1850: Denver, CO.

12 Immigration Certificate, Nat. Rec.-Pg.-13. 1854, Naturalization Service: McLean County Circuit Court, IL.

13 Kettelle, C., Rites of Marriage. 1857: Peoria, IL. p: 3919.

14 Wing, J., The Great Union Stock Yards of Chicago. 1865, Chicago: Religio-Philosophical Publishing Association.

15 Chanute, letter to Thomas C. Meyer, dated 6 May 1865, in Papers of Octave Chanute, Letterpress book No 2A, Manuscript Division, Library of Congress: Washington, DC.

16 Forney, M.N., "The Civil Engineers' Convention. Editorial Correspondence." Railroad Gazette, 1872, 8 June, Vol 16 no p: 244.

17 Reporter, "The Civil Engineers. Annual Convention of the American Society of Civil Engineers. Banquet at the Tremont House." Chicago Tribune, Chicago IL: 6 June 1872, p: 2.

18 Short, S., Locomotive to Aeromotive. Octave Chanute and the Transportation Revolution. 2011, Urbana, Chicago and Springfield: University of Illinois Press.

19 Chanute, O., Travel Diary of 1875, in Collection of papers and photos at the home of Joseph & Jean Hodges. Denver, CO.

20 Tenth Annual Report, in Aeronautical Society of Great Britain. 1875: Greenwich.

21 Ellis, T.G., "Rise and Progress in American Engineering." American Society of Civil Engineers, Proceedings, 1876, 13 June, Vol 2 no 6, p: 73–79.

22 Reporter, "The Flying Machine." Chicago Tribune, Chicago, IL: 8 July 1878, p: 2.

23 Reporter, "A Flying Machine." New York Times, New York, NY: 23 February 1878, p: 6.

24 Chanute, O., Mechanical Flight, in Papers of Octave Chanute. 1878, Manuscript Division, Library of Congress: Washington, DC.

25 Chanute, O., "Engineering Progress in the United States." American Society of Civil Engineers, Transactions, 1880, June, Vol 9 no 6, p: 217–258.

26 Reporter, "Octave Chanute discusses Aviation in 2009." St. Louis Post-Dispatch, St. Louis, MO: 10 October 1909, p: 1.

27 Welch, A., "Annual Address." American Society of Civil Engineers, Transactions, 1882, November, Vol 11 no 11, p: 153–180.

28 Chanute, O., "The Ethics of Consulting Practice." Engineering News, 1892, 10 November, Vol 28 no 46, p: 444–446.

29 McFarland, M., "Personal journal with a New Year's resolution." University of Notre Dame Archive, 1950, 19/20 September.

Chapter 2 (Page 17–72)

1 Chanute, O., The Effect of Invention upon the Railroad and other Means of Intercommunication, in American Patent Centennial Celebration. Proceedings and Addresses. 1891, Press of Gedney & Roberts Co.: Washington, DC. p: 161–173.

2 Chanute, O., "Engineering Progress in the United States." American Society of Civil Engineers, Transactions, 1880, June, Vol 9 no 6, p: 217–258.

3 Short, S., Locomotive to Aeromotive. Octave Chanute and the Transportation Revolution. 2011, Urbana, Chicago and Springfield: University of Illinois Press.

4 Reporter, "Illinois News — A great day in Rockford." Chicago Tribune, Chicago: 18 August 1855, p: 2.

5 Burnell, M., "Balloon Ascension." Peoria Weekly Republican, Peoria IL: 25 July, 1 August 1856, p: 2, 3.

6 Reporter, "Balloon Ascension." The Illinois Gazette, Lacon, IL: 2 August 1856, p: 2.

7 W.D.G., "Aerial Navigation." Scientific American, 1860, 18 August, Vol 3 no 8, p: 116.

8 Tredgold, T., Charter of the Institution of Civil Engineers, in ICE Archives. 1828: London.

9 Bescherelle, A., Histoire des Ballons et des Locomotives Aériennes, depuis Dédale jusqu'a Petin. L'Instruction popularisée par l'illustration, ed. L. Marescq & Company with Gustave Havard. 1854, Paris.

10 Reporter, "Camp Douglas. A Change in Commanders. Kite Flying and Rebel Messages." Chicago Tribune, Chicago, IL: 6 March 1864, p: 4.

11 C., "The True Theory of Flying." Scientific American, 1870, 27 August, Vol 23 no 9, p: 133.

12 Chanute, O., "Wind Pressure upon Bridges, Discussion of paper by C. Shaler Smith." American Society of Civil Engineers, Transactions, 1880, 15 December, Vol 10 no 5, p: 169–171.

13 Thurston, R.H., "Our Progress in Mechanical Engineering." American Society of Mechanical Engineers, Transactions, 1881, November, Vol 2 no 2, p: 425–453.

14 Chanute, O., "Experiments in Flying. An Account of the Author's own Inventions and Adventures." McClure's Magazine, 1900, June, Vol 15 no 2, p: 127–33.

15 Chanute, letter to Israel Lancaster, dated 26 April 1886, in Papers of Octave Chanute, Letterpress book No 7.

16 Chanute, O., Scientific Invention, in AAAS, Section D, Mechanical Science and Engineering. Proceedings. 1886: Buffalo, NY. p: 174–182.

17 Lancaster, I., "The Problem of the Soaring Bird." American Naturalist, 1885, November, December, Vol 19 no 11, 12, p: 1055–1058, 1162–1171.

18 Editor, "Proceedings, Section of Mechanical Science and Engineering." Science, 1886, 3 September, Vol 8 no 187, p: 215–217.

19 Editor, "Papers read before the American Association for the Advancement of Science, Buffalo Meeting, August 1886." Scientific American, 1886, 4 September, Vol 55 no 10, p: 154.

20 Reporter, "Editorial." The Courier, Buffalo, NY: 22 August 1886, p: 5.

21 Chanute, letter to F. Hobart, dated 27 May 1892, in Papers of Octave Chanute.

22 Wood, D.V., "The Philosophy of Soaring Birds." The Scientific American Supplement, 1886, 16 October, Vol 22 no 563, p: 8991.

23 Chanute, O., Langley's Contribution to Aerial Navigation, in Langley, Samuel Pierpont. Secretary of the Smithsonian Institution, 1887–1906. 1907, Smithsonian Institution: Washington DC. p: 30–35.

24 Verne, J., The Clipper of the Clouds (Original 1886 French title is Robur-le-Conquérant). 1887, London: S. Low, Marston, Searle & Rivington.

25 Chanute, O., "Evolution of the "two-surface" flying machine." Aeronautics (New York), 1908, September, October, Vol 3 no 3, 4, p: 9–10, 28–29.

26 Latimer, C., "Letter to the Editor." Railroad Gazette, 1888, 6 January, Vol 32 no 1, p: 11.

27 Chanute, letter to Charles Latimer, dated 14 January 1888, in Papers of Octave Chanute, Letterpress book 8, Manuscript Division, Library of Congress: Washington, DC.

28 Chanute, letter to Mrs. Eugene F. Falconnet, dated 7 March 1888, in Papers of Octave Chanute.

29 Chanute, O., "The Problem of Aerial Navigation. Campbell's Airship a Step Towards its Solution." Chicago Tribune, Chicago, IL: 28 December 1888, p: 1.

30 Reporter, "A Ship for the Skies. Is the Problem of Aerial Navigation Solved at Last?" Chicago Tribune, Chicago, IL: 23 December 1888, p: 2.

31 Chanute, letter to Matthias N. Forney, dated 10 November 1886, in Papers of Octave Chanute, Letterpress book No 7.

32 Chanute, O., "The Latest Rapid Transit Scheme." The Railroad and Engineering Journal, 1889, April, Vol 1 no 4, p: 199.

33 Chanute, letter to Charles W. Hastings, dated 2 December 1888, in Papers of Octave Chanute, Letterpress book No 8.

34 Chanute, letter to Robert Thurston, dated 4 February 1889, in Papers of Octave Chanute, Letterpress book 8.

35 Eiffel, G., Travaux Scientifique exécutés a la tour de trois cents métres de 1889 a 1900. 1900, Paris: L. Maretheuz, Impremeur.

36 Verne, J., From the Earth to the Moon direct in ninety-seven hours and twenty minutes: And a trip around it. 1874, New York: Scribner, Armstrong & Company.

37 Chanute, O., "Note sur la Résistance de l'Air aux Plans Obliques." L'Aéronaute, 1889, 1 August, Vol 22 no 9, p: 197–214.

38 Drzewiecki, S., "Les oiseaux considérés comme des aéroplanes animés. Essai d'une nouvelle théorie du vol." L'Aéronaute, 1889, October, Vol 22 no 10, p: 221–253.

39 Chanute, letter to Olie H. Landrath, dated 21 December 1889, in Papers of Octave Chanute.

40 Chanute, O., "Resistance of air to inclined planes in motion." AAAS Proceedings, Section D, Mechanical Science and Engineering, 1889, August, p: 198–199.

41 Chanute, O., "The Paris Exposition of 1889." Journal of the Association of Engineering Societies, 1890, July, Vol 9 no 7, p: 341–351.

42 Chanute, letter to Edward T. Jeffery, dated 30 September 1889, in Papers of Octave Chanute.

43 Reporter, "Chit-Chat from Paris. (Special Correspondence of the World)." The World, New York, NY: 29 November 1889, p: 2.

44 Mouillard, L.-P., The Empire of the Air. An ornithological essay on the flight of birds., in Annual Report of the Board of Regents of the Smithsonian Institution, showing the Operations, Expenditures, and Condition of the Institution to July 1892. 1893, Government Printing Office: Washington, DC. p: 397–463.

45 Mouillard, L.-P., "Gliding Flight." The Cosmopolitan, 1894, February, Vol 16 no 2, p: 459–466.

46 Chanute, letter to Louis-Pierre Mouillard, dated 22 October 1890.

47 Chanute, letter to Louis-Pierre Mouillard, dated 2 November 1894.

48 Chanute, O., "Progress in Flying Machines." The American Engineer and Railroad Journal, 1893, September, Vol 67 no 9, p: 445–446.

49 Wenham, F.H., "On Aerial Locomotion and the Laws by which Heavy Bodies Impelled through Air are Sustained." Transactions of the Aeronautical Society of Great Britain, 1866, 27 June, Vol 1, p: 10–40.

50 Chanute, letter to George P. Whittlesey, dated 12 December 1892, in Papers of Octave Chanute, Letterpress book No 11, Manuscript Division, Library of Congress: Washington, DC.

51 Editor, "The Attitude of the Patent Office toward Flying-Machine Inventors." AeronauticS, 1894, April, Vol 1 no 7, p: 86.

52 Mouillard, L.-P., Assignor of one-half to Octave Chanute, Means for Aerial Flight (No. 582,757), in United States Patent Office. 1897.

53 Chanute, letter to Louis-Pierre Mouillard, dated 24 May 1894.

54 Chanute, letter to Louis-Pierre Mouillard, dated 8 June 1895, in Papers of Octave Chanute, Letterpress book No 13, Library of Congress: Washington, DC.

55 Chanute, letter to Louis-Pierre Mouillard, dated 21 July 1895.

56 Chanute, letter to Louis-Pierre Mouillard, dated 31 December 1895.

57 Mouillard, L.-P., letter to O. Chanute, dated 5 January 1896, in Chanute-Mouillard correspondence on the subject of flight.

58 Thurston, R.H., "Our Progress in Mechanical Engineering. Annual Address." American Society of Mechanical Engineers, 1882, November, Vol 3 no 3, p: 425–452.

59 Chanute, O., "Aerial Navigation." Railroad and Engineering Journal, 1890, July-November Vol 64 no 7, 8, 9, 10, 11, p: 316–318, 365–367, 395–397, 442–444, 498–501.

60 Zahm, A.F., "Aerial Navigation." Journal of the Franklin Institute, 1894, October, November, Vol 138 no 4, 5, p: 265–287, 347–356.

61 Chanute, O., "Aerial Navigation." The Crank, 1890, May, Vol 4 no 8, p: 1, 2–6.

62 Chanute, O., "Motors for aerial machines." Scientific American Supplement, 1893, 18 February, Vol 35 no 894, p: 14281–14282.

63 Editor, "Progress in Flying Machines. Inventors of such structures no longer thought to be insane." Chicago Tribune, Chicago IL: 11 October 1891, p: 11.

64 Chanute, O., Aerial Navigation, in Modern Mechanism. 1892, Appleton's Cyclopedia of Applied Mechanics: New York. p: 1–9.

65 Chanute, O., "Progress in Aerial Navigation." Engineering Magazine, 1891, October, Vol 2 no 2, p: 1–13.

66 Chanute, O., "Proceedings of the Annual Meeting." Journal of the Western Society of Engineers, 1891, 7 January, Vol 10 no 1, p: 1–10.

67 Forney, M.N., "Editorial." Railroad and Engineering Journal, 1891, October, Vol 5 no 10, p: 433.

68 Langley, S.P., "Experiments in Aerodynamics." 1891.

69 Chanute, O., "On the Soaring of Birds." Railroad and Engineering Journal, 1891, March, Vol 5 no 3, p: 117–119.

70 Gibbs-Smith, C.H., Aviation. An Historical Survey from its Origins to the end of World War II. 1970, London: Her Majesty's Stationery Office.

71 Chanute, O., "Progress in Flying Machines." Railroad and Engineering Journal, 1891, October, Vol 5 no 10, p: 461–465.

72 Villeneuve, A.H., "Le Tableau d'Aviation de M. Dieuaide." L'Aéronaute, 1880, Vol 13 no 10, p: 241–245.

73 Tissandier, G., La Navigation Aérienne. L'Aviation et la Direction des Aérostats dans les temps anciens et modernes. 1886, Paris: Librairie Hachette et Cie.

74 Chanute, O., "Progress in Flying Machines. Aeroplanes." The Railroad and Engineering Journal, 1892, June, Vol 66 no 6, p: 270–273.

75 Chanute, letter to Francis H. Wenham, dated 13 September 1892, in Papers of Octave Chanute, Letterpress book No 11.

76 Phillips, H.F., "Experiments with currents of air." Engineering. An Illustrated Weekly Journal, 1885, 14 August, Vol 40, p: 160–161.

77 Chanute, letter to A. Goupil, dated 17 December 1892, in Papers of Octave Chanute, Letterpress book No 11.

78 Chanute, O., "Development and Future of Flying Machines." The City Club Bulletin, 1908, 18 November, Vol 2 no 15, p: 191–194.

79 Chanute, O., "Progress in Flying Machines (Otto Lilienthal)." The American Engineer and Railroad Journal, 1893, August, Vol 67 no 8, p: 395–398.

80 Zahm, A.F., "Flying Machine." Notre Dame Scholastic, 1882, 25 November, 2 December, 9 December, 16 December, Vol 16, 17 no 12–15.

81 Chanute, O. Proceedings of the International Conference on Aerial Navigation. in Columbian Exhibition. 1893. Chicago IL: The American Engineer and Railroad Journal.

82 Zahm, A.F., Stability of Aeroplanes and Flying Machines, in Proceedings of the International Conference on Aerial Navigation, O. Chanute, Editor. 1893, The American Engineer and Railroad Journal: New York, NY. p: 273–287.

83 Zahm, A.F., Pioneer in Aeronautics, in Zahm Papers. 1922, revised and updated in 1942: Notre Dame University Archive.

84 Bancroft, H.H., The Book of the Fair; an historical and descriptive presentation of the world's science, art, and industry, as viewed through the Columbian exposition at Chicago. 1893, Chicago, San Francisco: The Bancroft Company.

85 Chanute, O., "Committee Report on the International Engineering Congress. Proceedings, Western Society of Engineers." Journal of the Association of Engineering Societies, 1892, January, Vol 11 no 1, p: 104–107.

86 Chanute, O., Not Things, But Men. The World's Congress Auxiliary of the World's Columbian Exposition of 1893. Department of Engineering. General Division of Aerial Navigation. 1892: Chicago IL.

87 Zahm, A.F., Diary of conference on aerial navigation, in Albert F. Zahm Papers. 1893, Archives of the University of Notre Dame: Notre Dame, IN.

88 Chanute, letter to Carl E. Myers, dated 30 November 1892, in Papers of Octave Chanute.

89 Chanute, O., "Aerial Navigation Conference." The American Engineer and Railroad Journal, 1893, September, Vol 67 no 9, p: 416–417.

90 Langley, S.P., The Internal Work of the Wind, in International Conference on Aerial Navigation. Proceedings. 1893, American Engineer and Railroad Journal: Chicago IL. p: 66–104 (with discussions).

91 Wood, D., The Internal Work of the Wind, Discussion, in International Conference on Aerial Navigation. Proceedings. 1894, American Engineer and Railroad Journal: Chicago IL. p: 92–93.

92 Owen, H.S., "To Fly Through the Air. Practice is needed to make Aerial Navigation Practicable." The Evening Star, Washington, DC: 19 January 1894, p: 11.

93 Editor, "America's first glider club, from contemporary notes of Charles P. Steinmetz." The Sportsman Pilot, 1930, December, Vol 2 no 12, p: 26–28.

94 Brice, C.S., Aerial Navigation. Report of Committees of the Senate of the United States to accompany bill S. 1344. 1895, Government Printing Office: Washington, DC. p: 1–13.

95 Means, J., Senate Bill S.302. Fifty-Fourth Congress., in the Aeronautical Annual. 1896: Boston. p: 80–84.

96 Young, D., Chicago Aviation, an Illustrated History. 2003, De Kalb, IL: Northern Illinois Press.

97 Chanute, O., "Progress in Flying Machines. Conclusions." The American Engineer and Railroad Journal, 1894, January, Vol 68 no 1.

98 Parville, H.d., "L'Aviation. L'Homme Volant." La Science Illustrée, 1894, 3 February, Vol 13 no 323, p: 153–155.

99 Chanute, letter to Matthias N. Forney, dated 22 March 1894, in Papers of Octave Chanute.

100 Editor, "Literature: Book Reviews." The Electrical Engineer, 1894, 9 May, Vol 17 no 314, p: 411.

101 Editor, "New Publications received." The Internal Revenue Record and Customs Journal, 1894, 28 May, Vol 40 no 1489, p: 152, 168.

102 Hensley, S.a.J., The Unwelcome Assistant, Edward C. Huffaker and the Birth of Aviation. 2003, Johnson City, TN: The Overmountain Press.

103 Huffaker, E.C., Soaring Flight, in International Conference on Aerial Navigation. Proceedings. 1893, American Engineer and Railroad Journal: New York.

104 Chanute, letter to Edward C. Huffaker, dated 26 September 1893, in Papers of Octave Chanute.

105 Chanute, O., Soaring Flight, in Flying Machines, Construction & Operation. Experiments by the writer, introductory chapter by O. Chanute, W.J.a.T.H.R. Jackman, Editor. 1910, Charles C. Thompson Co: Chicago, IL. p: 179–190.

106 Huffaker, E.C., "My Four Years with Langley." The World News, 1919, p: 15–16.

107 Montgomery, J.J., Correspondence and other papers of and relating to Montgomery. 1958, University of North Carolina, Chapel Hill: Southern Historical Collection.

108 Montgomery, J.J., "The Origin of "Warping." Professor Montgomery's Experiments." Aeronautics, 1910, May, p: 63–64.

109 Chanute, letter to John J. Montgomery, dated 30 March 1894, in Papers of Octave Chanute, Letterpress book No 12.

110 Hargrave, L., letter to J. E. Watkins, Smithsonian Institution, Washington, DC, dated 8 December 1891: National Air and Space Museum archive.

111 Short, S., "Have a fly with me. Bold aeronauts meet and gleefully predicts." WW1 Aero, 2014, November, no 221, p: 13–26.

112 Forney, M.N., "Pilcher's Soaring Machine." American Engineer and Railroad Journal, 1895, August, Vol 69 no 8, p: 387.

113 Jarrett, P., Another Icarus. Percy Pilcher and the Quest for Flight. 1987, Washington, DC and London: Smithsonian Institution Press.

114 Chanute, O., Scrapbook of clippings, correspondence and photographs related to Pilcher's work in aeronautics. University of Chicago Library System. 1898.

115 Herring, letter to O. Chanute, dated 1 January 1895, in Research Notes collected by Sherwin Murphy, CrMs-171 Crerar Manuscript Collection. University of Chicago Library.

116 Chanute, letter to Augustus M. Herring, dated 31 December 1894, in Papers of Octave Chanute, Letterpress book No 13.

117 Chanute, letter to Samuel P. Langley, dated 10 May 1895, in Papers of Octave Chanute.

118 Herring, letter to O. Chanute, dated 25 May 1895, in Research Notes, collected by Sherwin Murphy.

119 Herring, A.M., Scrapbook, in Kroch Library, Cornell University. 1916: Ithaca, NY.

120 Chanute, letter to Samuel P. Langley, dated 15 February 1898, in Papers of Octave Chanute.

121 Marvin, C.F., "Special Contributions: A Weather Bureau Kite." Monthly Weather Review, 1895, November, Vol 23 no 11, p: 418–420.

122 Marvin, C.F., "Mechanics and equilibrium of kites." Monthly Weather Review, 1897, April, Vol 25 no 4.

123 Chanute, letter to Wilhelm Kress, dated 14 March 1895, in Papers of Wilhelm Kress, Technisches Museum Wien: Vienna, Austria.

Chapter 3 (Page 73–125)

1 Cayley, S.G.B., "On Aerial Navigation." A Journal of Natural Philosophy, Chemistry and the Arts, 1809, November, Vol 24 no 108, p: 164–173.

2 Reporter, "A Real Flying Machine at Last. The World successfully experiments with a machine that will actually fly over the tops of the houses and up and down the streets. Not a balloon, but a flying-machine with wings and propeller." The World, New York, NY: 4 August 1895, p: 21.

3 Lilienthal, O., Der Vogelflug als Grundlage der Fliegekunst. 1889, Berlin: R. Gaertner's Verlagsbuchhandlung.

4 Reporter, "Scientific Miscellany. The Problem of Flight." The Cortland Democrat, Cortland, NY: 7 February 1890, p: 1.

5 Wright Brothers, Editors. "He can half fly" (17 July) and "Needs more Wings" (26 July)." The Evening Item. The West Side Daily: 17, 26 July 1890.

6 Lilienthal, O., Flying-Machine (No. 544,816), in United States Patent Office. 1895.

7 Lilienthal, O., letter to Marie Lilienthal, dated 3 September 1893, in Otto Lilienthal's Flugtechnische Korrespondenz: Lilienthal Museum, Anklam, Germany.

8 Lilienthal, O., letter to H.-W. Moedebeck, dated 14 November 1893, in Gimbel Collection, XF-2–1 2462 1893/NOV. 14, Special Collections, McDermott Library, US Air Force Academy, CO 80840–6215.

9 Chanute, letter to Otto Lilienthal, dated 12 January 1894, in Papers of Octave Chanute, Letterpress book No 12, Manuscript Division, Library of Congress: Washington, DC.

10 Lilienthal, O., "Why is artificial flight so difficult an invention (Weshalb ist es so schwierig, das Fliegen zu erfinden). Translated from Prometheus, n.261 p: 7–10, January 1894." American Engineer and Railroad Journal, 1894, December, Vol 68 no 12, p: 575–576.

11 Schwipps, W., Die Schule Lilienthal's. Gesammelte Vorträge des Internationalen Symposiums im September 1991. 1992, Berlin, Germany: Museum für Verkehr und Technik.

12 Chanute, O., Recent Experiments in Gliding Flight, in Aeronautical Annual, J. Means, Editor. 1897, W. B. Clarke & Co.: Boston, Mass. p: 30–53.

13 Engineer, "Dr. Lilienthal." American Architect & Building News, 1895, 24 August, Vol 49 no 1026, p: 74.

14 Langley, letter to Augustus M. Herring, dated 6 August 1895, in Scrapbook of clippings, magazine articles by S. P. Langley, and material collected by Chanute, University of Chicago Library.

15 Runge, M.a.B.L., Erfinder Leben. 2005, Berlin, Germany: Berlin Verlag.

16 Jackman, W.J.a.T.H.R., Flying Machines, Construction & Operation. 2nd edition (1912) has several chapters added, including a "In Memoriam" of Chanute. ed. 1910, 1912, Chicago, IL: Charles C. Thompson Co.

17 Editor, "Flying Machines." Rome Daily Sentinel, Rome, NY: 6 March 1896, p: 2.

18 Reporter, "A Flying Machine at Last that Really Flies. A Journal Reporter Succeeds in Flying with One of Inventor Lilienthal's Flying Machines." The Journal, New York, NY: 3 May 1896, p: 17.

19 Reporter, "Journalism off its Feet. Reporters take wings to boom circulation. A flying machine tried on Staten Island with varying results, Sometimes the journalist ploughs the earth with his ear and sometimes alights on his neck." The Sun, New York, NY: 4 May 1896, p: 3.

20 Short, S., "Have a fly with me. Bold aeronauts meet and gleefully predicts." WW1 Aero, 2014, November, p: 13–26.

21 Langley, S.P., Story of Experiments in Mechanical Flight, in The Aeronautical Annual, J. Means, Editor. 1897, W. B. Clarke & Co.: Boston, Mass. p: 11–25.

22 Chanute, letter to J. Brown Goode, dated 10 April 1896, in Papers of Octave Chanute, Letterpress book No 14.

23 Wilson, R.F., "Sporting News of Interest. Recognition at last for father of aviation; Scientists to show that Langley's machine, built in 1896, can fly." The Pensacola Journal: 3 January 1913, p: 6.

24 Langley, S.P., The Internal Work of the Wind, in International Conference on Aerial Navigation. Proceedings. 1893, American Engineer and Railroad Journal: Chicago IL. p: 66–104 (with discussions).

25 Reporter, "Prof. Langley's Flying Machine. Alexander Graham Bell witnesses a trial and speaks enthusiastically of its success." The World, New York, NY: 13 May 1896, p: 7.

26 Chanute, letter to Augustus M. Herring, dated 28 May 1895, in Papers of Octave Chanute.

27 Langley, letter to O. Chanute, dated 23 December 1897, in Transcribed correspondence between Langley and Chanute. National Air & Space Museum Archives: Washington, DC.

28 Langley, letter to O. Chanute, dated 1 December 1897. National Air & Space Museum Archives: Washington, DC.

29 Talbert, A.E., "America's Air Navy of the Future." Interavia, 1949, September, Vol 4 no 9, p: 544–547.

30 Langley, letter to Robert H. Thurston, dated 9 May 1898. National Air and Space Museum.

31 Viereck, G.S., "What Life Means to Einstein. An Interview." The Saturday Evening Post, 1929, 26 October, p: 17, 110, 113–114, 117.

32 Chanute, O., Sailing Flight, in The Aeronautical Annual. 1896, 1897, W. B. Clarke & Co.: Boston. p: 60–76 (Part 1), p: 98–127 (Part 2).

33 Chanute, O., Soaring Machine (No. 582,718), in United States Patent Office. 1897.

34 Chanute, letter to William A. Glassford, dated 3 September 1895, in Papers of Octave Chanute.

35 Chanute, letter to Augustus M. Herring, dated 12 December 1895, in Papers of Octave Chanute.

36 Avery, W., "Some little success of the aeroplane in aerial navigation." The Cherry Circle, 1908, January, Vol 14 no 1, p: 36–43.

37 Chanute, O., Diary of the glides in 1896, in Octave Chanute Papers, Manuscript Division, Library of Congress. 1896: Washington, DC.

38 Reporter, "Will Sail in the Clouds. Octave Chanute Constructs a Peculiar Flying Machine. Inventor Tests Apparatus on the Shore of Lake Michigan. Fishermen View a Huge Kite Contrivance Flitting Through Space." Chicago Chronicle, Chicago, IL: 23 June 1896.

39 Reporter, "Men Fly in Midair." Chicago Tribune, Chicago, IL: 24 June 1896, p: 1.

40 Reporter, "Trying to Fly." Westchester Tribune, Chesterton, Westchester County, IN: 27 June 1896, p: 1.

41 Reporter, "O. Chanute and his Air Ship." The Kansas City Journal, Kansas City, MO: 25 June 1896, p: 5.

42 Manley, F., "Steal the Birds' Art." Chicago Record, Chicago IL: 29 June 1896, p: 2.

43 Chanute, letter to Charles H. Lamson, dated 13 October 1896. Also gives recipe for pyroxylene varnish, adapted from British Patent No.2249, dated 15 September 1860, in Papers of Octave Chanute, Letterpress book No 14.

44 Chanute, letter to Francis H. Wenham, dated 7 July 1896, in Papers of Octave Chanute.

45 Chanute, letter to Thomas Moy, dated 13 August 1897, in Papers of Octave Chanute.

46 Chanute, O., Improvements in and relating to Flying Machines (No. 13,372), in British Patent Office. 1898: Great Britain.

47 Chanute, letter to James Means, dated 7 July 1896, in Papers of Octave Chanute.

48 Editor, "His invention cost him his life." Chicago Tribune, Chicago, IL: 12 August 1896, p: 1.

49 Allport, W.H., "Octave Chanute and H. T. Ricketts — A Sidelight." Scientific American, 1911, 23 September, Vol 105 no 13, p: 275.

50 Chanute, O., "Evolution of the "two-surface" flying machine." Aeronautics (New York), 1908, September, October, Vol 3 no 3, 4, p: 9–10, 28–29.

51 Reporter, "Use Wings in Flight." Daily Inter Ocean, Chicago IL: 25 August 1896, p: 1.

52 Chanute, letter to Edward C. Huffaker, dated 30 September 1896, in Papers of Octave Chanute, Letterpress book No 14.

53 Manley, F., "Airship's Final Test." Chicago Record, Chicago IL: 28 September 1896.

54 Macbeth, H., "The Albatross, Queen of the Air." Times-Herald, Chicago, IL: 27 September 1896, p: 17, 18.

55 Spicer, S. The Octave Chanute Pages. Carr's Beach and the Carr Family of Miller. 1995 [cited 1995–2016; Available from: http://www.spicerweb.org/Chanute/Cha_index.aspx.

56 Bunting, H.S., "Go coasting in the air." Chicago Tribune, Chicago, IL: 8 September 1896, p: 1.

57 Editorial, "The Gull." Chicago Tribune, Chicago IL: 1 October 1896.

58 Reporter, "Will visit her subjects again. The Flying Machine." The Kansas City Journal, Kansas City, MO: 4 October 1896, p: 11, 19.

59 Butusov, W.P., Memorandum of Agreement, in Octave Chanute Papers, Manuscript Division, Library of Congress. 1896: Washington, DC.

60 Butusov, W.P., Soaring Machine (No. 606,187), in United States Patent Office. 1898.

61 Chanute, letter to Carl E. Myers, dated 10 November 1898, in Papers of Octave Chanute, Letterpress book No 16.

62 McGowan, P.M., "To soar at 1,000 feet." Times-Herald, Chicago, IL: 29 September 1897, p: 2.

63 Hodges, E. and P. I. Young, Papers of Octave Chanute, in Collection of papers and photos at the home of Joseph & Jean Hodges. 1850–1969: Denver, CO.

64 Chanute, letter to W. J. Lloyd (Associated News Bureau), dated 5 July 1896, in Papers of Octave Chanute, Letterpress book No 14.

65 Lloyd, W.J., "This is the latest Flying-Machine. Octave Chanute, a famous civil engineer, is the inventor, and he is now testing the contrivance among the sand hills of Indiana." New York Journal, New York, NY: 26 July 1896, p: 21.

66 Bunting, H.S., Primitive Birdmen, in Written for a writer's contest in the Atlantic Monthly, called "America's First". 1946: Naples, FL.

67 Reporter, "Ship Spoars in Air. Aeronauts Conduct a Successful Test at Millers, Ind." Chicago Chronicle, Chicago, IL: 11 September 1896, p: 1–2.

68 Chanute, O., "Experiments in Flying. An Account of the Author's own Inventions and Adventures." McClure's Magazine, 1900, June, Vol 15 no 2, p: 127–33.

69 Chanute, letter to James Means, dated 28 January 1897, in Papers of Octave Chanute, Letterpress book No 15.

70 Reporter, "See Airship or a Star. Vagrant of the Night Sky startles all but Prof. Hough. Appears all along from Evanston to South Chicago. Some observers declare they see wings. Astronomer says it is Alpa Orionis. Secretary of Aeronautic Association says he expected the travelers from Frisco on Sunday, but they're ahead of time." Daily Tribune, Chicago, IL: 10 April 1897, p: 1.

71 Reporter, "It Flies by Night. Everybody puzzling over the great Airship Mystery. Many say they have seen it. A reporter claims to have run it down. A Chicago paper gives the story a terse old fashioned title." Lockport Daily Journal, Lockport, NY: 10 April 1897, p: 3.

72 Chanute, O., "Notes of a Visit to the Keely Workshop." Engineering News, 1898, 20 December, Vol 40 no 26, p: 418.

73 Glassford, W.A., "Military Aeronautics." Journal of the Military Service Institution of the United States, 1896, May, Vol 19 no 7, p: 561–576.

74 Reporter, "Warfare in the Air. Proposed construction of a flying machine under Army auspices. It is to be on the aeroplane order. Study of Military aeronautics in all of the European armies. Superior to balloons." The Evening Star, Washington, DC: 6 July 1895, p: 18.

75 Lilienthal, letter to O. Chanute, dated 5 August 1895, in Papers of Octave Chanute, Manuscript Division, Library of Congress: Washington, DC.

76 Reporter, "As Engines of War. Relation of Kites to Future Military Operations." Chicago Tribune, Chicago, IL: 18 October 1896, p: 25.

77 Reporter, "Will be a kite's tail. Chief Signal Officer Maxfield of Chicago will ascend with a man-carrying kite." New York Herald, New York, NY: 4 October 1896, p: 10.

78 Chanute, O., Balloons in the War (unpublished manuscript), in Papers of Octave Chanute. 1899: at Regenstein Special Collections.

79 Reporter, "Neglect Balloons. Government overlooks a possibility in Warfare." Chicago Tribune, Chicago IL: 5 June 1898, p: 44.

80 Chanute, O., "Aviation." The Sibley Journal of Engineering, 1897, April, Vol 11 no 7, p: 266–268.

81 Means, D.J.H., James Means and the Problem of Manflight during the Period of 1882–1920. 1964, Washington, DC: Smithsonian Institution.

82 Chanute, O., "Experiments in Flying. An Account of the Author's own Inventions and Adventures." McClure's Magazine, 1900, June, Vol 15 no 2, p: 127–133.

83 Chanute, O., Notes, in Octave Chanute Papers, Manuscript Division, Library of Congress. 1896: Washington, DC. p: 2.

84 Bowers, A., Correspondence (2000–2008), in NASA Dryden Flight Research Center: Lancaster, CA.

85 Short, S., Birth of American Soaring Flight: A New Technology. American Institute of Aeronautics and Astronautics, 2005. 43(1): p: 17–28.

86 Macbeth, H., "How it Feels to Fly." Times-Herald, Chicago, IL: 8 September 1897, p: 2.

87 Chanute, letter to Albert F Zahm, dated 8 July 1908, in Papers of Octave Chanute, Letterpress book No 23.

88 Chanute, O., Aeronautics, in The New Volumes of the Encyclopaedia Britannica, T.S. Spencer, Editor. 1902, Adam & Charles Black: London, Edinburgh, New York. p: 100–104.

89 Herring, A.M., Scrapbook, in Kroch Library, Cornell University. 1916: Ithaca, NY.

90 Herring, letter to O. Chanute, dated 17 March 1901, in Exhibit 319. Herring-Curtiss Company vs Glenn Curtiss et al, Olin Library, Cornell University, Ithaca, NY: Steuben County, NY.

91 Chanute, letter to Augustus M. Herring, dated 24 March 1901, in Papers of Octave Chanute, Letterpress book No 18, Manuscript Division, Library of Congress: Washington, DC.

92 Chanute, letter to Ernest L. Jones, dated 22 October 1909, in Papers of Octave Chanute.

93 Dienstbach, C., "Herring's Work, with reproduction of an old print, and Herring correspondence at early date." American Aeronaut 1908, May, Vol 1 no 5, p: 154–159.

94 Zahm, A.F., "Invention of the 'Chanute Glider'." American Aeronaut, 1908, June, Vol 1 no 6, p: 250–251.

95 Kronfeld, R., On Gliding and Soaring. 1934, London: John Hamilton, Ltd.

96 Sweetland, R., "First American to Fly — Tells of 12-Winged Plane." Chicago Daily News, Chicago IL: 4 October 1940, p: 10.

97 Fowle, F.F., "Octave Chanute: Pioneer Glider and Father of the Science of Aviation." Indiana Magazine of History, 1936, September, Vol 32 no 3, p: 226–230.

98 Young, D., "The Wright Stuff. 100 years ago Octave Chanute paved the way for the first airplane flight in modern aviation." Sunday Chicago Tribune, Chicago, IL: 23 June 1996.

99 Herring, A.M., Recent advances toward a solution of the problem of the century, in Aeronautical Annual, J. Means, Editor. 1897, W. B. Clarke & Co.: Boston, Mass. p: 54–75.

100 Chanute, O., "Gliding Experiments." Journal of the Western Society of Engineers, 1897, December, Vol 2 no 5, p: 595–628.

101 Chanute, O., "American Gliding Experiments. Amerikanische Gleitflug-Versuche (translated by Rittmeister Warder)." Illustrirte Aeronautische Mittheilungen, 1898, January, Vol 2 no 1, p: 4–8, 9–12.

102 Chanute, O., "Some American Experiments." The Aeronautical Journal, 1898, January, Vol 2 no 5, p: 9–11, Frontis piece.

103 Chanute, O., "Gleitflugversuche in Nordamerika." Prometheus, 1898, September, Vol 9 no 458, p: 662–664.

Chapter 4 (Page 127–172)

1 Editor, "The Dream of Aerial Flight." The New York Times, New York, NY: 4 September 1901, p: 6.

2 Chanute, O., "Aerial Navigation." The Independent, 1900, 26 April, 3 May, Vol 52 no 2682, 2683, p: 1006–1007, 1058–1060.

3 Marvin, C.F., "Mechanics and equilibrium of kites." 1897, April, Vol 25 no 4.

4 Means, J., Boston Aeronautical Society. in The Aeronautical Annual. 1896, Boston: W. B. Clarke & Co.

5 Means, D.J.H., James Means and the Problem of Manflight during the Period of 1882–1920. 1964, Washington, DC: Smithsonian Institution.

6 Means, J., The Aeronautical Annual. 1895, Boston: W. B. Clarke & Co.

7 Reporter, "Will Men yet Fly? A convention of aeronauts to be held in Boston. Help expected from Congress." Los Angeles Herald, Los Angeles, CA: 21 March 1896, p: 7.

8 Schwipps, W., Otto Lilienthal's Flugtechnische Korrespondenz. 1993, Anklam, Germany: Herausgegeben und mit Anmerkungen versehen von Werner Schwipps im Auftrag des Lilienthal-Museums Anklam.

9 Chanute, letter to James Means, dated 3 January 1907, in Papers of Octave Chanute.

10 Eddy, W., "To the Editors of Aeronautics: A Simple Aeroplane." American Engineer and Railroad Journal, 1895, October, Vol 69 no 10, p: 484.

11 Wenham, F.H., "On Aerial Locomotion and the Laws by which Heavy Bodies Impelled through Air are Sustained." Transactions of the Aeronautical Society of Great Britain, 1866, 27 June, Vol 1 no p: 10–40.

12 Merrill, A.A., "Some simple experiments with Aero-Curves." The Aeronautical Journal, 1899, July, Vol 3 no 3, p: 65–67.

13 Merrill, A.A., The Great Awakening. The Story of the Twenty-Second Century. 1899, Boston: George Book Publishing Co.

14 Chanute, letter to Albert A. Merrill, dated 4 March 1902, in Papers of Octave Chanute, Letterpress book No 19.

15 Chanute, letter to Albert A. Merrill, dated 21 July 1902, in Papers of Octave Chanute.

16 Merrill, A.A., "New Air Balance and Small Wind Tunnel. California Institute of Technology." Aviation, 1925, 30 March, Vol 18 no 13, p: 580–581.

17 Reporter, "Prizes for Photographers. Cabot competition for photographs of soaring birds is continued. Why the photographs are desired. Professor Lilienthal's experiments." Boston Evening Transcript, Boston, Mass: 14 October 1896, p: 5.

18 Reporter, "New Flying Machine. Boston Man has some Ideas. He is ready to apply to one." The Oswego Daily Palladium, Oswego, NY: 11 April 1896, p: 10.

19 Schwipps, W.a.H.H., Flugpionier Gustav Weisskopf. Legende und Wirklichkeit. 2001, Oberhaching: Aviatik Verlag GmbH.

20 Cabot, letter to O. Chanute, dated 7 May 1897, in Papers of Octave Chanute, Letterpress book No 15.

21 Chanute, letter to Samuel Cabot, dated 11 May 1897, in Papers of Octave Chanute.

22 Cabot, letter to O. Chanute, dated 18 May 1897, in Papers of Octave Chanute.

23 Chanute, letter to Samuel Cabot, dated 5 January 1898, in Papers of Octave Chanute.

24 Cabot, letter to O. Chanute, dated 26 July 1897, in Papers of Octave Chanute.

25 Reporter, "Flying at Great Diamond. First use of Lilienthal's design in America. Mt. Charles H. Lamson soared against the wind. Interesting facts about it." Portland Daily Press, Portland, Maine: 23 October 1895, p: 8.

26 Reporter, "Personal." New York Tribune, New York: 25 October 1895, p: 6.

27 Lamson, C.H., Work on the Great Diamond, in The Aeronautical Annual. 1896, W. B. Clarke & Co: Boston. p: 133–137.

28 Reporter, "Make a test with Air Ships. Two inventors fly their machines over the lake in a stiff breeze." The Chronicle, Chicago, IL: 16 July 1896, p: 1.

29 Lamson, C.H., Scrapbook of clippings, in Collected by O. Chanute. University of Chicago Library System. 1896–1897.

30 Gould, G.M., "A Tale of a Kite. By the New Woman. She went skyward on inventor Lamson's new flyer and had a most exciting time of it." New York Journal, New York, NY: 11 July 1897, p: 13.

31 Lamson, C.H., Kite (No. 666,427), in United States Patent Office. 1901.

32 Chanute, letter to Charles H. Lamson, dated 8 February 1900, in Papers of Octave Chanute, Letterpress book No 18.

33 Herring, A.M., Recent advances toward a solution of the problem of the century, in Aeronautical Annual, J. Means, Editor. 1897, W. B. Clarke & Co.: Boston, Mass. p: 54–75.

34 Dees, P., The 100-Year Chanute Glider Replica, an adventure in education, in Society of Automotive Engineers. 1997, AIAA and SAE: World Aviation Congress, Anaheim, CA.

35 Short, S., Birth of American Soaring Flight: A New Technology. American Institute of Aeronautics and Astronautics, 2005. 43(1): p: 17–28.

36 Chanute, letter to Thomas Moy, dated 13 August 1897, in Papers of Octave Chanute, Letterpress book No 15.

37 Chanute, letter to James Means, dated 23 July 1898, in Papers of Octave Chanute, Letterpress book No 16.

38 Chanute, letter to Samuel P. Langley, dated 6 December 1898, in Papers of Octave Chanute, Letterpress book No 15.

39 Zahm, A.F., Aerial Navigation. A popular treatise on the growth of air craft and on aerial meteorology. 1911, New York and London: D. Appleton and Company.

40 Reporter, "Herring's Air Ship. Trial at St. Joseph, Mich. Inventor takes a flight of seventy-three feet against a twenty-five mile wind." Chicago Record, Chicago: 31 October 1898.

41 Reporter, "A Flying Machine. M. C. Arnot of this city is interested. This airship will fly." Daily Advertiser, Elmira, NY: 18, 24 November 1898.

42 Crouch, T.D., A Dream of Wings. Americans and the Airplane, 1875–1905. 1981, New York, London: W. W. Norton & Co.

43 Chanute, O., "American Gliding Experiments. Amerikanische Gleitflug-Versuche (translated by Rittmeister Warder)." Illustrirte Aeronautische Mittheilungen, 1898, January, Vol 2 no 1, p: 4–8, 9–12.

44 Chanute, O., "Conditions of Success in the Design of Flying Machines. Die Bedingungen des Erfolges im Entwurf von Flugapparaten (translated by Rittmeister Warder)." Illustrirte Aeronautische Mittheilungen, 1899, April, Vol 3 no 4, p: 37–41, 41–46.

45 Chanute, letter to Hermann W. L. Moedebeck, dated 22 July 1901, in Papers of Octave Chanute, Letterpress book No 18.

46 Wenham, F.H., letter to Chanute, dated 2 July 1900, in Papers of Octave Chanute, Manuscript Division, Library of Congress: Washington, DC.

47 Chanute, letter to Francis H. Wenham, dated 25 November 1900, in Papers of Octave Chanute, Letterpress book No 18.

48 Chanute, letter to Thomas Moy, dated 29 April 1901, in Papers of Octave Chanute, Letterpress book No 18.

49 Drzewiecki, S., "Methode Pour la Détermination des Éléments Mécaniques des Propulseurs Hélicoidaux." Bulletin de L'Association Technique Maritime, 1893, 15–16 December 1892, Vol 3 no p: 11–31.

50 Drzewiecki, S., "Des Hélices Propulsives." Bulletin de L'Association Technique Maritime, 1900, Vol 11 p: 89–116.

51 Reporter, "Zeppelin's Airship Trial." New York Times, New York: 4 July 1900.

52 Reporter, "Talks of Zeppelin's Airship." Chicago Tribune, Chicago, IL: 4 July 1900, p: 8.

53 Gilman, R.R.T.b.M.B.D., "Zeppelin in Minnesota. The Count's Own Story." Minnesota History, 1967, Summer, Vol 40 no p: 265–278.

54 Kelly, F.C., The Wright Brothers. A Biography Authorized by Orville Wright. 1943, New York, NY: Harcourt, Brace and Company.

55 Vernon, "The Flying Man, Otto Lilienthal's Flying Machine." McClure's Magazine, 1894, September, Vol 3 no 4, p: 323–331.

56 Wright Brothers, Editors. "He can half fly" (17 July) and "Needs more Wings" (26 July)." The Evening Item. The West Side Daily: 17, 26 July 1890.

57 W. Wright, letter to the Smithsonian Institution, dated 30 May 1899, in The Wilbur and Orville Wright Papers.

58 Wright, W., "First Rebuttal Deposition of Wilbur Wright. United States Circuit Court of Appeals for the second Circuit. The Wright Company, Complainant-Appellee, vs. The Herring-Curtiss Co. and Glenn H. Curtiss. Defendants-Appellants." Transcript of Record, 1912, 15 February, Vol 1, p: 473–495.

59 Chanute, O., Progress in Flying Machines. 1894, New York: American Engineer and Railroad Journal.

60 W. Wright, letter to O. Chanute, dated 13 May 1900, in The Wilbur and Orville Wright Papers.

61 Chanute, letter to Wilbur Wright, dated 17 May 1900, in Papers of Octave Chanute.

62 W. Wright, letter to Milton Wright, dated 3 September 1900, in The Wilbur and Orville Wright Papers.

63 W. Wright, letter to O. Chanute, dated 16 November 1900, in The Wilbur and Orville Wright Papers.

64 W. Wright, letter to O. Chanute, dated 12 May 1901, in The Wilbur and Orville Wright Papers.

65 Crouch, T.D., The Bishop's Boys. A Life of Wilbur and Orville Wright. 1989, New York, London: W. W. Norton & Comp.

66 W. Wright, letter to Reuchlin Wright, dated 3 July 1901, in Family Papers: Correspondence-Wright, Reuchlin. The Wilbur and Orville Wright Papers.

67 W. Wright, letter to Milton Wright, dated 26 July 1901, in Family Papers: Correspondence-Wright, Milton. The Wilbur and Orville Wright Papers.

68 Chanute, O., "Gliding Machines, the latest aeronautical experiments." The Illustrated Scientific News, 1903, February, p: 73.

69 Chanute, letter to W. Wright, dated 23 August 1901, in Papers of Octave Chanute.

70 Wright, W. Subject File: Chanute, Octave — Photographs, Kitty Hawk, North Carolina, Originals. The Wilbur and Orville Wright Papers 1901–1902.

71 Wright, W., "Some Aeronautical Experiments." Journal of the Western Society of Engineers, 1901, December, Vol 6 no 6, p: 490–510.

72 Kochersberger, K., "An evaluation of the Wright 1902 glider using full scale wind tunnel data." AIAA - 0096, 2003, 6–9 January.

73 K. Wright, letter to Milton Wright, dated 25 September 1901, in Family Correspondence: Wilbur and Orville Wright Papers.

74 Chanute, letter to W. Wright, dated 27 November 1901, in Papers of Octave Chanute.

75 Chanute, O., "Address of Mr. Chanute, the Retiring President." Journal of the Western Society of Engineers, 1902, 7 January, Vol 7 no 1, p: 1–9.

76 Spratt, letter to O. Chanute, dated 20 December 1898, in Papers of Octave Chanute.

77 Chanute, letter to George A. Spratt, dated 24 December 1898, in Papers of Octave Chanute, Letterpress book No 16.

78 Chanute, letter to George A. Spratt, dated 15 October 1899, in Papers of Octave Chanute.

79 Wyman, L. Flying Manuscripts. Unfolding History. Manuscripts at the Library of Congress., 2021, 4 November.

80 Chanute, letter to George A. Spratt, dated 8 February 1900, in Papers of Octave Chanute.

81 Chanute, letter to George A. Spratt, dated 4 July 1901, in Papers of Octave Chanute.

82 Chanute, letter to George A. Spratt, dated 21 January 1902, in Papers of Octave Chanute, Letterpress book No 19.

83 Huffaker, E.C., Experiments with gliding models conducted by E. C. Huffaker for O. Chanute, in Collected by O. Chanute. 1899, University of Chicago.

84 Chanute, letter to Edward C. Huffaker, dated 7 February 1901, in Papers of Octave Chanute, Letterpress book No 18.

85 Chanute, letter to Edward C. Huffaker, dated 1 June 1901, in Papers of Octave Chanute.

86 Chanute, letter to W. Wright, dated 29 June 1901, in Papers of Octave Chanute.

87 Chanute, letter to Edward C. Huffaker, dated 3 July 1901, in Papers of Octave Chanute.

88 Huffaker, E.C. Diary (July-August 1901) kept for Mr. O. Chanute at Kill Devil Hills near Kitty Hawk, NC, of experiments with gliding models. The Wilbur and Orville Wright Papers, 1901.

89 Randers-Pehrson, N.H., Pioneer Wind Tunnels, in Smithsonian Miscellaneous Collections. 1935, Smithsonian Institution: Washington, DC. p: 1–24 (with four plates).

90 Zahm, A.F., "New Methods of Experimentation in Aerodynamics." Science, 1902, 29 August, p: 342–343.

91 Anderson, J.D., Jr, Infancy of Aerodynamics. To Lilienthal and Langley (Chapter 4), in A History of Aerodynamics and Its Impact on Flying Machines. 1997, Cambridge University Press: Cambridge, United Kingdom.

92 Chanute, letter to Hermann W. L. Moedebeck, dated 15 February 1902, in Papers of Octave Chanute, Letterpress book No 19.

93 W. Wright, letter to George Spratt, dated 12 November 1902, in General Correspondence.

94 K. Wright, letter to Milton Wright, dated 12 October 1901, in Family Papers: Correspondence.

95 Chanute, letter to Samuel P. Langley, dated 16 October 1901, in Papers of Octave Chanute, Letterpress book No 18.

96 Chanute, letter to Samuel P. Langley, dated 12 November 1901, in Papers of Octave Chanute.

97 Chanute, letter to W. Wright, dated 10–12 November 1901, in Papers of Octave Chanute, Letterpress book No 18.

98 W. Wright, letter to O. Chanute, dated 14 November 1901, in The Wilbur and Orville Wright Papers.

99 Chanute, letter to Hermann von Schrenk, dated 19 November 1904, in Papers of Octave Chanute.

100 Chanute, letter to W. Wright, dated 5 May 1902, in Papers of Octave Chanute, Letterpress book No 19.

101 W. Wright, letter to O. Chanute, dated 2 June 1902, in The Wilbur and Orville Wright Papers.

102 Langley, S.P., letter to Chanute, dated 21 October 1902, in Papers of Octave Chanute.

103 Chanute, letter to Augustus M. Herring, dated 30 May 1902, in Papers of Octave Chanute, Letterpress book No 19.

104 Chanute, letter to George A. Spratt, dated 17 September 1902, in Papers of Octave Chanute.

105 Chanute, letter to Charles H. Lamson, dated 14 October 1902, in Papers of Octave Chanute.

106 Chanute, letter to Samuel P. Langley, dated 21 October 1902, in Papers of Octave Chanute.

107 W. Wright, letter to O. Chanute, dated 3 November 1902, in The Wilbur and Orville Wright Papers.

108 Chanute, letter to Hermann W. L. Moedebeck, dated 21 October 1902, in Papers of Octave Chanute, Letterpress book No 19.

109 Chanute, letter to B. Baden-Powell, dated 9 December 1902, in Papers of Octave Chanute, Letterpress book No 19.

110 Irish, W.E., "Editorial." The Aeronautical World, 1902, 1 August, Vol 1 no 1, p: 3–6.

111 Jackman, W.J., "Has secret of equilibrium in riding the air. Octave Chanute solves a puzzling problem. Turns his attention now to the construction of motors, going to study foreign experiments." Sunday Times-Herald, Chicago, IL: 30 July 1899.

112 Chanute, O., "Experiments in Flying. An Account of the Author's own Inventions and Adventures." McClure's Magazine, 1900, June, Vol 15 no 2, p: 127–33.

Chapter 5 (Page 173–215)

1 Pawlowski, F.W., "The Evolution of Airplane Wing-Trussing." SAE Bulletin, Proceedings at the first Aeronautic Session, 1916, February, Vol 11 no 5, p: 516–528.

2 Editor, "Variétés: La Locomotion aérienne." Revue Scientifique (Revue Rose), 1901, 1 June, Vol 15 no 22, p: 689–691.

3 Lilienthal, G., letter to Ferdinand Ferber, dated 21 September 1901, in archive, Otto Lilienthal Museum: Anklam, Germany.

4 Chanute, letter to Ferdinand Ferber, dated 24 November 1901, in Papers of Octave Chanute, Letterpress book No 18.

5 Chanute, letter to Wilbur Wright, dated 7 March 1903, in The Wilbur and Orville Wright Papers.

6 Ferber, F., letter to O. Chanute, dated 7 September 1903, Chanute Papers.

7 W. Wright, letter to O. Chanute, dated 19 September 1903, in The Wilbur and Orville Wright Papers.

8 Carlier, C., "Ferdinand Ferber et l'Aviation." Presses Universitaires de France, 2003, Vol 1 no 209, p: 7–23.

9 Chanute, O., Travel Diary, in Collection of papers and photos at the home of Joseph & Jean Hodges. 1902–3: Denver, CO.

10 Blanc, E., "L'oeuvre posthume de Mouillard. "Nous étions sur ie point d'abandonner nos travaux, ont déclaré les Wright eux-mêmes, lorsque 'L'Empire de l'Air' tomba entre nos mains. Nous avons continué." Les Ailes, 1947, 27 September, Vol 27 no 1131, p: 5–7.

11 Ferber, F., "Expériences D'Aviation." L'Aerophile, 1903, February, Vol 11 no 2, p: 36–40.

12 Chanute, O., "Gliding Machines, the latest aeronautical experiments." The Illustrated Scientific News, 1903, February, p: 73.

13 Nimführ, R., "Die neusten Fortschritte in der praktischen Fliegekunst." Illustrierte Zeitung, Leipzig, Germany: 5 March 1903.

14 Beschu, F.L., "Locomotion Aérienne en Amérique." Le Monde Illustré, 1903, 28 March, p: 299.

15 Chanute, O., "The first flying model of Wilhelm Kress." Aeronautics, 1915, 30 May, Vol 16 no 6, p: 83, 85.

16 Moedebeck, H.W.L., Taschenbuch zum praktischen Gebrauch für Flugtechniker und Luftschiffer, unter Mitwirkung von Ingenieur O. Chanute, Dr. R. Emden und anderen Mitarbeitern. 2nd ed. 1904, Berlin W, Germany: W. H. Kühl.

17 McFarland, M., ed. The Papers of Wilbur and Orville Wright, including the Chanute-Wright letters and other papers of Octave Chanute. 1953, McGraw-Hill Book Comp., Inc.: New York, Toronto, London.

18 Archdeacon, E., "M. Chanute á Paris." La Locomotion, 1903, 11 April, Vol 30 no 80, p: 225–227.

19 Chanute, O., "La Navigation Aérienne aux Etats-Unis." L'Aérophile, 1903, August, Vol 11 no 8, p: 171–183.

20 Besançon, G., "Diner-Conférence du 2 Avril 1903, M. Chanute á l'Aéro-Club." L'Aérophile, 1903, April, Vol 11 no 4, p: 81–86.

21 Bacon, G., Memories of Land and Sky. 1928, London: Methuen & Co., Ltd.

22 Bacon, G., "Pigs That Fly. How 'Blindfold Pigs' indicate Character." The Strand Magazine, 1912, December, Vol 44 no 264, p: 733–738.

23 Reporter, "Correspondence." The Automotor Journal, 1903, 2 May, Vol 8 no 18, p: 462.

24 Chanute, O., "Experiments in Flying. An Account of the Author's own Inventions and Adventures." McClure's Magazine, 1900, June, Vol 15 no 2, p: 127–33.

25 Blanchet, G., "La pratique du vol plané." L'Aérophile, 1904, March, Vol 12 no 3, p: 60–63.

26 Chanute, letter to Victor Tatin, dated 16 August 1903, in Papers of Octave Chanute, Letterpress book No 19.

27 Blanchet, G., "La pratique du vol plané (The practice of gliding)." L'Aérophile, 1904, March, Vol 12 no 3, p: 60–63.

28 Voisin, G., Men, Woman and 10,000 Kites. 1963, London: Putnam & Company, Ltd.

29 Voisin, G., Mes 10,000 Cerfs-Volants. 1960, Paris: La Table Ronde.

30 Jaubert, M.J., "L'Aviation au Commencement de 1904 (Aviation at the beginning of 1904)." Revue Scientifique (Revue Rose) 1904, 2 April, Vol 5 no 14, p: 421–425.

31 Peyrey, F., L'Idée Aérienne. Les Oiseaux Artificiels. 1909, Paris: H. Dunod et E. Pinat, Editeurs.

32 Chanute, O., Artificial Flight. Part 3, in Pocket-Book of Aeronautics, H.W.L. Moedebeck, Editor. 1907, Whittaker & Co: London.

33 Esnault-Pelterie, R., "Expériences d'Aviation exécutés en 1904, en vérification de celles des fréres Wright. Conférences faite de 5 Janvier 1905 a l'Aéro-Club de France." L'Aérophile, 1905, June, Vol 13 no 6, p: 132–138.

34 Editor, "The Mission of Santos-Dumont." The New York Times, New York, NY: 22 April 1902.

35 Reporter, "By airplane 45 miles hourly. German inventors pushing experiments in sky sailing." The New York Press, New York, NY: 7 October 1907, p: 2.

36 Holland, J.P., "Flying Machines not Impossible. Feasibility of the achievement proved in theory by some curious figures. Nuts for Edison to crack. Conditions to be observed by these who address themselves to the problem of the future." The New York Herald, New York, NY: 23 November 1890, p: 23.

37 Reporter, "All to be Flying in Twelve Months. John B. Holland of submarine fame says he has solved difficult problem. He has applied for a patent upon a flying machine and will soon test it publicly." The Philadelphia Inquirer, Philadelphia, PA: 12 December 1905.

38 Reporter, "Holland as an Aeronaut." New York Daily Tribune, New York, NY: 22 July 1904.

39 Bryan, G.H., "History and Progress of Aerial Locomotion." Proceedings of the Royal Institution of Great Britain, 1901, 8 February, Vol 16 no 95, Part 3, p: 487–492.

40 Bryan, G.H.a.W.E.W., "The Longitudinal Stability of Aerial Gliders." Proceedings of the Royal Society of London, 1904, 24 February, Vol 73 no 489, p: 100–116.

41 W. Wright, letter to O. Chanute, dated 11 December 1902, in The Wilbur and Orville Wright Papers.

42 Reporter, "Langley's Flying Machine Makes its First Flight. Performance was a Disappointment. Went 500 Yards, then Sank in the River." Syracuse Journal, Syracuse, NY: 8 August 1903, p: 5.

43 Reporter, "Flying Machine Fiasco. Prof. Langley's Airship Proves a Complete Failure." New York Times, New York: 8 October 1903.

44 Reporter, "Aerodrome did a flipflop." The Sun, New York, NY: 9 December 1903, p: 1.

45 Editor, "The Langley Aerodrome." The New York Times, New York, NY: 10 December 1903, p: 8.

46 Reporter, "Seek Flying Machines' Cost. House tables resolution asking for expense of Prof. Langley's experiments." New York Times, New York, NY: 28 January 1904, p: 5.

47 Chanute, O., Langley's Contribution to Aerial Navigation, in Langley, Samuel Pierpont. Secretary of the Smithsonian Institution, 1887–1906. 1907, Smithsonian Institution: Washington DC. p: 30–35.

48 Chanute, letter to Wilbur Wright, dated 30 June 1903, in Papers of Octave Chanute, Letterpress book No 19.

49 W. Wright, letter to O. Chanute, dated 24 July 1903, in The Wilbur and Orville Wright Papers.

50 Chanute, letter to W. Wright, dated 27 July 1903, in Papers of Octave Chanute, Letterpress book No 19.

51 Chanute, O., "L'Aviation en Amérique." La Revue Générale des Sciences, 1903, 30 November, Vol 14 no 22, p: 1133–1142.

52 Combs, H., We certainly have been 'Jonahed' this year, in Kill Devil Hill. Discovering the Secret of the Wright brothers. 1979, Houghton Mifflin Company: Boston. p: 236–243.

53 W. Wright, letter to O. Chanute, dated 16 October 1903, in The Wilbur and Orville Wright Papers.

54 Editorial, "Performances of the Wright brothers." New York Times, New York, NY: 20 December 1903.

55 Herring, letter to Wright Brothers, dated 26 December 1903, in The Wilbur and Orville Wright Papers.

56 W. Wright, letter to O. Chanute, dated 8 January 1904, in The Wilbur and Orville Wright Papers.

57 Chanute, letter to W. Wright, dated 14 January 1904, in Papers of Octave Chanute, Letterpress book No 20.

58 M, "Editorial: The Mission of Santos-Dumont." The New York Times, New York, NY: 22 April 1902.

59 Archdeacon, "La machine volante des freres Wright." L'Aérophile, 1904, January, p: 16–18.

60 Moedebeck, H.W.L., "Gleitflugwettfahrten. Editorial." Illustrierte Aeronautische Mitteilungen, 1904, April, Vol 8 no 4, p: 142–143.

61 Chanute, O., "Development and Future of Flying Machines." The City Club Bulletin, 1908, 18 November, Vol 2 no 15, p: 191–194.

62 Chanute, O., "Aerial Navigation." Popular Science Monthly, 1904, March, Vol 64 no 25, p: 385–393.

63 Magruder, W.T., "Aeronautics. Section D — Mechanical Science and Engineering." Science, 1904, 4 March, Vol 19 no 479, p: 367.

64 Serrell, E.W., "Discussion and Correspondence: A Flying Machine in the Army." Science, 1904, 24 June, Vol 19 no 495, p: 952–955.

65 Editor, "Aeronautic Contests." New York Daily Tribune, New York: 26 January 1902, p: 8.

66 Bartow, C.A., "The Aerial Sweepstakes at the St. Louis Fair." The Amsterdam Daily Democrat, Amsterdam, NY: 12 April 1902, p: 16.

67 Chanute, letter to Hermann W. L. Moedebeck, dated 14 October 1904, Papers of Octave Chanute.

68 Chanute, O., Aerial Navigation, in Modern Mechanism. 1892, Appleton's Cyclopedia of Applied Mechanics: New York. p: 1–9.

69 Chanute, O., Means for Aerial Flight (No. 834,658), in United States Patent Office. 1906. p: 4.

70 Chanute, letter to W. Wright, dated 5 September 1904, in Papers of Octave Chanute, Letterpress book No 20.

71 Avery, W., letter to O. Chanute, dated 9 September 1904: Papers of Octave Chanute.

72 Avery, W., letter to Percy Hudson, dated 24 January 1905, in Papers of Octave Chanute.

73 Avery, W., letter to Octave Chanute, dated 9 September 1904: Papers of Octave Chanute.

74 Chanute, letter to William Avery, dated 23 September 1904, in Papers of Octave Chanute.

75 Reporter, "Airship Contest Time Extended." St. Louis Republic, St. Louis, MO: 1 October 1904, p: 5.

76 Reporter, "L'Aeroplane Chanute a Saint-Louis." La Locomotion Automobile, 1904, 27 October, Vol 12 no 42, p: 685–687.

77 Reporter, "Aeroplane makes three flights. William Avery's flying machine glides successfully at the World's Fair Stadium." The St. Louis Republic, St. Louis, MO: 8 October 1904, p: 1.

78 Kruckman, A., "Nerve of Aeronaut averts a serious fall. Sudden gust disturbs aeroplane, but steerman's coolness rights it again at the Fair." St. Louis Republic, St. Louis, MO: 11 October 1904, p: 3.

79 Avery, W., "Some little success of the aeroplane in aerial navigation." The Cherry Circle, 1908, January, Vol 14 no 1, p: 36–43.

80 Chanute, letter to William Avery, dated 16 October 1904, in Papers of Octave Chanute, Letterpress book No 20.

81 Chanute, letter to William Avery, dated 23 October 1904, in Papers of Octave Chanute.

82 Short, S., "The First American Aeroplane Export. Octave Chanute's 1904 Glider at the Threshold of Powered Flight." AIAA SciTech Forum, 2016, January.

83 Editor, "Where were the Airships?" The Technical World, 1905, January, Vol 2 no 5, p: 589.

84 Chanute, letter to William T. Magruder, dated 15 December 1904, Papers of Octave Chanute.

85 Reporter, "Wright Brothers Fly. Their New Machine Said to Have Successful Test." The Evening Star, Washington, DC: 27 May 1904, p: 1.

86 W. Wright, letter to O. Chanute, dated 18 September 1904, in The Wilbur and Orville Wright Papers.

87 Root, A.I., "Our Homes." Gleanings in Bee Culture, 1909, 15 August, Vol 37 no 16, p: 516.

Chapter 6 (Page 217–258)

1 Verne, J., The Clipper of the Clouds (Original 1886 French title is Robur-le-Conquérant). 1887, London: S. Low, Marston, Searle & Rivington.

2 Chanute, letter to W. G. Raymond, dated 10 May 1904, in Papers of Octave Chanute, Letterpress book No 20.

3 Chanute, letter to George W. Saathoff, dated 16 November 1905, in Papers of Octave Chanute, Letterpress book No 21.

4 Reporter, "Urbana Installs New President." Chicago Tribune, Chicago, IL: 19 October 1905, p: 7.

5 Chanute, letter to Henrietta M. Belcher, dated 15 May 1905, in Papers of Octave Chanute, Letterpress book No 20.

6 Reporter, "Lawyer hopes to fly within two weeks. Ludlow says he has sailed 25 yards in the air. Lot at Riverside Avenue and Ninetieth Street the scene of his experimental flights." New York Times, New York, NY: 6 July 1905, p: 9.

7 Reporter, "Two boys run away with big balloon and land in safety. Toledo, O., youngsters steal wild ride through clouds and escape without injury after being half frozen." The Daily Tribune, Chicago, IL: 22 September 1902, p: 1.

8 Chanute, letter to A. Roy Knabenshue, dated 19 October 1902, in Papers of Octave Chanute, Letterpress book No 19.

9 Kruckman, A., "Many thousands watched flight of airship. Giving young navigator an ovation at finish. Crowds wild as airship skims over New York." The New York World, New York, NY: 21 August 1905, p: 1.

10 Knabenshue, R., Pioneers in Aviation. Unpublished manuscript., in IAS Papers, assembled by Lester Gardner. 1942, Library of Congress. Manuscript Division.

11 Chanute, letter to William B. Stout, dated 7 June 1907, in Papers of Octave Chanute, Letterpress book No 22.

12 Stout, W.B., So away I went! 1951, Indianapolis, IN: Bobbs-Merrill Publisher.

13 Lesh, L.J., "Flying, as it was. Memories of Octave Chanute." The Sportsman Pilot, 1938, 15 May, Vol 13 no p: 18, 36–37.

14 Chanute, letter to O. Wright, dated 19 May 1907, in Papers of Octave Chanute, Letterpress book No 22.

15 Reporter, "Airships in Chicago. The Aero Club gives its first exhibition." Popular Mechanics, 1907, August, Vol 9 no 8, p: 859–860.

16 Lesh, L.J., "Some Preliminary Experiments in Flying." Scientific American, 1907, 19 October, Vol 97 no 16, p: 272–273.

17 Teale, E.W., Dune Boy. The early years of a naturalist. Illustrated by Edward Shenton. Lone Oak Edition ed. 1957, New York: Dodd, Mead & Company.

18 Wild, H.B., "Flying with the Pioneers. Thrilling events in the air are told in this second part of "My Forty Years of Flying"." Popular Science Monthly, 1931, January, Vol 118 no 1, p: 40–42, 131–132.

19 Short, S., "The Curious History of the "Chanute" Sailplane." Time Flies. Aviation's Golden Age., 2021, February, Vol 1 no 1, p: 2–25.

20 Wright, O.a.W., Flying-Machine (No. 821,393), in United States Patent Office. 1906: United States.

21 Hill, T.A., "Status of the Wrights' suit." Aeronautics, 1909, October, Vol 5 no 4, p: 122–123, 164.

22 Chanute, letter to W. Wright, dated 26 December 1904, in Papers of Octave Chanute, Letterpress book No 20.

23 W. Wright, letter to Chanute, dated 28 May 1905, in The Wilbur and Orville Wright Papers.

24 W. Wright, letter to Ferdinand Ferber, dated 4 November 1905, The Wilbur and Orville Wright Papers.

25 W. Wright, letter to O. Chanute, dated 31 January 1906, in The Wilbur and Orville Wright Papers.

26 Chanute, letter to W. Wright, dated 7 January 1906, in Papers of Octave Chanute, Letterpress book No 21.

27 Moedebeck, H.W.L., Taschenbuch zum praktischen Gebrauch für Flugtechniker und Luftschiffer, unter Mitwirkung von Ingenieur O. Chanute, Dr. R. Emden und anderen Mitarbeitern. 2nd ed. 1904, Berlin W, Germany: W. H. Kühl.

28 Reporter, "The flight of a flying machine. Was in the air twenty-five minutes Thursday afternoon near Simms Station." Dayton News, Dayton, OH: 6 October 1905.

29 Chanute, letter to W. Wright, dated 21 December 1905, in Papers of Octave Chanute, Letterpress book No 21.

30 Chanute, O., "Wiener Flugtechnischer Verein. Letter from O. Chanute to the Directors." Illustrierte Aeronautische Mitteilungen, 1906, March, Vol 10 no 3, p: 142–143.

31 W. Wright, letter to O. Chanute, dated 20 November 1905, in The Wilbur and Orville Wright Papers.

32 Wenham, F.H., letter to O. Chanute, dated 8 December 1905, in Papers of Octave Chanute.

33 Chanute, letter to Hermann W. L. Moedebeck, dated 1 May 1906, in Papers of Octave Chanute.

34 Besançon, G., "L'Aéroplane Wright. Nouveaux documents. Appareil à vendre. Le pour et le centre. Un moyen de savoir." L'Auto, 1905, 1 December, Vol 6 no 1873, p: 5.

35 Reporter, "The Conquest of the air. A speed of thirty-five miles an hour attained by the brothers Wright." The Car, a Journal of Travel by Land, Sea, and Air, 1906, 21 February, Vol 16 no 196, p: 8.

36 Reporter, "New Aero Club Told How to Fly." New York Herald, New York: 15 November 1905.

37 Chanute, letter to Augustus Post, dated 9 January 1906, in Papers of Octave Chanute, Letterpress book No 21.

38 Chanute, letter to Aero Club of America, dated 18 December 1905, in Papers of Octave Chanute, Letterpress book No 21.

39 Reporter, "France to own first real airship in the world." New York Herald, New York: 1906, p: 5

40 Reporter, "Want to Fly? Go to Aero Exhibit. There you will have choices of so many designs they'll leave you up in the air." Morning Telegraph, Saturday, New York, NY: 13 January 1906, p: 5.

41 Beach, S., "The Aero Club of America's Exhibit of Aeronautical Apparatus." Scientific American, 1906, 27 January, Vol 94 no 4, p: 93–94.

42 Chanute, letter to R. Rathbun, dated 12 December 1905, in Papers of Octave Chanute.

43 Dienstbach, K., "Die erste aeronautische Ausstellung in Amerika (vom 13. bis zum 23. Januar 1906 in New York)." Illustrierte Aeronautische Mitteilungen, 1906, March, April, Vol 10 no 3, 4, p: 264–270, 304–308.

44 Reporter, "Bold Aeronauts meet and gleefully predict." New York Times, New York, NY: 19 January 1906.

45 Myers, C.E., "A visit to the first show of the Aero Club of America." Scientific American Supplement, 1906, 17 February, Vol 111 no 1572, p: 25193–25194.

46 Crouch, T., Discussion on "What were the first ten items in the Smithsonian Air and Space Museum?" by Rebecca Maksel (NASM blog of 23 August 2017). Washington, DC.

47 Gueydan, P., letter to Pearl I. Young, dated 9 March 1964, in Collection of papers and photos at the home of Joseph & Jean Hodges: Denver, CO.

48 Beach, S., "The Wright Aeroplane and its Fabled Performances." Scientific American, 1906, 13 January, Vol 94 no 2, p: 40.

49 Chanute, O., "Chanute on the Wright Brother's Achievement on Aerial Navigation." Scientific American, 1906, 14 April, Vol 94 no 15, p: 307.

50 W. Wright, letter to O. Chanute, dated 10 October 1906, in The Wilbur and Orville Wright Papers.

51 Chanute, letter to W. Wright, dated 15 October 1906, in Papers of Octave Chanute.

52 W. Wright, letter to O. Chanute, dated 2 March 1906, in The Wilbur and Orville Wright Papers.

53 Chanute, O., Langley's Contribution to Aerial Navigation, in Langley, Samuel Pierpont. Secretary of the Smithsonian Institution, 1887–1906. 1907, Smithsonian Institution: Washington DC. p: 30–35.

54 Reporter, "The Motor Car of 1907. Seventh Annual Exhibition of the Automobile Club of America. Grand Central Palace crowded with enthusiasts." The New York Press, New York, NY: 2 December 1906, p: 1.

55 Reporter, "Wright Brothers at the Aero Show. Men said to be the first to fly. Say they owe their lives to caution. Admit air is dangerous. Ohio Experimenters decline to discuss their negotiations with France." New York Herald, New York, NY: 8 December 1906.

56 Reporter, "He Tells of a Real Airship." The Kansas City Star and The Kansas City Times, Kansas City, MO: 11 October 1906, p: 2.

57 Reporter, "Said Mr. Chanute." Chanute Daily Tribune, Chanute, KS: 21 December 1906.

58 Interview, "Five Years' Study of Flying Machines — Interview with Octave Chanute." New York Herald, New York: 19 May 1907, p: 5.

59 Morse, S., "Say Wright's in Aeroplane have flown miles. Neighbors of Ohio Inventors testify to wonderful flights in airship. Evolutions under perfect control. Brothers guard secret. Refuse to make public test, but insist that they have solved the aerial problem." The New York Herald, New York, NY: 22 November 1906, p: 1, 4, 10.

60 W. Wright, letter to O. Chanute, dated 1 December 1906, in The Wilbur and Orville Wright Papers.

61 Chanute, letter to Charles R. Flint, dated 26 December 1906, in Papers of Octave Chanute, Letterpress book No 21.

62 Reporter, "Flyers or Liars." New York Herald, Paris Edition, Paris: 10 February 1907.

63 Bell, A.G., "The Tetrahedral Principle in Kite Structure." National Geographic Magazine, 1903, June, Vol 14 no 6.

64 Editorial, "Problems of the Air. Aerial Flight Sure to Come — Interview with Alexander Graham Bell." New York Herald, New York, NY: 29 April 1907.

65 Reporter, "Professor Bell is honored by Oxford. Says the Germans are behind in the race of inventors." The Evening Star, Washington, DC: 3 May 1907, p: 23.

66 Short, S., "From kite to glider to powered flight: the AEA gets into the air." WWI Aero, 2013, May, No 215, p: 14–25.

67 Roseberry, C.R., Glenn Curtiss, Pioneer of Flight. 1972, Garden City, New York: Doubleday & Company, Inc.

68 Curtiss, G.H., "An Airship Chauffeur." American Aeronaut and Aerostatist, 1908, January, Vol 1 no 3, p: 27–28.

69 Reporter, "Announcement." The Dayton Herald, Dayton, OH: 21 November 1907, p: 2.

70 Reporter, "New Airship Flies. Selfridge Aerodrome sails steadily for 319 feet at 25 to 30 miles an hour. First public trip of heavier-than-air car in America." The Washington Post, Washington, DC: 13 March 1908, p: 1.

71 Moedebeck, H.W.L., Pocket-Book of Aeronautics, in collaboration with O. Chanute and others. Authorized English Translation. 1907, London: Whittaker & Co.

72 Chanute, letter to Hermann W. L. Moedebeck, dated 15 April 1907, in Papers of Octave Chanute, Letterpress book No 22.

73 Chanute, O., The Wright Brothers' Motor Flyer, in Navigating the Air. 1907, Aero Club of America. Doubleday, Page & Company: New York.

74 Chanute, O., "Conditions of Success with Flying Machines." American Magazine of Aeronautics, 1907, July, Vol 1 no 1, p: 7–9.

75 Chanute, O., "Pending European Experiments in Flying." American Aeronaut and Aerostatist, 1907, October, Vol 1 no 1, p: 13–16.

76 Jones, E.L., "The Scientific American Flying Machine Trophy." American Magazine of Aeronautics, 1907, August, p: 11–12.

77 Moore, W.L., "International Aeronautical Congress, President's Address." American Magazine of Aeronautics, 1907, November, Vol 1 no 11, p: 19–23.

78 Reporter, "An Aerial Battleship. Admiral Chester says aeroplane is future's warrior." The Evening Star, Washington, DC: 10 June 1907, p: 7.

79 Glassford, W.A., "Our Army and Aerial Warfare." American Magazine of Aeronautics, 1908, January, Vol 2 no 1, p: 18–20.

80 Reporter, "Air ship fleet is planned for Army. General Allen tells Aeronautical Congress. Secretary Taft will spend $200,00." New York Herald, New York, NY: 29 October 1907, p: 1.

81 Les Freres Voisin- Ingenieurs-Constructeurs- demeurant a Billancourt (Seine), Q.d.P., dune part, et Monsieur August Euler - demeurant a Frankfort a/Main, Contract Agreement. 1910: Billancourt (Seine).

82 Chanute, letter to William A. Glassford, dated 31 July 1907, in Papers of Octave Chanute, Letterpress book No 22.

83 Squier, G.O., "Present Status of Military Aeronautics." Flight, 1909, 1 May, Vol 1 no 21, p: 251–252.

84 Clark, P.W.a.L.A.L., George Owen Squier: U.S. Army Major General, Inventor, Aviation Pioneer, Founder of Muzak. 2014, Jefferson, NC: McFarland & Company Inc.

85 Chanute, letter to W. Wright, dated 4 December 1907, in Papers of Octave Chanute.

86 Chanute, letter to W. Wright, dated 15 December 1907, in Papers of Octave Chanute.

87 W. Wright, letter to O. Chanute, dated 1 January 1908, in The Wilbur and Orville Wright Papers.

88 Chanute, letter to W. Wright, dated 4 January 1908, in Papers of Octave Chanute.

89 Archdeacon, E., "Diner Mensuel du 7 Novembre 1907." L'Aérophile, 1907, November, Vol 15 no 11, p: 307–308.

Chapter 7 (Page 259–293)

1 Voisin, G.e.C., "Ou est née l'aviation? En France. Les Freres Voisin defendent la cause du genie national." Le Matin, Paris: 5 September 1908, p: 1–2.

2 Reporter, "C'est un Français Qui a Créé L'Aviation. Le Matin, Paris: 8 November 1908.

3 Chanute, letter to F. Crowninshield, dated 12 March 1906, in Papers of Octave Chanute, Letterpress book No 21.

4 Farman, H., "Farman's own story of his aeroplane flight." The New York Times, New York, NY: 2 February 1908, p: 2.

5 Reporter, "Airship without a gas bag." The Sun, New York, NY: 24 December 1907, p: 7.

6 Reporter, "Three aeroplanes for Signal Corps. Of forty-one bidders only three followed the Army's instructions." The New York Times, New York, NY: 9 February 1908, p: 5.

7 Roseberry, C.R., Glenn Curtiss, Pioneer of Flight. 1972, Garden City, New York: Doubleday & Company, Inc.

8 Reporter, "Langley Machine. Army would continue experiments with aeroplane." The Paducah Evening Sun, Paducah, KY: 1 August 1908, p: 7.

9 Reporter, "It will never fly. Langley Aerodrome will not be tried again. Dr. Cyrus Adler says so. Reporters get first close look at "the Buzzard." Its mechanism is a marvel." The Evening Star, Washington, DC: 29 July 1908, p: 1.

10 Chanute, letter to Charles D. Wolcott, dated 30 November 1908, in Papers of Octave Chanute.

11 W. Wright, letter to O. Chanute, dated 16 January 1908, in The Wilbur and Orville Wright Papers.

12 Chanute, O., "Recent Aeronautical Progress in the United States." The Aeronautical Journal, 1908, July, Vol 12 no 47, p: 52–55.

13 Reporter, "Reporter who witnessed 1906 [1908] flight still lives. Arnold Kruckman, now 72, recalls events." The Evening Telegram, Herkimer-Ilion, NY: 18 February 1953, p: 11.

14 Chanute, O., "The Wright Brothers' Flights." The Independent, 1908, 4 June, Vol 64 no 3105, p: 1287–1288.

15 Hargrave, L., letter to Octave Chanute, dated 22 February 1908, in Papers of Octave Chanute.

16 Chanute, letter to Laurence Hargrave, dated 11 June 1908, in Papers of Octave Chanute.

17 Editorial, "For a School of Aeronautics." The New York Times, New York, NY: 19 January 1908.

18 Means, J., "The Value of the Motorless Glider." Aeronautics, 1908, March, Vol 2 no 3, p: 4–6.

19 Reporter, "Aero Club takes wing on a "Glider". Members secretly test simple flying device on Long Island estate. They try short flights. Other tests to be made after their machine has been equipped with a motor." The New York Times, New York, NY: 13 April 1908.

20 Reporter, "See big races in New York skies. Members of the Aero Club of America actively promoting plans for aeronautics. Conference held here. Messrs Means, Rotch and Chanute are at work devising plans and raising the money. Testing ground projected. Some very successful glides reported to have taken place on Los Island with a new aeroplane." New York Herald, New York, NY: 19 April 1908, p: 3.

21 Chanute, O., "Future Uses of Aerial Navigation." Aeronautics, 1908, June, Vol 2 no 6, p: 15–16.

22 Jones, E.L., "Aero Club of America." Aeronautics, 1908, May, Vol 2 no 5, p: 38.

23 Reporter, "New Aeronautic Club. Society not satisfied with work of old Organization." New-York Daily Tribune, New York, NY: 11 June 1908, p: 4.

24 Iliaco, A.C., "Aeronautics Hereabout. Cooperation of brain, wealth and work needed in America." The Sun, New York, NY: 30 May 1908, p: 2.

25 Reporter, "We'll All Fly Soon. Farman, sky pilot, predicts future of aeroplane. Heavier-than-air machines safer than automobiles and a whole lot faster. Aeronaut talks of the problem of aerial navigation and growing knowledge of it." The Evening Star, Washington, DC: 27 July 1908, p: 4.

26 Reporter, "Farman here to fly. English aviator says aeroplanes will be safer than autos." New York Daily Tribune, New York, NY: 27 July 1908, p: 10.

27 Curtiss, G.H.a.A.P., The Curtiss Aviation Book. 1912, New York: Frederick A. Stokes Company.

28 Editor, "Another Guaranteed Flying Machine in America." Aeronautics, 1908, June, Vol 2 no 6, p: 57.

29 Munn, C., "The Scientific American Aeronautic Trophy." Scientific American, 1907, 20 April, Vol 46 no 16, p: 326.

30 Reporter, "American Aeroplane. First public demonstration successful on Lake Keuka. Flies 319 feet easily. Design of Lieut. Selfridge. Slight accident brings it to the ice again. Prof. Alex. Graham Bell one of the backers of the entreprise. W. Baldwin aviator." The Evening Star, Washington, DC: 13 March 1908, p: 12.

31 Selfridge, T.E., Notes by Thomas E. Selfridge, from September 24, 1907, to July 24, 1908, in Alexander Graham Bell Papers, Library of Congress: Washington, DC.

32 Reporter, "Airship Mail Carriers. Expert predicts their use on short routes. Limits passenger to five. Octave Chanute, noted for aerial experiments, says he

does not see how heavier-than-air machines can carry more. Will be valuable for reconnaissance in war and for exploration." The Washington Post, Washington, DC: 2 July 1908, p: 15.

33 Reporter, "Aeronauts to see airship tests. Members of Aero Club of America will witness contest for trophy." New York Daily Tribune, New York, NY: 3 July 1908, p: 4.

34 Editor, "Won the Trophy. Curtiss' name goes down in History. First official flight of Aerodrome in America." The Hammondsport Herald, Hammondsport, NY: 8 July 1908, p: 8.

35 O. Wright, letter to Glenn H. Curtiss, dated 20 July 1908, in General Correspondence: Wilbur and Orville Wright Papers.

36 M. Wright, letter to W. Wright, dated 13 June 1908, in Family Correspondence: Wilbur and Orville Wright Papers.

37 K. Wright, letter to Wilbur Wright, dated 2 July 1908, in Family Correspondence: Wilbur and Orville Wright Papers.

38 O. Wright, letter to W. Wright, dated 19 July 1908, in Family Correspondence: Wilbur and Orville Wright Papers.

39 Editor, "Paris, May 30." New York Times, New York, NY: 31 May 1908.

40 Reporter, "Continental Notes and News. Wilbur Wright Testing his Aeroplane." The Autocar, 1908, 15 August, Vol 20 no p: 261.

41 Chanute, letter to Paul Renard, dated 22 November 1908, in Papers of Octave Chanute, Letterpress book No 23.

42 Breyer, V., "Chez le Premier 'Homme Oiseau.' Wilbur Wright bat tous ses records… de conversation." L'Auto, 1908, 18 October, p: 1.

43 W. Wright, letter to Reuchlin Wright, dated 3 July 1901, in Family Papers: Correspondence-Wright, Reuchlin.

44 W. Wright, letter to O. Chanute, dated 10 November 1908, in The Wilbur and Orville Wright Papers.

45 Reporter, "Baldwin Talks Shop. Believes airship of future is not yet conceived, thousands will plan it. Lieut. Lahm is inspecting his gas machine to see that it answers specifications." The Evening Star, Washington, DC: 25 July 1908, p: 5.

46 Reporter, "Baldwin Ship has Test. Dirigible Balloon for Army Use Makes Good Showing in First Flight." Brooklyn Daily Eagle, NY: 5 August 1908, p: 6.

47 Reporter, "Baldwin Ship Passed. Army Board finds balloon up to specifications. Speed trial postponed, Curtiss wants to make improvements. Aeronaut Herring Witness. Looks over experiments at Fort Meyer and discusses marvels of own solution of air travel." The Sunday Star, Washington, DC: 9 August 1908, p: 5.

48 Reporter, "Baldwin Airship flies two hours. Big dirigible meets every test and is accepted for United States Army, Under perfect control. Average speed 21 miles." The New York Times, New York, NY: 16 August 1908.

49 Reporter, "Awaiting Conditions. Airship flight again delayed by the wind. Orville Wright hopes to fly plane before nightfall. How the Skyflyer flies." The Evening Star, Washington, DC: 3 September 1908, p: 13.

50 Reporter, "Wright keeps on breaking records. Aeroplanist stays up over 70 minutes in flight at Fort Myer. Goes 39.55 miles an hour." Chicago Daily Tribune, Chicago, IL: 12 September 1908, p: 2.

51 Reporter, "Orville Wright, in his Aeroplane, carries two men for nine minutes." The New York Herald, New York, NY: 13 September 1908, p: 1.

52 Reporter, "Flying Machine. Orville Wright makes another record." The Troy Times, Troy, NY: 12 September 1908, p: 1.

53 Satzman, C.M., C. S. Wallace, F. P. Lahm, The First United States Army Aircraft Report (September 1908), in War Department Office of the Chief Signal Officer, Aeronautical Division. 1909: Washington, DC.

54 Reporter, "Wright will live; Selfridge is dead. Flying machine experiments are not to be abandoned. No setback for tests." The Evening Star, Washington, DC: 18 September 1908, p: 1, 2, 4.

55 Chanute, letter to Katharine Wright, dated 29 September 1908, in Papers of Octave Chanute.

56 Reporter, "The Men Behind Orville Wright." The Sunday Star, Washington, DC: 27 September 1908, p: 3.

57 K. Wright, letter to W. Wright, dated 24 September 1908, in Family Correspondence: Wilbur and Orville Wright Papers.

58 Chanute (?), letter to Major George 0. Squier at Fort Myer, dated 26 September 1908, on Wright brothers stationary.

59 Herring, A.M., letter to O. Chanute, dated 3 December 1898, in Papers of Octave Chanute.

60 Editor, "The Aeronautic Society's First Exhibition." Scientific American, 1908, 14 November, Vol 98 no 20, p: 338.

61 Reporter, "Aeroplane fleet predicted soon. Immense activity is noted among inventors of apparatus for conquest of air." New York Herald, New York, NY: 2 August 1908, p: 3.

62 Editors, "Comet like transit of the future in the Ocean of the skies. As foreseen by eminent aeronautical authorities, scientists and inventors. What will be the ultimate type of airship? (p1); Facts and Fancies of the Conquest of the Skies, (p: 2); Solving the Secrets of Aerial Flight (p: 3)." The New York Herald, New York, NY: 30 August 1908, p: 1–3.

63 Schukking, W.H., "Een Nederlandsch zweeftoestel van het stelsel Wright." Orgaan van de Nederlandsche Vereeniging voor Luchtvaart, 1908, 15 September, Vol 1 no 7, p: 73–77.

64 Reporter, "Army plane falls at Aldershot again and is wrecked." Evening Star, Washington, DC: 16 October 1908, p: 1.

65 Reporter, "200 foot fall ends Capt. Cody's career. American Cowboy-Aviator and passenger killed at Aldershot, England. Showed great perseverance in developing the science of aeronautics." The Sun, New York, NY: 8 August 1913, p: 9.

66 Rives, M.G., Rapport sur le premier salon de l'aéronautique. Grand Palais, Paris. 1908, Paris: Automobile-Club de France.

67 Reporter, "An aeronautic salon. Small aeroplane $1,000. Popularity of airships shown by 120,000 visitors a day." New York Daily Tribune, New York, NY: 10 January 1909.

68 Special Cable, "French Aeronaut crushed to death. Ferber's machine, in alightening, turns somersault and he is pinned under motor. Had just predicted that more lives would be sacrificed to the Science of Aviation." The New York Times, New York, NY: 23 September 1909, p: 2.

69 Chanute, O., "Captain Ferber Killed in a Fall." Aeronautics, 1909, November, Vol 5 no 5, p: 187.

70 Short, S., "The First American Aeroplane Export. Octave Chanute's 1904 Glider at the Threshold of Powered Flight." AIAA SciTech Forum, 2016, January.

71 Claessens, J.-L.P. Le plus vieux des Musées Aéronautiques. 2015 [cited 2015; Available from: <www.pyperpote.tonsite.biz/listinmae/index.php?option=com_content&view=article&id=473&Itemid=72>.

72 Verneuil, M., "Aéronautique: Le musée de l'Aéronautique a Chalais-Meudon pres de Paris." Le Génie Civil, 1930, 20 September, Vol 97 no 12.

73 Reporter, "You may order your balloon on Broadway. Want a balloon?" The Evening Telegram, New York, NY: 11 December 1908, p: 7, 10.

74 Reporter, "Opens store to sell airships. Ready to take orders for aerial craft." Los Angeles Sunday Herald, Los Angeles, CA: 20 December 1908, p: 1, Section 5.

75 Chanute, O., "Development and Future of Flying Machines." The City Club Bulletin, 1908, 18 November, Vol 2 no 15, p: 191–194.

Chapter 8 (Page 295–337)

1 Jones, E.L., "Side Lights on the Future of Flying." Aeronautics, 1909, February, Vol 4 no 2, p: 84–85.

2 Reporter, "Orville Wright here, Aero Men greet him. Ohio Flyer welcomed by New Yorkers on eve of departure for Europe." The Thrice-A-Week World, New York, NY: 6 January 1909, p: 3.

3 Reporter, "Wrights' sister in first flight. Makes seven minute trip in aeroplane with her brother Wilbur, Orville looking on. Makes three ascents today, in one Comte de Lambert goes up, the comtesse de Lambert being taken up in another." New York Herald, New York, NY: 10 February 1909.

4 Kruckman, A., "Wrights come home honored, left obscure. Wealthy and learned men will greet Ohio brothers first to fly." St. Louis Post-Dispatch, St Louis, MO: 9 May 1909, p: 15

5 Chanute, letter to W. Wright, dated 19 May 1909, in Papers of Octave Chanute.

6 Chanute, O., "Review of Scientific Books: Artificial and Natural Flight." Science, 1909, 27 August, Vol 30 no 765, p: 282–283.

7 Chanute, O., "Soaring Flight, how to perform it." Aeronautics, 1909, April, Vol 6 no 4, p: 134–137. Article was republished in August 1914.

8 Slater, A.E., "The 'Mystery' of Soaring Flight." Weather, 1955, September, Vol 10 no 9, p: 298–303.

9 Means, J., "Grist for the mathematical mill." Aeronautics, 1909, March, Vol 4 no 3, p: 101–102.

10 Zahm, A.F., "Comments on Mr. James Means' article on "Grist for the mathematical mill"." Aeronautics, 1909, March, Vol 4 no 3, p: 103.

11 Chanute, letter to James Means, dated 26 March 1909, in Papers of Octave Chanute, Letterpress book No 23.

12 Chanute, letter to Chas. Walcott, dated 22 February 1910, in Papers of Octave Chanute, Letterpress book No 24.

13 Editor, "Another Guaranteed Flying Machine in America." Aeronautics, 1908, June, Vol 2 no 6, p: 57.

14 Reporter, "First aeroplane just purchased in this country. Stimulated by World's Fulton Flight, Aeronautical Society of New York buys one of G. H. Curtiss for $5,000." The Thrice-A-Week World, New York, NY: 3 March 1909.

15 Reporter, "Aeroplane Trust to be formed. G. H. Curtiss the Central Figure. The Herring and Curtiss Aeroplanes and Curtiss Motorcycle Interests involved." The Hammondsport Herald, Hammondsport, NY: 10 March 1909, p: 4.

16 Herring, A.M., Anti-drip nose for pitchers (No.377,109), in United States Patent Office. 1891: United States. p: 1–2.

17 Chanute, O.a.A.M.H., Improvements in or relating to means and appliances for effecting aerial navigation (No. 15,221), in British Patent Office. 1897, Communicated to Thomas Moy, England.

18 Reporter, "Octave Chanute discusses Aviation in 2009." St. Louis Post-Dispatch, St. Louis, MO: 10 October 1909, p: 1.

19 Roseberry, C.R., Glenn Curtiss, Pioneer of Flight. 1972, Garden City, New York: Doubleday & Company, Inc.

20 Reporter, "Another aerial record broken. Several contests are planned for next fortnight. Curtiss makes spectacular flight in St. Louis." The Cairo Bulletin, Cairo, IL: 10 October 1909, p: 1.

21 Reporter, "Chicago Sees Curtiss Fly." Chicago Tribune, Chicago, IL: 17 October 1909, p: 1.

22 Chanute, O., "Curtiss Flies in Chicago." Aeronautics, 1909, December, Vol 5 no 6, p: 216.

23 Reporter, "Chicago has become a vast testing ground for amateur aviators. One hundred and more young men from all walks of life, are now building flying machines with which they hope to leave the earth and soar at will through the atmosphere, bidding defiance to the treacherous lake breezes and equaling the best records of Wright, Curtiss, Bleriot, et al." Chicago Sunday Tribune, Chicago, IL: 18 September 1910, p: 3–4.

24 Curtiss, letter to Alexander G. Bell, dated 27 October 1909, in The Papers of Alexander Graham Bell, Manuscript Division, Library of Congress: Washington, DC.

25 Chanute, letter to Emerson R. Newell, dated 14 October 1909, in Papers of Octave Chanute, Letterpress book No 24.

26 Chanute, O., "Recent Progress in Aviation and Chronology of Aviation." Journal of the Western Society of Engineers, 1910, April, Vol 15 no 2, p: 111–147.

27 Chanute, O., "Recent Progress in Aviation. The Present State of the Art. Chronology of Aviation." Scientific American Supplement, 1910, 23 July (part 1); 30 July (part II); 6 August (part III); 106–108 (part IV), Vol 120 no 1803, 1804, 1805, 1806 p: 56–58, 72–74, 88–90.

28 Moore, W.L., "International Aeronautical Congress, President's Address." American Magazine of Aeronautics, 1907, November, Vol 1 no 11, p: 19–23.

29 Ferber, F., "The Aviation World. Our Paris Correspondent writes." American Aeronaut, 1908, June, Vol 1 no 6, p: 215–224.

30 Reporter, "'Sea Gull' Blériot describes flight. French Aviator, first to cross English Channel, tells of every sensation." The Daily Tribune, Chicago, IL: 26 July 1909, p: 1, 5.

31 Chanute, letter to William A. Glassford, dated 8 August 1909, in Papers of Octave Chanute, Letterpress book No 24.

32 Reporter, "Not yet war engines." The New York Times, New York, NY: 26 July 1909.

33 Chanute, O., "First Steps in Aviation and Memorable Flights." Aeronautics, 1909, January, Vol 4 no 1, p: 24.

34 "Chanute Compiles List showing Aviation Steps. Noted expert gives particulars of first and most memorable subsequent flights, from the primitive machine to the wonderful inventions of today." The Philadelphia Inquirer, Philadelphia, PA: 10 January 1909, p: 11.

35 Kruckman, A., "Herring Aeroplane still a Mystery." The World, New York, NY: 16 May 1909.

36 Dickinson, J.M., Report of The Board of Ordnance and Fortification to the Secretary of War, in Annual Reports, War Department. June 30, 1909, Government Printing Office: Washington, DC.

37 Reporter, "Das erste Heim für Flugmaschinen und Gleitflieger in Deutschland." Deutsche Zeitschrift für Luftschiffahrt, 1909, July, Vol 13 no 14, p: 374–377.

38 Short, S., "Gliders at the ILA'09." Bungee Cord, 2012, Fall, Vol 38 no 3, p: 15.

39 Hirth, H., Meine Flugerlebnisse. 1915, Berlin W 35: Verlag Gustav Braunbeck, GmbH.

40 Wright, O.a.W., Flying-Machine (No. 821,393), in United States Patent Office. 1906: United States.

41 "Affidavit of Wilbur Wright." United States Circuit Court, Western District of New York, 1909, 18 September.

42 Reporter, "Big men of finance back of the Wrights. Shoots, Cornelius Vanderbilt, August Belmont, and Howard Gould among directors. Corporation expects machine will come into general use soon, standing fast for patent rights." The New York Times, New York, NY: 23 November 1909.

43 Reporter, "Wright brothers form combination to control aviation in this country and Canada. Capital placed at $1,000,000 and no stock on Sale. Monopoly of air current. Company to take over Wright patents and to prosecute all infringements." The Evening Star, Washington, DC: 23 November 1909, p: 11.

44 W. Wright, letter to O. Chanute, dated 6 December 1909, in The Wilbur and Orville Wright Papers.

45 Chanute, letter to Ernest L. Jones, dated 29 June 1909, in Papers of Octave Chanute, Letterpress book No 23.

46 Chanute, letter to Ernest L. Jones, dated 26 August 1909, in Papers of Octave Chanute, Letterpress book No 24,.

47 Hill, T.A., "Status of the Wrights' suit." Aeronautics, 1909, October, Vol 5 no 4, p: 122–123, 164.

48 Kruckman, A., "Wrights begin patent fight on Tuesday. World-wide interest in the case against Glenn Curtiss, which comes up in United States Circuit Court, Buffalo. Many will watch fight from different angles." New York World, New York, NY: 12 December 1909.

49 W. Wright, letter to Arnold Kruckman, dated 21 December 1909, in General Correspondence: Wilbur and Orville Wright Papers.

50 Kruckman, A., "Wrights open court fight on Rival Air Pilots. Sue for infringement of patents against Curtiss and Herring-Curtiss Company given anopen hearing." The Thrice-a-Week World, New York, NY: 16 December 1909.

51 Wright, W., "What Mouillard Did." Aero Club of America Bulletin, 1912, April, Vol 1 no 3, p: 3–4.

52 Merrill, A.A., "An Analysis of Mouillard's Claim Twelve." Aeronautics, 1913, May, Vol 12 no 5, p: 169.

53 Chanute, O., "Progress in Flying Machines." Railroad and Engineering Journal, 1891, October, Vol 5 no 10, p: 461–465.

54 Mouillard, L.-P., Assignor of one-half to Octave Chanute, Means for Aerial Flight (No. 582,757), in United States Patent Office. 1897.

55 Chanute, letter to Paul Renard, dated 22 November 1908, in Papers of Octave Chanute, Letterpress book No 23.

56 Editorial, "The Selden Patent." St. Louis Post-Dispatch, St. Louis, MO: 12 January 1911.

57 Anderson, J.D., Jr., Historical Note: The Development of Flight Controls (Chapter 7.21) in Principals of Stability and Control (Chapter 7), in Introduction to Flight. 2004, McGraw-Hill: New York.

58 Spratt, letter to W. Wright, dated 29 September 1909, in Wilbur and Orville Wright Papers.

59 W. Wright, letter to George Spratt, dated 16 October 1909, in Wilbur and Orville Wright Papers.

60 Spratt, letter to O. Chanute, dated 14 December 1909, in Papers of Octave Chanute.

61 Spratt, letter to Orville Wright, dated 27 November 1922, in The Wilbur and Orville Wright Papers.

62 Wright, O., The Early History of the Airplane. 1922, The Dayton-Wright Airplane Co.: Dayton, OH. p: 1–8: The Wright Brothers' Aeroplane; pp 9–15: How we made the first flight; pp 16–24: Some Aeronautical Experiments.

63 Chanute, letters to Emerson R. Newell and Victor Lougheed, dated 5 January 1910, in Papers of Octave Chanute, Letterpress book No 24, Manuscript Division, Library of Congress: Washington, DC.

64 Chanute, O. Proceedings of the International Conference on Aerial Navigation. 1893. Chicago IL: American Engineer and Railroad Journal.

65 Reporter, "After the Wrights. An attack probable upon their airship patent. Prior claim to be filed. Prof J. J. Montgomery's developments in 1903. Flexing wrong experiments. Inventor declares he will prevent monopoly in aviation, not one man's work." The Sunday Star, Washington, DC: 17 April 1910, p: 5.

66 Editor, Wright vs. Paulhan, Extracts from Affidavits and Judge Hand's decision in the case of the Farman and Blériot aeroplanes. Scientific American Supplement, 1910. 69(1785, 1786): p: 182–193, 198.

67 Chanute, letter to Israel Ludlow, dated 25 February 1910, in Papers of Octave Chanute, Letterpress book No 24.

68 Chanute, letter to Israel Ludlow, dated 18 March 1910, in Papers of Octave Chanute.

69 Jones, E.L., "The Wright Suits." Aeronautics, 1910, August, Vol 7 no 2, p: 60.

70 Reporter, "Masters of the Air. To fly slow is now the airproblem. Lilienthal, who says Wrights owe him royalties, thinks he has solved it." The Thrice-A-Week World, New York, NY: 22 February 1910, p: 3.

71 W. Wright, letter to Arnold Kruckman, dated 4 March 1912, in General Correspondence: Wilbur and Orville Wright Papers.

72 Fanciulli, J.S., "Smithsonian the tomb of airships that failed to fly. Unique collection of aeroplanes and other air craft long forgotten. But work of early inventors has aided the present day tests. Wrights visited Institute and examined old types of gliders, helicoptres and monoplanes. Late models are patterned somewhat on inventions tested many years ago. Aviators never able to plan their machines like birds." The Sunday Star, Washington, DC: 22 August 1909, p: 6.

73 Wright, W., "Otto Lilienthal." Aero Club of America Bulletin, 1912, September, Vol 1 no 8, p: 19–20.

74 McFarland, M., ed. The Papers of Wilbur and Orville Wright, including the Chanute-Wright letters and other papers of Octave Chanute. 1st edition ed. 1953, McGraw-Hill Book Comp., Inc.: New York, Toronto, London.

75 Chanute, letter to W. Wright, dated 28 November 1906, in Papers of Octave Chanute.

76 W. Wright, letter to O. Chanute, dated 1 December 1906, in The Wilbur and Orville Wright Papers.

77 Bell, A.G., "Aerial Locomotion." Proceedings of the Washington Academy of Sciences, 1907, 4 March, Vol 8 no p: 407–448.

78 Kruckman, A., "Dr. Chanute denies Wright Flying Claim." New York World, New York, NY: 17 January 1910.

79 Reporter, "Pioneer aviator praises Wrights. Octave Chanute, who gave start in Aeronautics, regrets Patent Suits." Philadelphia Ledger, Philadelphia, PA: 17 January 1910.

80 Birdman, "Chatter of the Man-Birds." The Philadelphia Inquirer, Philadelphia, PA: 27 November 1910, p: 8.

81 Kruckman, A., "Chanute deplores Wright patent suits. Early experimenter who turned his data over to them thinks they will check progress. Engineer who perfected the bi-plane glider thinks there would be enough for all." The New York Times, New York, NY: 23 January 1910.

82 Reporter, "The Wrights' Suits for Infringement. Chanute, pioneer glider, who gave Wrights their start, is in favor of public flights. Wing warping idea is not wholly original with Wright brothers, he says. Sorry lawsuits are started." The Hammondsport Herald, Hammondsport, NY: 26 January 1910, p: 8.

83 Editor, "Progress in Aeronautics. What Chanute has done for flying. The interesting story, told by himself, of his connection with the Wright brothers and their work. His ideas as to the future of the aeroplane" Boston Evening Transcript, Boston, Mass: 5 February 1910, p: 10.

84 Goldstone, L., Birdmen. The Wright Brothers, Glenn Curtiss, and the Battle to Control the Skies. 2014, New York: Ballantine Books.

85 Chanute, letter to George A. Spratt, dated 25 January 1910, in Papers of Octave Chanute.

86 Chanute, letter to W. Wright, dated 23 January 1910, in Papers of Octave Chanute.

87 W. Wright, letter to O. Chanute, dated 29 January 1910, in The Wilbur and Orville Wright Papers.

88 W. Wright, letter to O. Chanute, dated 28 April 1910, in The Wilbur and Orville Wright Papers.

Chapter 9 (Page 339–374)

1 Reporter, "Curtiss Wins the World's $10,000 prize, his daring flight smashes all records. Start from Albany on flight and finish on Governor's Island." The Thrice-A-Week World, New York, NY: 30 May 1910, p: 1, 4.

2 Editorial, "The Listener." Boston Evening Transcript, Boston, MA: 27 September 1911, p: 17.

3 Prandtl, L., Betrachtungen über das Flugproblem, in Denkschrift der ersten Internationalen Luftschiffahrts-Ausstellung (ILA) zu Frankfurt a/M 1909. 1909, Julius Springer: Berlin. p: 140–150.

4 W. Wright, letter to Colonel William A. Glassford, dated 30 November 1910, in The Papers of Wilbur and Orville Wright, including the Chanute-Wright letters and other papers of Octave Chanute, edited by Marvin McFarland: Washington, DC.

5 A. Flier, "A Lesson in Flying for our Brooklyn Boys." Brooklyn Citizen, Brooklyn, NY: 24 July 1910.

6 Ogilvie, A., letter to Wright brothers, dated 1 September 1909, in General correspondence with the Wright Brothers.

7 Clarke, T.W.K., "The Wright Glider as Made by Clarke." Flight, 1909, 18 September, 25 September, Vol 38, 39 no 52, 1, p: 568–571, 585–588.

8 Clarke, T.W.K., "Gliding as a sport and as an aid to flight." Aeronautics, 1911, August, Vol 9 no 2, p: 43–47.

9 Voisin, C.a.G., "The Practice of Aviation. Building a Chanute glider." Aeronautics. Supplement to "Knowledge & Illustrated Scientific News", 1907–1908, December through February, Vol 1 no 1, 2, 3, p: 1–2; 5–6; 9.

10 Cierva, J.d.l.a.D.R., Wings of Tomorrow. The story of the Autogiro. 1931: Brewer, Warren & Putnam.

11 Reporter, "Gliding at Narrabeen. Sensational Incidents. An Australian-Built Machine." Sydney Morning Herald, Sydney, NSW: 7 December 1909, p: 3.

12 Young, D., Chicago Aviation, an Illustrated History. 2003, De Kalb, IL: Northern Illinois Press.

13 Bates, C., "How to Make a Glider." Popular Mechanics, 1909, April, Vol 11 no 4, p: 386–388.

14 Waterman, W.D.w.J.C., Waldo: Pioneer Aviator. A personal history of American aviation, 1910–1944. 1988, Carlisle, Massachusetts: Arsdale, Bosch & Co.

15 Reporter, "American Flies With Smallest Novel Aeroplane. M B Sellers has been flying in Kentucky for more than a year in craft smaller than Santos-Dumont's. Uses less power than other aviators. Successfully invents automatic balancing device that does not infringe Wright Patents." The-Thrice-A-Week World, New York, NY: 17 and 18 April 1910, p: 5.

16 Kruckman, A., "American flies with smallest novel aeroplane. M. B. Sellers has been flying in Kentucky for more than a year in craft smaller than Santos-Dumont's. Uses less power than other aviators. Sellers, whose novel aeroplane

with an automatic balancing device is attracting wide attention. Successfully invents automatic balancing device that does not infringe Wtight patents." The World, New York, NY: 17 April 1910.

17 Kruckman, A., "American flies with smallest novel aeroplane. M. B. Sellers has been flying in Kentucky for more than a year in craft smaller than Santos-Dumont's." The Thrice-A-Week World, New York, NY: 18 April 1910, p: 3.

18 E. Husting, letter to Paul A. Schweizer, dated 14 April 1985, in Schweizer Papers, National Soaring Museum: Elmira, NY.

19 Slawson, H.H., "Octave Chanute's Model Gliders Found." Aero Digest, 1928, January, Vol 12 no 1, p: 58.

20 Chanute, letter to John A. Schnaane, dated 21 October 1909, in Papers of Octave Chanute, Letterpress book No 24, Manuscript Division, Library of Congress: Washington, DC.

21 Boyd, A.C., Some Memories of my Father, in Collection of papers and photos at the home of Joseph & Jean Hodges. A later manuscript contains an account of his trip to New Orleans. 1915, 1918: Denver, CO.

22 Aitken, W.H., "How to glide. Demonstrator at Philadelphia Country Club." Aeronautics, 1909, June, Vol 6 no 6, p: 183.

23 Reporter, "This young man monopolizes an aero profession. Willian H. Aitken is the first man to earn his living by teaching budding aeronauts to glide. Popular priced airships for lean pocketbooks." The World, New York, NY: 31 October 1909.

24 Reporter, "Night School for Flyers; Pupils Guaranteed to Rise. YMCA in New York plans to give practical instructions in aeronautics. Experts in demand." The Chicago Sunday Tribune: 5 September 1909, p: 6.

25 Short, S., "90 Years Ago. SOARING: Germany's Gift to Sporting America. Flying planes without motor." American Aviation Historical Society Journal, 2018, Summer, Vol 63 no 2, p: 130–139.

26 Barnaby, R.S., Gliders and Gliding. Design principles, structural features, and operation of gliders and soaring planes. 1930, New York: The Ronald Press Company.

27 Cox, A., letter dated 14 August 1952 to the Joint Committee, Congress of the United States, Office of the Architect of the Capitol: Washington, DC.

28 Doolittle, J.w.C.V.G., I could never be so lucky again. 1991, New York: Bantam Books.

29 Reporter, "Medals for the Wrights. President presents gifts of Aero Club today. After official honors in Washington the two navigators will go to Fort Myer to resume trials of aeroplane for Army use. Octave Chanute's experiments recalled (by Maj. Menoher?)." The Evening Post, New York, NY: 9 June 1909, p: 3.

30 Clark, E.B., "First U. S. Army Plane Decade ago. General Menoher's telegram to Orville Wright. Wright brothers are considered developers of the airplane, but always gave full credit to Langley, Chanute and other pioneers." Daily Sentinel, Rome, NY: 17 September 1919, p: 5.

31 Kruckman, A., "Wright Brothers at last will get Honor at Home. President Taft on next Thursday in the White House will present the Aero Club's medals to the aeronauts. Notable gathering to witness the ceremony. Famous scientists and high officials to be guests. Aeroplane flight a feature." The Thrice-A-Week World, New York, NY: 6 June 1909.

32 Reporter, "Wrights receive Nation thanks. President tells aviators that dawn of flight's age is here; Not for war alone. Brothers one day lions. Orville resumes flights June 21 in type of aeroplane wrecked in Fort Myer trials." Chicago Tribune, Chicago, IL: 11 June 1909.

33 Chanute, letter to Chas. Walcott, dated 27 January 1909, in Papers of Octave Chanute, Letterpress book No 23.

34 Reporter, "Presentation of the Langley Medal to the Wright Brothers." Science, 1910, 4 March, Vol 31 no 792, p: 334–337.

35 W. Wright, letter to O. Chanute, dated 2 March 1906, in The Wilbur and Orville Wright Papers.

36 C., "The True Theory of Flying." Scientific American, 1870, 27 August, Vol 23 no 9, p: 133.

37 Ferber, F., L'Aviation. Ses Débuts - Son Développement. De Crete a Crete, De Ville a Ville, De Continent a Continent, ed. E. Berger-Levrault & Cie. 1908, Paris and Nancy, France.

38 Chanute, letter to George A. Spratt, dated 15 January 1909, in Papers of Octave Chanute.

39 Chanute, letter to George A. Spratt, dated 23 June 1909, in Papers of Octave Chanute.

40 Spratt, letter to O. Chanute, dated 13 August 1909, in Papers of Octave Chanute.

41 Lorin, R., "Aviateurs Contemporains. Propulsion par Reaction Directe et son application a l'aviation. Nouvelles considerations." L'Aérophile, 1910, 15 July, Vol 18 no 14, p: 322–325.

42 Lorin, R., L'air et la vitesse: vues nouvelles sur l'aviation 1919. Librairie Aéronautique. 1919, Paris.

43 "Aeronautics at Columbia. Growing interest shown." Columbia Spectator, 1908, 4 November, Vol 52 no 36, p: 7.

44 Editor, "A course in Aerial Engineering." Cornell Alumni News, 1909, 24 November, Vol 12 no 9, p: 97.

45 Editor, "Editorial on Scientific Achievements." The Sibley Journal of Engineering, 1910, December, Vol no 3, p: 123.

46 Jackman, W.J.a.T.H.R., Flying Machines, Construction & Operation. 2nd edition (1912) has several chapters added, including a "In Memoriam" of Chanute. ed. 1910, 1912, Chicago, IL: Charles C. Thompson Co.

47 Brown, E.C., '12 "Aeronautics at Harvard." The Harvard Illustrated Magazine, 1910, March, Vol 11 no 6, p: 189–190.

48 Reporter, "Boston Y.M.C.A. will open aerial school next Tuesday night. H. Helm Clayton, Dean of the faculty, will have able assistants for regular educational course. Covers wide field." Christian Science Monitor, Boston, MA: 20 November 1909, p: 4.

49 Lanza, G., "Departments of Mechanical Engineering and Applied Mechanics." Bulletin of the Massachusetts Institute of Technology, 1911, January, Vol 46 no 2, p: 76–78.

50 Ovington, E.L., "The Modern Aeroplane. The plain facts about aviation and aeroplanes told by licensed aviator and expert performer, who is also a skilled engineer." Science Conspectus, 1912, April, Vol 2 no 5, p: 125–133.

51 Reporter, "Circumstances that are destined to make Chicago the Aeronautical Center of America." Chicago Tribune, Chicago, IL: 18 July 1909, p: 4–5.

52 Avery, W., "Some American Gliding Experiments." The Aero, 1911, 22 February, Vol 4 n 92 p: 158–159.

53 Reporter, "Professor of Aeronautics. Paris establishing post. Wanted Wright to take it. Chicago Man second choice." The Sun, New York, NY: 26 January 1909, p: 6.

54 Reporter, "Dr. Chanute Refuses." The Times-Democrat, New Orleans: 3 February 1909, p: 1.

55 Marchis, M.L., Leçons sur La Navigation Aérienne (Ballons sphériques., Aérostation militaire, Aérostation scientifique, Aéronautique maritime, Ballons dirigeables. Université de Bordeaux. Faculté des Sciences. 1903–1904, Paris: Vve Ch. Dunod.

56 Chanute, letter to Lucien Marchis, dated 25 August 1906, in Papers of Octave Chanute.

57 Reporter, "Aviation meet is now assured. Wright brothers and Aero Club reach an agreement. Inventors to be protected." The New York Times, New York, NY: 10 April 1910.

58 Jones, E.L., "News in General. A. C. A. Recognizes Wright Patent." Aeronautics, 1910, May, Vol 6 no 5, p: 173.

59 Editor, "The Open Door of Aviation." Scientific American, 1910, 26 June, Vol 102 no 26.

60 Chanute, O., "Book Review: Board-D, Vehicles of the Air." Chicago Tribune, Chicago, IL: 24 June 1910, p: 1–2.

61 Means, J., Epitome of the Aeronautical Annual. 1910, Boston: W. B. Clarke & Co.

62 Manly, C.M., "Tribute to the Herald. Letter to the Editor." New York Herald, New York, NY: 8 September 1911, p: 8.

63 Chanute, O., Travel Diary, in Collection of papers and photos at the home of Joseph & Jean Hodges. 1910: Denver, CO.

64 Special Cable, "Aviation as a Sport in Chicago. Chanute says in Paris wealthy men are building machines." Daily News Company, Chicago, IL: 1910.

65 Reporter, "Un précurseur de l'aviation. M. Chanute, de passage à Paris, a donné ses impressions à un rédacteur du Petit Journal." Le Petit Journal, Paris, France: 4 June 1910, p: 1.

66 Reporter, "Aviators Gather for Belmont Flight." New York Times, New York, NY: 9 October 1910, p: 2.

67 Chanute, C.D., "Letter to Lee S. Burridge, The Aeronautic Society, dated 5 December 1910." Aeronautics, 1911, January, Vol 8 no 1, p: 4.

Chapter 10 (Page 375–386)

1 Cronquist, D., "President's Message." ASCE Los Angeles Section, 2020, February, Vol 61 no 2, p: 1.

2 Zahm, A.F., "Octave Chanute, his work and influence in aeronautics." Scientific American, 1911, 13 May, Vol 104 no 19, p: 463, 488.

3 Jones, E.L., "Editorial - News in general." Aeronautics, 1911, January, Vol 8 no 1, p: 29.

4 Jones, E.L., "The life and work of Octave Chanute." Aeronautics, 1911, January, Vol 8 no 1, p: 3, 4, 35.

5 Chanute, O., "President American Society of Civil Engineers." Engineering News, 1891, 23 May, Vol 25 no 23, p: 496, full page portrait facing page 496.

6 Means, J., "Octave Chanute's Work in Aviation." Aeronautics, 1911, January, Vol 8 no 1, p: 3.

7 Wright, W., "The Life and Work of Octave Chanute." Aeronautics, 1911, January, Vol 8 no 1, p: 4.

8 W. Wright, letter to Colonel William A. Glassford, dated 30 November 1910, in The Papers of Wilbur and Orville Wright.

9 Maxim, H., "In Honor of Octave Chanute." Aircraft, 1911, February, Vol 1 no 12, p: 432–433.

10 Editor, "Topics of the Times. Aviation and its accidents." The New York Times, New York, NY: 16 January 1911.

11 Editor, "The Patent Situation settled. Manufacturers Aircraft Association, Inc. Organized." Aviation and Aeronautical Engineering, 1917, 1 August, Vol 3 no 1, p: 23, 43.

12 Bates, O., M.W.S.E., "Fundamentals. The Essential Principles of Engineering Practise, Part I and Part II." Scientific American Supplement, 1912, 27 April, 4 May, p: 266–267, 274–275.

13 Post, A., "The Soaring Airplane. The art of riding the air in a motorless flight may soon be stolen from the birds." The Sun, New York, NY: 17 December 1921, p: 9.

14 Brennan, G.A., The Wonders of the Dunes. 1923, Indianapolis, IN: The Bobbs-Merrill Company.

Index